Allan Barker has been extremely successful in SCCA B Production racing with his Q-jet equipped small-block engine.

## Why a Book on Rochester Carburetors

It is a frustrating experience to seek information on any subject and not be able to draw on a reliable source. Rochester Products Division of General Motors (RPD) markets more carburetors than any other supplier in the world. They also print manuals, training books and parts reference guides by the thousands. A number of independent technical book writers contribute to the library of information published yearly. All of these are designed to cover general service and maintenance procedures. This book was written to provide you with knowledge far beyond that obtainable from manuals. . .and much of it has never been presented anywhere else. We have gone to considerable effort to gather materials that venture into new depths of Rochester Carburetor know-how. Every attempt has been made to illustrate the versatility of these excellent carburetors for street use and for racing applications, too.

### PERFORMANCE ENTHUSIASTS

We know a great number of people want accurate and tested information on using Rochester Carburetors for many varied applications.

Some car buffs participate in drag races, run slaloms, road race, go desert racing or do any number of things besides just drive down the street.

We have gathered speed secrets from winners who use Rochester Carburetors and combined it with our wide and varied background to provide a special section on High Performance With Rochester Carburetors.

You will find a lot of tips on special-performance carburetors. We tried to incorporate answers to all of the questions which enthusiasts ask again and again at technical sessions, or at competitive events. Hopefully, we have dispelled some of the rumors, myths and half-truths which have become part of the romance of carburetors.

That line of Monojets looks endless. And, when you consider that the line runs day in and day out, perhaps it really is endless.

Here a Monojet is being given its computer-controlled audit inspection. Arrow indicates air cylinder used to close vacuum tube in throttle body during test.

G.M. test driver monitoring tape-controlled instruments as he runs through emission test schedule on General Motors Desert Proving Grounds dynamometer rolls.

NASCAR photographer and officials scrutinize Corvair prior to its record gas mileage run at the 1965 Pure Oil Performance trials. Vince Piggins, Manager of the Chevrolet Production Promotion Department and Doug Roe, driver of the entry stand at the rear observing. Vehicle was equipped with a pair of Rochester H carburetors.

Foreman advises operator on techniques for assembling 1-1/2-inch Oldsmobile carburetor.

## EMISSIONS

One of the most important factors in vehicle manufacturing today is emission. Hopefully we can illustrate for you what has really taken place in the past five years and some of what is in the future for emission devices and associated tuning for low emissions. The carburetor plays a very important role in this task of reducing emissions. How does this affect you? Some comments and hopefully enlightening facts lie within these pages.

## DRIVEABILITY AND ECONOMY

A good portion of this book deals with the Rochester Quadrajet (Q-jet) carburetor. This unit has undoubtedly had more design and test consideration directed toward driveability and economy than any other single carburetor model in the world. We will take you into many driving situations and relate each one to the specific portion of the carburetor which serves that need.

This book has been written with the intention of benefitting and improving your knowledge of carburetion and how it applies to your driving.

Photos and illustrations have been used freely to describe design features and operation of these carburetors. Some photos and graphs show how special parts can be used to provide improved performance in specific applications. . .and how the stock parts should be checked or blue printed to serve you best.

This book is not a substitute for factory service and overhaul manuals which provide essential basic tuning information for particular vehicles, together with extensive repair information for the specific carburetors used in those

Top/Just prior to clean-air inspection an operator uses a mirror to check both sides of carb to ensure that correct parts are installed all around.

Center/Doug Roe instructing Maurine Smith and Earl Reed, 1964 Mobile Gas Economy Run drivers, on how to drive for maximum economy. The vehicle simulator helped relate driver actions to engine and carburetor function.

Bottom/The famous "Baja Boot" in action. The builder and designer, Vic Hickey, says, "For off-road racing, if it doesn't have a Q-jet for carburetion, I won't touch it."

applications. It does not provide complete adjustment details for Rochester carburetors because these vary widely from vehicle to vehicle. Also with the ever-changing emission requirements, each carburetor for a specific automobile is factory adjusted to provide emission performance legally required for that car.

A carburetor is the most worked-on and often the least-understood component on a vehicle.

Co-author Doug Roe has more than two decades of carburetor-development experience, starting in the early 50's with Stromberg Carburetor, a Division of Bendix Aviation. He spent four years at their Elmira, N.Y. Eclipse Plant working at road and laboratory testing. He then worked four years with Rochester Products Division (RPD) of General Motors in their test laboratory and as a field representative to Chevrolet Motor Division in Michigan. A transfer to Chevrolet Motor Division provided 12 more years of concerted effort in engine-carburetor development and test programs.

H.P. Books' Publisher Bill Fisher, editor and co-author of this book, has been writing automotive technical books for 26 years. His concerted efforts towards creating carburetor books started in 1965.

RPD's product-information people have been most helpful in providing technical materials related to their carburetors.

To further round out the research for this book we have interviewed many automotive-oriented people including GM training instructors.

A vast amount of work, research and recall has made this book so you the reader will better understand Rochester carburetors — and carburetion in general.

**Where to begin?** At the beginning, you say? Right! Chapter one discusses the fuel and air requirements of an engine and introduces the standard way to specify carburetor air-flow rates: cubic feet per minute (CFM) along with how to figure it.

And away we go. Into how carbs work and the families of Rochester carbs with names like Monojet, 4GC, H, and Quadrajet—known to its fans as Q-jet. Why families? Because each of these carbs has been used on so many different engines and applications that lots of the little parts are different and some of the big ones. The Q-jet is so good it has been built in over 200 different versions. The book tells you all the important stuff about the differences even down to details such as models with more leverage to help the float close the needle-valve.

One problem is, you can't talk about carburetors without using carb language—unless you use baby talk, and that's not what you came here for. So the book uses carb words and expressions some of which you may already use, and some strangers. Such as the word *signal* which to a carb engineer means the amount of pressure-drop in the venturi which signals the fuel supply to dump in the correct amount of gasoline. Such signals are also used to control the distributor and some emission-control devices.

If you find a few unfamiliar expressions at the beginning, please hang in there. This is a big book and it all gets sorted out for you, a piece at a time. With definitions, some calculations, descriptions, practical examples, drawings, and a lot of pictures. Pretty soon you'll be speaking carburetor right along with us.

# Engine Requirements

**L**earning what an engine requires in the way of air flow and air/fuel ratio puts your tuning or vehicle way out front of the competition. Whether you work on engines to make them run quicker, faster—or merely "sweeter," you'll get more performance after you have fully digested this chapter on engine requirements, plus those which follow on basic principles and high-performance tuning.

The purpose of the carburetor on the gasoline engine is to meter, atomize and distribute the fuel throughout the air being pulled into the engine. All of these three functions must be performed by the carburetor over a wide range of engine-operating conditions, varying engine speeds, loads and operating temperatures. Manifolding and inlet-mixture temperatures are also influential factors in good engine operation as you will see.

You do not need to worry about tough mathematical formulas because we have made the whole idea of carburetion dirt-simple. After all, you want your tuning or building to make engines run better, so we've left the complex calculations for the engineers who designed the Rochester carburetor.

So, what's the first step? Well, we have to learn what the engine needs so we can feed it correctly.

It may be said that the carburetor has one other function to perform, controlled by an outside force—the driver. This carburetor function—regulating the amount of gasoline mixture which flows to the engine—gives the driver control of the engine speed and load.

No matter how fast or at what load the engine operates, the carburetor must do its job automatically.

The carburetor must determine the air flow to the engine and the load at which the engine is operating. Then it must meter and mix thoroughly the correct amount of gasoline with the intake air stream in ac-

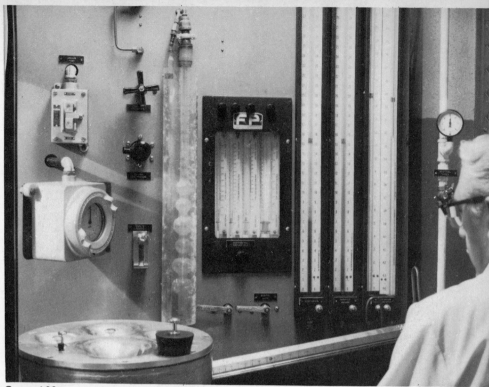

General Motors test engineers use many types of accurate fuel-flow and air-flow meters on carburetor flow boxes and in conjunction with engine dynamometers.

cordance with that particular engine-operating condition. You can see that the carburetor is a very intricate device and one which, viewed as a whole, is extremely complicated. We'll study it one phase at a time so it is easy to understand.

No one component on a vehicle is affected by as many elements as the carburetor. Its basic function of metering fuel and air in correct proportions is often disrupted by varying temperatures, air density, vehicle movements, seasonal fuel-blending changes, etc. In short, a carburetor is a very sensitive and delicate instrument with many variables affecting its operation.

## METERING

An engine's performance is dependent on the air and gasoline mixture supplied to it. Engines must operate over such a wide range of speed and the engine sizes vary so greatly, it means little to say an engine uses so many pounds of fuel per hour. However, if an engine is said to have

used so much gasoline while consuming so much air, this is a significant statement. The weight ratio of air consumed per fuel burned in a given amount of time is termed the air/fuel, or mixture ratio. Using the term air/fuel ratio, two radically different engines may be compared as to their fuel requirements.

We can say, for a given fuel there is only one ideal air-fuel ratio where there will be just enough fuel to react with all the air available in the burning process. This results in a maximum amount of energy being produced. If the air-fuel ratio varies from this value, less energy will be given off per pound of mixture. This "best energy" air-fuel ratio for gasoline-air mixtures is approximately 15 pounds of air per pound of fuel.

## AIR-FLOW REQUIREMENTS

Because the air an engine consumes has to come in through the carburetor, knowing how much air the engine can consume will help

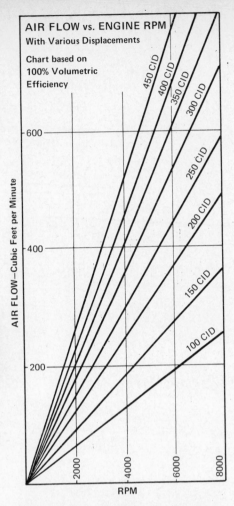

AIR FLOW vs. ENGINE RPM
With Various Displacements
Chart based on 100% Volumetric Efficiency

**Pick Cubic-Inch Displacement, estimate your maximum RPM, read your air-flow requirement, then correct air flow by Volumetric Efficiency as described in text. This chart is for four-cycle engines. A two-stroke requires twice as much air flow.**

---

**Carburetor flow rating in CFM is a better** way to state carburetor capacity than using the older method of comparing venturi sizes. This is because venturi size does not accurately represent the actual flow capacity of the carburetor, especially where one or more booster venturis reduce the effective opening and increase the restriction or pressure drop across the carburetor.

---

NOTE: 1728 cubic inches per cubic foot. Displacements in cubic centimeters can be divided by 16.4 to convert them to cubic inches for use in these formulas.

us to select the correct carburetor size.

How big should the carburetor be? Most likely, if you were to guess—or tour the hot rod shops where super carburetors are on display—you would end up with one too large for the application. The Rochester Q-jet, for instance, was designed to cover a broad range of engine sizes and performance levels. **Engine displacement**—expressed in cubic inches (CID or Cubic Inches Displacement) or cubic centimeters (cc).

**Type of engine**—two-cycle or four-cycle. Two-cycle engines have an intake stroke every revolution. Four-cycle engines have an intake stroke every other revolution.

**Maximum RPM**—the peak RPM which the engine will "see." In this area you must be absolutely honest and realistic without kidding yourself. "Dream" figures will only lead you into the trap of getting a carburetor which is too big, causing problems which are discussed throughout this book.

Once you know the engine size, type and RPM, figuring the air-flow requirement in cubic feet per minute (CFM) gets real easy.

**For 2-Cycle Engines**

$$\frac{CID \times RPM \times Volumetric\ Efficiency}{1728} = CFM$$

**For 4-Cycle Engines**

$$\frac{CID}{2} \times \frac{RPM}{1728} \times Volumetric\ Efficiency = CFM$$

**Example:** 350 cubic inch engine
7000 RPM maximum
Assume volumetric efficiency of 1 (100%)

$$\frac{350\ CID \times 7000\ RPM}{2 \times 1728} \times 1 = 709\ CFM$$

A volumetric efficiency of 100% or 1 is not usually attainable with a naturally aspirated (un-supercharged) engine. Thus, the engine in question will not flow 709 CFM of air. Let's talk about volumetric efficiency for a few minutes so we can see how it affects the air-flow requirement.

## VOLUMETRIC EFFICIENCY

Volumetric efficiency indicates how well the engine breathes. The better the "breathing ability" —the higher the volumetric efficiency. We should pause here to point out that *volumetric efficiency* is really an incorrect description of what is measured. But the term has been in use for so many years now that there's no real reason to try to change the usage to the correct term, *mass efficiency*.

Volumetric efficiency is defined as the ratio of the *actual* mass (weight) of air which is taken into the engine as compared to the mass which the engine displacement would *theoretically* take in if there were no losses. This ratio is expressed as a percentage. It is quite low at idle and low RPM because the "pump" or engine is being throttled by throttle-blade position.

$$V.E. = \frac{Actual\ mass\ of\ air\ taken\ in}{Mass\ of\ air\ which\ could\ be\ taken\ into\ perfect\ engine.}$$

Volumetric efficiency reaches a maximum at a speed close to where maximum-torque wide-open throttle occurs and then falls off as engine speed is increased to peak RPM.

Actually, the volumetric-efficiency curve closely follows the torque curve.

An ordinary low-performance engine has a V.E. of about 75% at maximum speed; about 80% at maximum torque. A high-performance engine has a V.E. of about 80% at maximum speed; about 85% at maximum torque. An all-out racing engine has a V.E. of about 90% at maximum speed; about 95% at maximum torque. A highly tuned intake and exhaust system with efficient cylinder-head porting and a camshaft which takes full advantage of the engine's other equipment can provide such complete cylinder filling that a V.E. of 100%—or slightly higher—may be obtained at the speed matching the system's tuned point or range.

Let's go back to our example of the 350 CID engine we calculated as flowing 709 CFM of air (at standard temperature and pressure) at 100% volumetric efficiency. If this is a high-performance engine with a maximum 80% V.E., then the flow becomes 709 CFM x 0.80 = 567 (at standard temperature and pressure).

Because the mass of air which is ingested is directly related to the density of the air, volumetric efficiency can be expressed as a ratio of the density achieved in the cylinder versus the inlet density, or

$$\eta = \gamma_{cyl} / \gamma_i$$

Ideal mass flow for a four-cycle engine is calculated by multiplying

$$\frac{RPM}{2} \times CID \times \gamma_i$$

where $\gamma_i$ is the inlet air density.

Air density varies directly with pressure, therefore, the lower the pressure, the less dense the air. Similarly, at altitudes above sea level, pressure drops, reducing power because the density is reduced.

Tables which relate air density to pressure (correct barometer) and temperature are available. Or you can use this formula:

$$\gamma = \frac{1.326P}{t + 459.6}$$

where

$\gamma$ = density in lbs/cubic foot

P = absolute pressure in inches of mercury (Hg) read directly off of barometer (not corrected)

t = temperature in degrees Fahrenheit at induction-system inlet

Actual mass flow into an engine can be measured as the engine is running. A laminar-flow unit or other gas-measuring device such as a calibrated orifice or a pitot tube can be used.

Actual mass flow is usually lower than ideal in a naturally aspirated engine because air becomes less dense as it is heated in the intake manifold. Absolute pressure (and density) also drop as the mixture travels the path from the carburetor inlet into the combustion chamber. This further reduces the mass of the charge reaching the cylinder.

The greater the pressure drop across or through the carburetor the lower the density can be inside the manifold and in the combustion chamber. If the carburetor is too

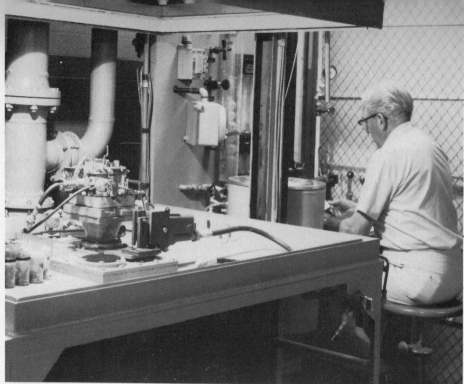

Carburetor flow box at Rochester Products' Arizona Desert Facility. The director of this hot-weather-oriented facility, Mr. Robert Prior, is flowing a late-model Quadrajet.

Air/fuel ratios for each model must be met during inspection or the carburetor is rejected for repair. The operator puts the unit on the fixtures and punches a button. The computer programs the entire inspection including making adjustments by use of positioned electric motors (arrows).

small, the pressure drop at wide-open throttle will be greater than the desired 1-inch Hg (for high performance) and power will be reduced because the mixture will not be as dense as is required for full power.

Selection of carb size (diameter of the air passage) gets you into two conflicting requirements. You probably already know that the air-passage tapers down inside the carb; this causes reduced pressure and the restriction is called a venturi. Carburetor theory treats this pressure drop separately and the venturi pressure reduction is not lost to the engine. The pressure comes back up as soon as the air passage is widened—either in the carburetor or later in the manifold or even in the cylinder.

The conflict is that pressure is also lost by friction between the air and the sidewalls of the air passages. This drop is gone forever and cannot be restored after it is lost.

A small-venturi carburetor generates a larger "signal" from the pressure drop in the venturi and can turn on the main fuel system at a lower air flow or lower RPM than would a larger venturi. The smaller venturi causes more friction losses however and that hurts flow, which means less WOT power.

Carburetors are flow-rated wide open and the CFM rating is that amount of air flow which causes some specified pressure drop due to air friction through the carburetor. Rochester has rated their four-barrels at a drop of 1.5 inches of mercury (same as barometric pressure readings) and other carbs at 3.0 inches of mercury in a chart on page 178. If you are not familiar with barometer readings, an inch of mercury is equivalent to an atmospheric pressure of about 0.5 psi (pounds per square inch).

To keep volumetric efficiency as high as possible, we would like to keep the size of the carburetor up, hence the pressure drop down. *The limiting factor is at the other end of the flow curve.* Will the carburetor be able to meter fuel correctly at low air flows?

## FUEL REQUIREMENTS

Fuel requirements relate to the air-flow requirement because fuel is consumed in proportion to the air being taken in by the engine. Fuel flow is typically stated in lbs./hr. and sometimes in lbs./HP hr., termed "specific fuel consumption."

First, let's look at the wide-open-throttle full-power fuel requirement for the same engine which needed 709 CFM of air at 7,000 RPM at 100% volumetric efficiency. To get the air flow into lbs./hr.:

$$CFM \times 4.38 = Air\ Flow\ lbs./hr.$$

Taking this a step further, multiply by the fuel/air ratio, which we will assume to be a typical full-power ratio of 0.077 lbs. fuel/ lbs. air (air/fuel ratio of 13 : 1).

$$CFM \times 4.38 \times F/A = Fuel\ Flow\ lbs./hr.$$

$$709 \times 4.38 \times 0.077 = 239\ lbs./hr.\ @\ 7,000\ RPM$$

This is the maximum amount of fuel which the engine can consume with an air flow of 709 CFM. However, this amount will never be reached because, like air flow, fuel flow

**NOTE:** 4.38 is a factor for 60°F. at standard barometric pressure (one atmosphere).

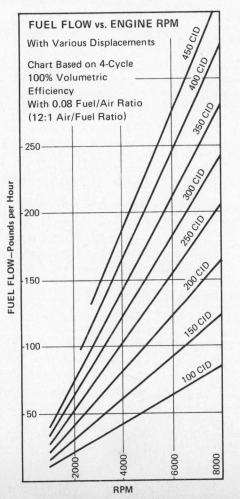

FUEL FLOW vs. ENGINE RPM
With Various Displacements
Chart Based on 4-Cycle
100% Volumetric Efficiency
With 0.08 Fuel/Air Ratio
(12:1 Air/Fuel Ratio)

FUEL FLOW—Pounds per Hour

RPM

450 CID
400 CID
350 CID
300 CID
250 CID
200 CID
150 CID
100 CID

must be reduced by volumetric efficiency, assuming $\eta = 0.80$:

$$CFM \times 4.38 \times F/A \times \eta = Fuel\ Flow\ lbs./hr.$$

$$709 \times 4.38 \times 0.077 \times 0.80 = 191\ lbs./hr.\ @\ 7,000\ RPM$$

And, the fuel flow will be less when the engine is turning fewer RPM. For instance, the 191 lbs./hr. @ 7,000 RPM would be 1/2 that at 3,500 RPM and 1/4 that at 1,750 RPM . . . assuming wide-open throttle in each instance.

An accompanying chart shows the fuel flow for various engine sizes over typical RPM ranges. If you don't happen to have this book along with you when you need to estimate fuel requirements for fuel pump and fuel line selection, just remember that wide-open-throttle full power typically requires 0.5 lb. fuel per HP every hour. Thus, a 300 HP engine needs 300 x 0.5 = 150 lbs./hr., or 150/6.0 lbs. per gallon = 25.0 gallons per hour.

The preceding discussion is related to the maximum amount of fuel that the engine could be expected to consume under full-power, wide-open-throttle conditions. This knowledge will help you in selecting fuel-line sizes, fuel-pump capacity and the required tank capacity to get your vehicle where you want it to go within the limitations of fuel stops. A lot of high-performance enthusiasts blame the carburetor for leaning out when the real problem is one of inadequate fuel supply. You'll find more about this subject in the fuel supply system chapter.

Now let's look at the fuel requirements for other conditions.

## EQUIVALENT RATIO TABLE

| A/F (Air/Fuel) | F/A (Fuel/Air) | A/F (Air/Fuel) | F/A (Fuel/Air) |
|---|---|---|---|
| 22:1 | 0.0455 | 13:1 | 0.0769 |
| 21:1 | 0.0476 | 12:1 | 0.0833 |
| 20:1 | 0.0500 | 11:1 | 0.0909 |
| 19:1 | 0.0526 | 10:1 | 0.1000 |
| 18:1 | 0.0556 | 9:1 | 0.1111 |
| 17:1 | 0.0588 | 8:1 | 0.1250 |
| 16:1 | 0.0625 | 7:1 | 0.1429 |
| 15:1 | 0.0667 | 6:1 | 0.1667 |
| 14:1 | 0.0714 | 5:1 | 0.2000 |

**Stoichiometric or chemically-correct mixture**—This is a mixture proportioned so all of the fuel burns with all of the air—actually all of the oxygen in the air—so that none of *either* is left over. The exhaust would contain only carbon monoxide, water and nitrogen as shown by the chemical equation:

$$Fuel + Air \longrightarrow CO_2 + H_2O + N_2$$

This would be a fuel/air ratio of approximately 0.068, or an air/fuel ratio of 14.7 : 1.

The actual ratio at which this occurs with an ideal set of conditions varies with the fuel's molecular structure. Gasolines will vary somewhat in structure, but not significantly. Fuels other than gasoline require different ratios for the ideal or stoichiometric condition. Alcohol, for instance, has a lower heat content (calorific value) than gasoline and requires a 0.14 F/A ratio (7.15 : 1 air/fuel) for its ideal burning condition. This is so much more fuel volume than required for gasoline that most carburetors cannot be used with alcohol, even when highly modified. The passages in the carburetor are actually smaller than would be required to meter the fuel correctly.

**Maximum power**—Maximum power requires an excess of fuel to make sure that all of the air is consumed. The reason for the excess fuel is that mixture distribution to the various cylinders and fuel-air mixing are seldom perfect. When all of the air enters into the combustion process, more heat is generated and heat means pressure that the engine converts to work.

The reaction looks like this:

$$Fuel + Air \longrightarrow CO_2 + H_2O +$$
$$+ CO + HC$$
$$+ N_2$$

where

| | | |
|---|---|---|
| $CO_2$ | = | carbon dioxide |
| $H_2O$ | = | water |
| $CO$ | = | carbon monoxide |
| $HC$ | = | unburned hydrocarbons |
| $N_2$ | = | nitrogen |

The fuel excess usually amounts to 10 to 15%, giving fuel/air ratios 0.075 to 0.080 (13.3 to 12.5 : 1 air/fuel ratios). Sometimes, an excess of fuel beyond that which produces maximum power is used for internal cooling of the engine. From a pollu-

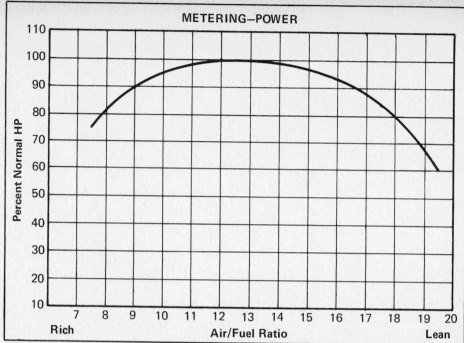

Figure 1 illustrates a power curve of a typical engine relative to various air-fuel mixtures. Note how the curve is quite flat between air/fuel ratios 11 and 14. This says an engine will produce good horsepower in a fairly broad range, but if you are looking for that last horsepower or a tenth of a second better performance in the quarter mile, the band you seek is even narrower. This optimum ratio varies from engine to engine and even within the same engine as RPM changes. The very precise art of finding this narrow band and then metering to stay within it throughout the power range is called "super tuning." It gets you trophies when competition is close.

tion standpoint, the unburned hydrocarbons and carbon monoxide are an undesirable by-product.

If excess gasoline is used the ratio becomes less than 15 to 1, and though all the air is utilized in the burning process, unburned gasoline will remain in the exhaust. Twelve to one is generally a good maximum power mixture because the horsepower will be good and the richer mixture aids combustion-chamber cooling to help avoid detonation and preignition. At about a 7 to 1 ratio the mixture no longer burns and black smoke and extreme sluggishness exist. Thus, from approximately 11 to 1 on down to rich-misfire—or at any time the air-fuel ratio varies outside the range of approximately 7 to 1 to 20 to 1—the engine will stall due to excessive richness or leanness.

In actual practice it has been proved that a 15 to 1 ratio supplied to an engine does not necessarily provide maximum efficiency for a particular engine. If there is less-than-perfect atomization and

distribution, there may not be a 15 to 1 ratio in all cylinders even when the carburetor supplies this ratio. Only by careful testing of a given engine design can the best carburetor air-fuel ratio be found for that engine.

The average automobile owner wants to get as many miles per gallon as he can. He also wants to get as much power as possible for some driving conditions. Carburetor designers have made a compromise with the consumer to give him some of each. When the engine is at or near wide-open throttle, the designers have increased the richness of the mixture supplied by the carburetor. When the driver is operating the engine at less than wide-open throttle, he is not interested in getting maximum power out of the engine. Consequently, for part-throttle operation a leaner mixture is supplied until a point is reached where either surge is objectionable or the miles per gallon is maximized. Generally these two conditions (lean surge and best economy) occur at nearly the same mixture ratio.

The carburetor is then so designed that when maximum power of an engine is desired the throttle is opened wide so the engine can draw a maximum amount of air into its cylinders. Theoretically the carburetor should add enough gasoline so, even with imperfect distribution, every particle of this air will be used in combustion. Beyond this point adding more fuel could not result in added energy liberation or more power. This rich mixture—called the maximum-power ratio—is about 12 or 13 to 1. The leanest cylinder often dictates the required carburetor metering, so other cylinders may run with a slightly over-rich mixture.

**Maximum economy**—If more air is admitted so that the A/F ratio becomes greater than 15 to 1, excess air will be left after the burning process is completed but all gasoline would be utilized. This is said to be a lean ratio and you can feel it with your sense of touch as engine surge. If the air-fuel ratio exceeds about 20 to 1, the mixture ratio becomes too lean to burn. Heavy surge, hesitation and backfire accompanies ratios nearing 20 to 1.

For maximum economy at part throttle the designer is interested in having every particle of fuel finding the correct amount of air so complete combustion of the fuel will occur. Maximum economy requires an excess of air to ensure that all of the fuel is consumed, giving a reaction that looks like this:

$$Fuel + Air \longrightarrow CO_2 + H_2O + N_2$$
$$+ \text{ small amounts of}$$
$$CO + HC + O_2$$

In any mixture which is leaner than stoichiometric under heavy loads, there is sufficient heat to cause any free oxygen to be combined with the remaining nitrogen to produce oxides of nitrogen ($NO_x$). This is one of the undesirable emission products.

At high engine speeds and low loads, mixtures approaching 0.055 F/A (18 : 1 A/F) are sometimes approached when seeking peak economy. However, this is on the borderline where fuel/air begins to become unstable in its ability to burn.

**Idle**—At very small throttle openings or closed throttle, the picture changes again. Here very little fresh air and fuel get to the engine. Also, because the pressure is ex-

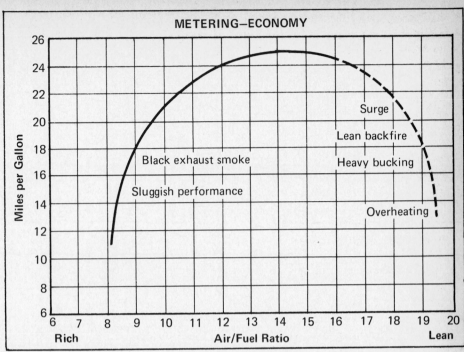

METERING—ECONOMY

Let's assume your engine is capable of giving you 25 miles per gallon at highway cruising speeds—not flat out. You will only get that good mileage if the carburetor is metering fuel correctly. This chart shows best economy at an air/fuel ratio of around 14 to 1 (this will vary among engines). If your carb gives the engine an A/F ratio of 10 or 11, bad things happen as shown. Poor economy, poor performance. If you or your carb try to make it run too lean, the engine will do different bad things for a while, then it may give up entirely due to overheating.

tremely low in the cylinders, all exhaust gases do not leave the combustion chamber. Therefore, a 15 to 1 ratio might leave the carburetor, but in the combustion chamber it is diluted with residual exhaust gases until it becomes leaner than 20 to 1 and the engine stalls. The carburetor has to supply a rich mixture—as rich as 10–13 to 1— so that with dilution the resulting combustion-chamber-mixture ratio will be nearly 15 to 1 and the engine will run correctly. There are many variables from engine to engine so an idle fuel metering adjustment screw is provided to set this idle ratio. This is another area in which emission regulations have changed things dramatically. The carburetor manufacturer narrows the requirements of specific engines with extensive testing and builds in only minor adjustment capabilities for use by the tune-up mechanic. This is often done by idle-limiter devices on the idle-mixture-adjustment screw.

Dilution of the idle intake charge is caused by the high manifold vac-

uum and by opening of the intake valve prior to the time that the piston has reached top dead center. When the intake valve opens, some of the exhaust gases are sucked into the intake manifold by the pressure difference, assisted by the still-upward-rising piston. These dilute and effectively lean the charge. Also, some of the exhaust gas stays in the cylinder clearance volume. When the piston descends on the intake stroke, the initial charge which is drawn in consists largely of exhaust gas, with a small proportion of fresh air/fuel mixture. As the piston completes its intake stroke, the percentage of fresh air/fuel mixture is usually somewhat higher and it is this portion of the inlet charge which burns well enough to supply power for idling the engine.

In the process of dilution, some of the fuel molecules combine (or line up, as the chemists say) with the exhaust molecules and some of the fuel molecules line up with the oxygen in the air. To make sure that the mixture is combustible, the mixture is made 10 to 20% richer than stoichiometric to offset that part of the fuel which combined with the exhaust gas. The richer mixture also helps to offset distribution problems. It also provides additional emission problems in that rich idle mixtures generate large amounts of carbon monoxide (CO).

**Cold starting**—Starting a cold engine requires the richest mixtures of all because slow cranking speeds provide air velocity which is too low for vaporization to occur. And, both the fuel and the manifold are cold, so there is no help from either of those areas in causing the fuel to vaporize so that it will be well distributed. Thus, a lot of excess fuel is required for starting. Typical air/fuel ratios are 10:1 to 5:1 (1.0 to 2.0 fuel/air ratios). The air/fuel ratio being supplied is not that which is getting to the cylinders. Distribution problems, depositing of liquid fuel on the walls of the manifold and on the ports in the cylinder heads all rob fuel from the mixture so that the mixture which actually reaches the cylinders in a vaporized form is probably in the 16.6:1−10:1 air/fuel ratio range (0.06−0.10 fuel/air).

Once the engine fires, speed goes up, velocity through the carburetor improves so that there is better vaporization and the fuel which has been deposited on the manifold walls in a liquid state vaporizes. As this happens, the mixture is gradually leaned out to 8.3:1−7.2:1 A/F (0.12−0.14 F/A) in the cylinders—still 80 to 120% richer than nominal. When the engine warms up, normal operating mixtures can be used.

## UNIFORM DISTRIBUTION

Now that we have looked at the engine's air and fuel requirements, let's consider another important engine requirement: uniform or equal air/fuel mixture distribution to the various cylinders. Uniform distribution helps every phase of engine operation. Good distribution also means the mixture is evenly distributed throughout the combustion chamber of each cylinder. Non-uniform (unequal) distribution causes a loss of power and efficiency, and may increase emissions.

Uniform distribution relies on a number of factors: good atomization of the fuel by the carburetor, use of fuel with the correct volatility for seasonal temperature variations, utilizing appropriate tech-

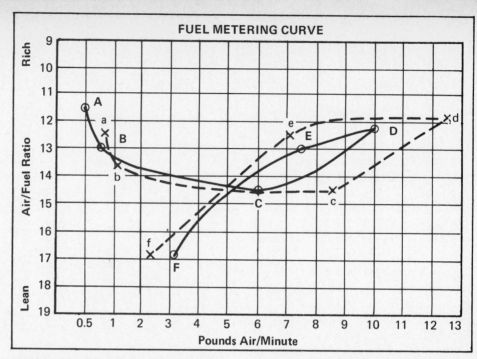

Because the average owner is interested in both power and economy, the carburetor must be metered to supply the best mixture for both conditions. Fortunately, similar air/fuel ratios are required for most engines to serve the needs of the various operating ranges (idle, part-throttle, power and wide-open throttle).

The drawing above compares the approximate delivery characteristics for a typical single-barrel carburetor as used on the Vega four-cylinder engine (solid line) with a carburetor used on Chevrolet 6's in the early 1960's. Note that the two carburetors serve slightly different air-flow ranges and the earlier unit provides slightly different (non-emission-controlled) air/fuel ratios.

The throttle is gradually opened from A(a) to D(d) and held wide open from D(d) to E(e) to F(f). The idle mixture at point a is a little leaner on the larger engine which has a higher air flow (A/F 12.4 at 0.8 pounds of air per minute). Points c and d reflect a vacuum-operated power system in the larger carburetor.

Referring again to the solid line for the modern-day carburetor: Point A represents idle at about 0.5 lb. air/min. at an A/F of about 11.5. Points B−C show the part-throttle range extending from about 1.0 to 6 lbs. air/min. at A/F ratios 13 −14.4. Light-throttle accelerations and steady throttle speeds depend on this carburetion area.

Point D is wide-open throttle, showing a requirement for 10 lbs.air/min. when the engine is turning high RPM. Note the A/F ratio of about 12.2. Line D−E represents an engine at wide-open throttle (WOT) progressively slowing down because of a steep grade.

As the engine slows down D−E−F the air/fuel ratio leans out because less air is being pulled through the carburetor. Even though the throttle is wide open, air speed and venturi signal applied to the discharge nozzle are diminishing to the point where fuel feed gradually stops. As the engine approaches point F, fuel flow has literally ceased with the engine faltering and finally stopping. This is known as "lugging to stall." The A/F ratio at this point is about 17.0 at 3 lbs.air/min.

These curves representing air/fuel ratios at various air-flow rates are only obtainable with elaborate equipment. We've shown them here to give you a better picture of what your carburetor does under the stated conditions.

niques to ensure vaporization of the fuel, and correct manifold design. This latter subject is covered in detail in a separate chapter later in the book. We will discuss the other items after we have established the importance of uniform distribution.

**Uniform distribution at idle**—As discussed in the preceding fuel-requirements section, idle mixture is typically 10 to 20% richer than ideal to offset dilution by exhaust gas. Idle mixtures with 1% CO (a common emission setting) require that all cylinders receive approximately the same idle mixture in terms of air/fuel ratio. Distribution problems can cause the use of an idle mixture too rich to pass emission regulations.

**Uniform distribution for economy**—Maximum economy, as previously discussed, requires an A/F ratio of approximately 15.5:1. If one cylinder receives a lean charge due to a mixture-distribution problem, this can increase the amount of fuel required for a given power output. To avoid misfiring in the lean cylinder, the other cylinders will have to be operated at mixtures richer than the desired 15.5:1 to bring the lean one up to the 15.5:1 figure.

All cylinders should be operating at the same A/F because lean cylinders produce more oxides of nitrogen ($NO_x$) if they are still firing. And, if they are so lean that misfiring is occurring, unburned hydrocarbons will be emitted. Rich-running cylinders will provide excess CO and unburned hydrocarbons. In either case, undesirable effects include decreased economy and increased emissions.

**Uniform distribution for power**—In the fuel-requirements section we discussed the need for A/F ratios of approximately 13.3 : 1 or 0.075 A/F to produce maximum power. If one cylinder runs lean, an excessively rich mixture must be supplied to the other cylinders to bring the lean cylinder up to the correct ratio so that it will not run into detonation. The rich cylinders will waste a lot of fuel and thus fuel consumption will be increased to maintain a specified power level. Running some cylinders rich reduces their power output because all cylinders must have the same A/F to get the best possible power out of the engine.

Unequal distribution not only affects fuel consumption and power—it also means that ignition timing can only be a rough approximation or compromise of what

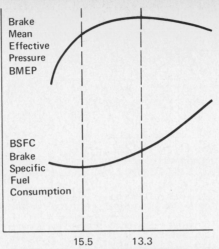

Brake mean effective pressure (BMEP) of engine operating with perfect distribution, showing relationship of A/F ratio supplied by carburetor. For maximum power, about 13.3–15.5:1 A/F gives lowest possible fuel consumption for that output. For maximum economy, 15.5:1 A/F gives lowest possible fuel consumption consistent with a lower, best-economy power output.

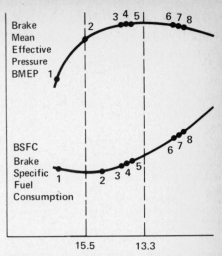

Same curve as at left with numbers representing cylinders receiving various A/F ratios at maximum-power output with non-perfect distribution. All cylinders are sharing the same carburetor and manifold. 6, 7, 8 with too-rich A/F produce less power and consume excess fuel. 3, 4, 5 receiving correct A/F for best power consume least fuel for that output. 1, 2 have lean A/F ratio. A richer mixture would have to be supplied to ALL cylinders to bring 1 & 2 to the flat part of the curve so as to avoid detonation. The net result: only cylinders 1 & 2 would operate at peak power with minimum fuel consumption for that output. 3, 4, 5, 6, 7, 8 would produce less power than if the engine had perfect distribution—and consume more fuel. If the curve is assumed to represent distribution during economical part-throttle operation, cylinder 1 receiving a lean A/F will operate at high fuel consumption due to misfiring. The other cylinders will operate at ratios above the correct 15.5:1 A/F.

could be used if all cylinders received equal A/F mixtures. The advance which can be used is directly related to the A/F available in the cylinder. Advance is typically limited to that which will not produce knock in the leanest cylinder. This limits the power which can be developed from those cylinders which have been richened to compensate for unequal distribution.

**Checking for correct distribution**—Automobile engine designers—and designers of carburetion systems, too—use a number of methods to check whether the various cylinders in an engine are receiving a uniform mixture. It should be noted that all of these methods require the use of a chassis or engine dynamometer. Generally speaking, these methods are not often used by individual tuners because of the cost and complexity of the equipment which is required. In many instances combinations of these methods are used. Distribution is checked by:

1. Chemically analyzing exhaust-gas (combustion products) samples from the individual cylinders at various operating modes: idle, cruise, acceleration, maximum power and maximum economy. There is a definite relationship between A/F and the chemical components in the exhaust gas, as described and pictured in the emissions chapter. By using the very accurate analytical equipment developed for emission studies, exact determinations of A/F ratio are used to check distribution. The availability of such equipment has made chemical analysis the method preferred by automotive engineers for distribution studies.

2. Measuring exhaust-gas temperature (EGT) with thermocouples inserted into the exhaust manifold or header at each cylinder's exhaust-port. The designer looks for EGT peaks as various main-jet sizes are tried. If all cylinders were brought to the same temperature, one cylinder (or more) might be 200° below its peak (and therefore below its peak output) because all cylinders do not produce the same EGT due to differences in cooling, valves, porting and ring sealing. Perfect distribution would place all cylinders at their peak EGT's with the same jet—*an ideal situation which is rarely achieved.*

3. Studying the brake specific fuel consumption to determine how closely the A/F requirements of a particular carburetor/manifold combination relate to ideal A/F ratios at various operating conditions.

INTAKE STROKE

FUEL PARTICLES    LIQUID FUEL ON WALLS

**When the mixture is only partially vaporized, liquid particles tending to cling to the manifold walls — or avoiding sharp turns into a cylinder may cause some cylinders to run lean and some rich. In this example, the center cylinder will tend to be leaned and the end cylinder will tend to receive a rich mixture. Where possible, it is best to divide the mixture before a change in direction occurs. Liquid particles are relatively heavier than the rest of the mixture and tend to continue in one direction. A fully vaporized mixture promotes good distribution in every instance.**

Any great variance is cause for suspecting that a distribution problem exists.

4. Observing combustion temperatures at various operating conditions.

**FACTORS AFFECTING DISTRIBUTION**
**Atomization & Vaporization**—Before gasoline can be burned, it must be vaporized. Vaporization changes the liquid to a gas state and this change only occurs when the liquid absorbs enough heat to boil. For example, a tea kettle changes water to water vapor (steam) by transferring heat into the water until the boiling point is reached. At this time the water changes into steam and the steam enters the atmosphere as water vapor (a gas, really). A temperature-pressure relationship controls the boiling point of any liquid. In the case of water, 212°F. is the sea-level boiling point. At higher altitudes, due to lowered atmospheric pressure, water boils at lower temperatures. Remember this relationship as we continue because it is important in understanding what happens in the functioning of the carburetor and manifold.

A common example is a cigarette lighter which works

fine indoors—but cannot be lit after it has been outside in a colder temperature for awhile. At warmer temperature, the fuel vaporizes easily and thus ignites easily. Evaporation is slower and less complete at lower temperatures, hence ignition becomes difficult.

The higher the temperature—the better the vaporization. There will be some sacrifice in top-end power when the mixture is heated sufficiently to ensure that all of the fuel will be vaporized. This is due to the reduction in charge density which occurs when the mixture is heated to this point. In most passenger-car engines, the loss of power is more than offset by the smoother running gained at part throttle. No manifold heat is used in a racing engine because a cold, dense mixture produces more power and has less tendency to detonate than a warm mixture.

Not only is it necessary to vaporize fuel before it can be expected to burn—it is essential to vaporize it to aid in distribution. It is much easier to distribute vaporized fuel in an air/fuel mixture than it is to distribute liquid fuel. It should be noted that some liquid fuel is nearly always present on the manifold surfaces. The only time that the manifold surfaces will be "dry" is during conditions of high manifold vacuum, which promotes vaporization.

Gasoline in the carburetor is discharged into the air stream as a spray, which is atomized (torn or sheared into fine droplets) into a mist. At this point it will be helpful if the fuel mist can be vaporized (changed to a gas form). Pressure in the intake manifold will be much lower than that of atmosphere (except at wide-open throttle) and this considerably lowers the boiling point of the gasoline. At the reduced pressure, some of the fuel particles are vaporized as they absorb heat from the surrounding air. And, some of the fuel particles are vaporized when they come in contact or close proximity to the hot spot on the manifold floor.

It must be understood that it is comparatively easy to get fair atomization of fuel, but getting *perfect* atomization is next to impossible. The carburetor-designer gets best results by locating the fuel-delivery nozzle at the point where air has the highest velocity. Rapidly moving air passing the slowly moving gasoline breaks the fuel into a spray of small particles. A second method of atomizing the fuel is to introduce air into the gasoline channel before the gasoline gets to the nozzle. This places air bubbles in the gasoline which produces a great amount of turbulence as they pass out the nozzle. This turbulence helps to break up the otherwise solid stream of liquid fuel into a fine spray. Look at the carburetor closely and you will see the air bleeds and metering orifices used to route air to the gasoline prior to its entering the venturi from the discharge nozzle.

Imperfect vaporization may occur if the mixture velocity is too low, if the manifold or incoming air is cold, if manifold vacuum is low (higher pressure), and if the fuel volatility is too low for the ambient temperature. Vaporization is also affected by manifold design and the carburetor size (flow capacity). With the same engine speed, a large carburetor has less velocity through its venturi/s, hence less pressure drop and a greater tendency for the fuel to come out of the discharge nozzle in liquid blobs or large drops which are not easily vaporized. In the case of a two- or four-barrel carburetor with progressively operated secondary throttles, small primary venturis aid vaporization by giving a higher velocity through the venturi and hence a greater pressure drop to ensure good vaporization. The design of the carburetor also affects the process. Multiple venturis (main venturi plus one or more boost venturis) and good emulsioning action in the main well (from which fuel is drawn) aid in vaporizing the fuel.

Manifold design also enters into the vaporization picture because the size of the passages affects the mixture velocity and heating. If the mixture travels slowly, some of the liquid particles of fuel may deposit onto the manifold walls before they have a chance to vaporize. The hot-spot size and location and the surface area inside the manifold dramatically influence vaporization.

Near-perfect distribution and best performance is only attained if the fuel is in the gaseous state rather than separated fuel and air. This is because gasoline in droplet form obtains momentum from the air stream in the manifold so that it becomes reluctant to turn corners. On the other hand, air is receptive to sudden changes of direction and consequently goes around corners easily to fulfill cylin-

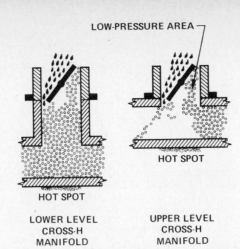

LOW-PRESSURE AREA

HOT SPOT

HOT SPOT

LOWER LEVEL CROSS-H MANIFOLD

UPPER LEVEL CROSS-H MANIFOLD

**Cross section through V-8 engine manifold of Cross-H or two-level type shows how riser height can affect distribution. Throttle shown in worst position causing greatest effect on distribution.**

der demand . . . sometimes leaving the gasoline droplets on the manifold surfaces or at the corners.

So if the fuel and air mixture leaving the carburetor is not correctly atomized, cylinder efficiency will vary.

When vaporization is poor, an excess of liquid fuel gets into the cylinders. Because it does not completely burn due to the lack of time available for evaporation and burning, it is expelled as unburned hydrocarbons. Some of the excess fuel washes the oil off the cylinder walls to cause rapid wear. And a portion of the liquid fuel drains past the rings into the crankcase where it dilutes the oil.

Evaporation of fuel typically requires 140 BTU's per pound and this amount of heat must be supplied for every pound of fuel that is to be consumed in the engine—if complete evaporation or vaporization of the fuel is to be achieved.

Exhaust-heated hot spots are typically small areas just under the area which is fed by the carburetor. The ends of the manifold are not usually heated. The size is kept as small as possible, consistent with the needs for flexible operating and smooth running. By keeping the spot fairly small, the manifold automatically cools off as RPM is increased. The

large amount of fuel being vaporized at high speed extracts heat from the manifold—often making it so cold that water condenses on its exterior surfaces. Although most passenger-car manifolds heat the mixture with an exhaust-heated spot, some manifolds are water-heated by the engine coolant. Cars equipped with emission controls often heat the incoming air by passing the air over the exhaust manifold on its way to the air-cleaner inlet.

Excepting racing intake systems, intake manifolds are compromise devices. Their shape, cross-sectional areas, and heating arrangements accomplish the necessary compromises between good mixture distribution and volumetric efficiency over the range of speeds at which the engine will be used. If only maximum or near-maximum RPM is being used, high mixture velocity through the manifold will help to ensure good distribution and will help to vaporize the fuel—or at least hold the smaller particles of fuel in suspension in the mixture. At slower speeds, the use of manifold heat becomes essential to ensure that the fuel is vaporized. If heat is not used, the engine will become rough running at slower speeds and distribution problems will be worsened.

**Fuel composition**—The more volatile the fuel, the better it will vaporize. All fuels are blends of hydrocarbon compounds and additives. The blending is typically accomplished to match the fuel to the ambient-temperature and altitude conditions. Thus, fuel supplied in the summertime has a higher boiling point than that which is available in winter. Fuel volatility is rated by a number known as "Reid Vapor Pressure." In Detroit, for example, the Reid number varies from 8.5 in the summer to 15 in winter. Fuel blending is always a compromise made on the basis of estimates of what the temperature will be when the gasoline is used and a sudden temperature change—such as a warm day in winter—usually causes a rash of vapor-lock and hot-starting problems.

**Carburetor placement**—The location of the carburetor on the manifold in relation to the internal passages can drastically affect distribution. If the geometric layout of the manifold places the carburetor closer to one or more cylinders, this can create problems. Considerations of carburetor location become especially important with multiple carburetors.

**It's a combination**—The engine designer checks power, economy and mixture distribution with the carburetor, air cleaner and manifold installed. Sometimes, the physical positioning of the air cleaner or a connecting elbow atop the carburetor is so critical that turning it to a different position can create distribution problems.

**Throttle-plate angle**—A directional effect is given to the fuel when the throttle is partially open. This effect becomes especially apparent when the manifold has little or no riser between the carburetor and the manifold passages.

Throttle-plate angle affects distribution in the upper level of a cross-H (or two-level) manifold (for a V-8 engine) more than it does the lower level because the longer riser into the lower level has a straightening effect on the mixture. This points up one of the advantages of high-riser designs which allow better straightening of the mixture flow with less directional effects at all throttle openings. High-riser designs typically offer better cylinder-to-cylinder distribution of the mixture. While height is usually limited by hood clearance, tests have shown definite improvements in distribution at part throttle with risers 1-1/2 to 2 inches high prior to opening into the manifold branches.

**Mixture speed & turbulence**—Mixture velocities and turbulence within the manifold definitely affect vaporization and hence distribution. More details are in the manifold chapter. Turbulence in the combustion chamber helps to prevent stratification of the fuel and promotes rapid flame travel.

**Time**—The volatility of the fuel and the heat available to assist in vaporization are especially important when you consider the tiny amount of time available for vaporization of the atomized fuel supplied by the carburetor. Unlike water in a tea kettle which can be left on the stove until it boils—fuel must be changed to the vapor state in less than about 0.008 second in a 12-inch-long manifold passage at a mixture velocity of approximately 125 feet per second. This assumes an engine speed of 5000 RPM and a cam with approximately 270° duration.

## DISTRIBUTION SUMMARY

To summarize the factors which can aid distribution:
1. Vaporize as much fuel as possible in the

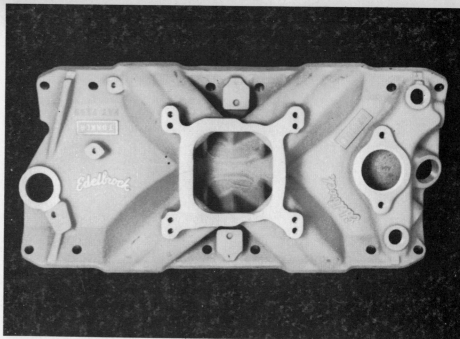

Edelbrock Torker manifold is a single-plane aftermarket item designed to use high stream velocities to ensure good distribution, even with carburetors set purposely lean for emission-controlled engines. Another manifold of this type was introduced as this book was being taken to the printer. It is the Edelbrock Streetmaster manifold which works to improve torque at speeds up to 4500 RPM—exactly the ticket for street and dual-purpose cars. It uses the stock Q-jet and has provisions for the EGR valve.

manifold so that a minimum of liquid fuel gets into the cylinders.
2. Use fuels with the correct volatility for the ambient-temperature and altitude conditions.
3. Ensure that velocities are kept high in the manifold by using the smallest passages consistent with the desired volumetric efficiency.
4. Provide good atomization in the carburetor through careful selection of venturi size (flow capacity).
5. Avoid manifold construction which causes fuel to separate out of the mixture due to sharp turns and severe changes in cross-sectional area.
6. Provide sufficient turbulence in the manifold to ensure that the fuel and air are kept well mixed as they travel to the cylinder.
7. Provide manifold heating and/or heating of the incoming air to ensure that fuel is well vaporized.

Obviously not all of these factors can be controlled by an individual. But, a perusal of them should bring out a few important points such as avoiding carburetors which are too large for the engine. Use fuel which is correct for the season, i.e., don't try to race with fuel you bought in a different season or another locality. And, for engines being operated on the street, use a heated manifold with the smallest passages consistent with the desired performance.

# How Your Carburetor Works

**U**nderstanding how your carburetor works is the key to getting top performance from your engine/carburetor combination. And, it's also the quick way to get more performance with less work because you will know what is happening and be able to do your tuning in a meaningful way without wasting time on a lot of cut-and-try changes which could wreck your carburetor and/or leave your combination in a worse state of tune than it was when you started. Also, once you know what happens inside of a carburetor, you will be able to make the correct choice of carburetor for your engine without having to rely on "experts" or rumors about the current "hot setup."

There's really nothing "tricky" about how your carburetor works. And, there is no "black magic" which makes one work differently from another. Whether you have a one-, two-, or a four-barrel carburetor, all operate essentially the same as all others. There are minor differences, but these are related to the way which the same kinds of systems were built into a particular carburetor. Carburetors are really very simple and similar devices, as you will see. Just as all four-cycle engines—from one-lung lawnmowers to Chrysler hemis—operate with the same principles of intake, compression, power and exhaust, all carburetors operate similarly when you compare them.

Regardless of how many venturis (barrels) your carburetor has, let's keep things really simple by starting out with a one-barrel carburetor, just as we would learn about engines by examining what goes on in a single cylinder. After you have become completely familiar with the one barrel—then we'll proceed to the two- and four-barrel units. Don't jump ahead because you have to learn what goes on in one barrel before we start talking about more complex carburetors which are literally several one-barrel carburetors built into a single unit.

The carburetor is constructed to do several jobs:
1. It controls engine input and therefore controls the power output.

**Knowing how it works is the key to fixing it**—This section explains in simple terms how each system of a carburetor operates. It will help you to analyze engine malfunctions too often blamed on the carburetor . . . and to pinpoint the causes and make needed fixes.

There is test/analysis equipment readily available for scrutinizing the electrical components. You can get the distributor's mechanical and vacuum-advance operation checked on a machine. The fuel pump and restrictions of fuel flow up to the carburetor can easily be checked with inexpensive equipment and so on.

Beyond a visible leak, how do you check a carburetor? The manufacturer uses flow equipment, but you will probably never have access to $100,000—$200,000 flow equipment, so you must go another route. Combustion analyzers are a definite guide but they have their own built-in limitations. Ofttimes the repairman must rely on experience based on four of his natural senses: touch, sight, smell and hearing. All too often, if your experience is limited, you end up buying a replacement carburetor that you don't need at all. Put your senses to work and study these systems to know what you should be aware of.

2. It mixes air and fuel in the correct proportions for engine operation.
3. It vaporizes the air/fuel mixture to put it in a homogeneous state for combustion.

To satisfy the range of requirements of engines and drivers, several different systems of operation must be employed. We will use a RPD single-bore carburetor for an example as it is the simplest and yet represents the typical modern carburetor used in current automotive industry. In another chapter we explain the systems of a Q-jet carburetor. The several different systems used in the carburetor are: inlet (float), idle, part-throttle, power, accelerator pump and choke.

Let's look at these basic systems one at a time and see what parts do each job. These show the relationships involved and how they all work together. We will look at the inlet system first.

## Float System

**Description of Operation**—The float (inlet) system consists of three major items:
1. Fuel bowl
2. Float
3. Inlet valve (needle and seat).

Fuel for the basic metering systems of the carburetor is stored in the fuel bowl. The fuel-inlet system must maintain the specified fuel level as the basic fuel-metering systems are calibrated to deliver the correct mixture only when the fuel is at this level. The correct fuel level also greatly affects fuel handling, which is the carburetor's ability to withstand maneuvering: quick accelerations, turns, and stops.

The amount of fuel entering the bowl through the fuel-inlet valve is determined by the space (flow area) between the movable needle and its seat and by fuel-pump pressure. Movement of the needle in relation to its seat is controlled by the float which rises and falls with fuel level. As fuel level drops, the float drops, opening the needle valve to allow fuel to enter the bowl. When the fuel reaches a specified level, the float moves the needle to a position where it restricts the flow of fuel, admitting only enough fuel to replace that which is being used. Any slight change in the fuel level causes a corresponding movement of the float, opening or closing the fuel-inlet valve to restore or maintain the correct fuel level.

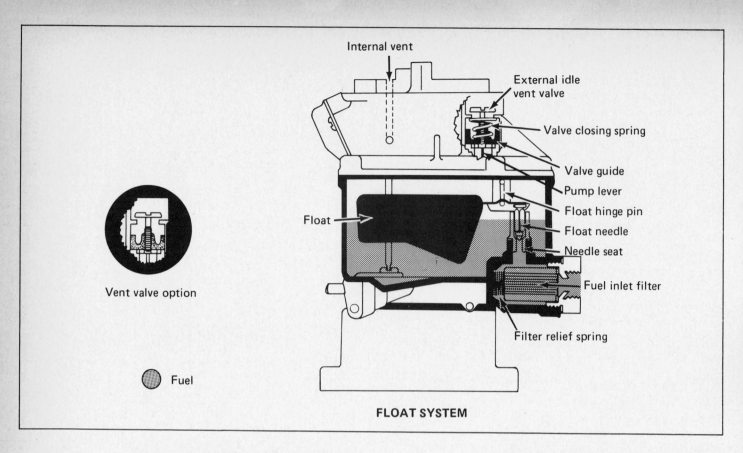

Internal vent

External idle vent valve

Valve closing spring

Valve guide

Pump lever

Float hinge pin

Float needle

Needle seat

Fuel inlet filter

Filter relief spring

Float

Vent valve option

● Fuel

**FLOAT SYSTEM**

## EFFECTS OF INLET SYSTEM CHANGES

| CHANGE | EFFECT |
|---|---|
| High float level | Raises fuel level in bowl. Speeds up main system "start up" because less depression is required at venturi to start fuel "pullover." Increases fuel consumption. May cause fuel spillage through discharge nozzles and/or vent into carbure- tor air inlet on quick stops or turns—causing engine to run erratically or stall. Increases tendency for the carburetor to "percolate" or to "boil over"—a condition in which fuel is pushed by rising vapor bubbles out of the discharge from the main well in a hot-soak situation. Also, when the vehicle is parked on a hill or side slope, high float level may cause spillage through the vent or main system at an inclination angle less than the design specification. |
| Low float level | Lowers fuel level in bowl. Delays main-system "start-up" because more depression is required at venturi to start fuel "pull-over." This delay may cause flat spots or "holes." May expose main jets in hard maneuvering, causing "turn cutout." Can lessen maximum fuel flow capability. |
| Bumper spring (float assist spring) too strong (too heavy) | Causes low fuel level. Inlet valve closes prematurely because of added closing force of closing spring. |
| Bumper spring too weak (too light) | Causes high fuel level. Inlet valve closes after correct fuel level has been reached because additional float displacement is required to compensate for weaker bumper spring force. |
| High fuel pressure | Will raise fuel level approximately 0.020-inch per psi fuel- pressure increase. Of course this varies to some degree with buoyancy, leverage, etc. |
| Low fuel pressure | Will lower fuel level. |
| Larger inlet-valve seat | Will raise fuel level. |
| Smaller inlet-valve seat | Will lower fuel level. |

NOTE: Changing almost anything in the fuel-inlet system makes it necessary to reset the float to obtain the correct fuel level.

## DESIGN FEATURES/Fuel Inlet System

**Fuel bowl**—The fuel bowl or float chamber is a reservoir which supplies all fuel to the idle, main, pump and power systems at a vented bowl pressure and height. Vented bowl (reference) pressure and height means that the main, power, idle and accelerating-pump systems are all calibrated to provide correct mixture when the fuel level is at this height and bowl-vent pressure (not fuel-pump pressure).

The bowl also acts as a vapor separator. A vent in the bowl con-nects to the inlet air horn. Vapors which may have been entrapped in the fuel as it was pumped from the tank escape through this vent so only very minor pressure build-up can exist in the bowl. On pre-emission-control models there was often a mechani-cal vent valve on the float bowl. At curb idle or when the engine was stopped, this external vent released fuel vapors into the engine compartment. The vent maintains constant near-atmospheric pressure on the fuel bowl. With this system restricted air cleaners (dirty) have a significant effect on metering. By connecting the vent to the inlet air horn, the fuel bowl is vented to clean air which is obtained through the air cleaner. And, because the vent "sees" the same pressure as the inlet, air bleeds and idle system, the mixture-changing effects of a dirty air cleaner are largely eliminated. This is provided no external venting accompanies it. Dirty air cleaners restrict the air flow, creating a lower absolute pressure to the carburetor, which causes a power loss. If the fuel bowl were not properly vented to the inlet air horn, a dirty air cleaner would richen the mixture.

The preferred location for a vent is near the center of the fuel bowl—high enough so that fuel will not slosh into the air horn during hard stops or maneuvering. Additionally, some 1970 and later cars have a vent which connects the float bowl to a charcoal canister when the engine is turned off. This part of the emission-control system collects escaping fuel vapors which are generated when heat from the engine block heats the carburetor, causing gasoline in the fuel bowl to "boil" off. The same canister also collects fuel-tank vapors via plumbing from tank vents. Vapors collected in the charcoal canister are sucked back into the intake manifold through a system of valves when the engine is running.

The vent to the vapor canister can be vacuum operated, operated by vapor pressure generated within the bowl, and/or mechanically actuated at off-idle.

On some pre-emission-control carburetors a bimetal actuated vent released fuel vapors to the atmosphere when temperature exceeded 75°F.

The fuel bowl is an integral part of the main carburetor body casting.

Bowl size or capacity is important. It must contain sufficient reserve fuel to allow good response when accelerating from stop after a hot engine has been idling or stopped. Hot fuel from the fuel pump may be delivered intermittently as spurts of liquid fuel and pockets of fuel vapor. Thus the bowl must be large or designed to provide all fuel for the metering systems until the pump is purged and able to deliver liquid. Rochester Monojet and 2G carburetors have a good fuel-bowl capacity for their air-fuel delivery capabilities. The Quadrajet has a small bowl for its size but its shape and location combined with excellent metering orifice location causes it to be exceptional in nearly all areas of performance.

**Float**—A float on a hinged lever operates the inlet valve so that fuel enters the bowl

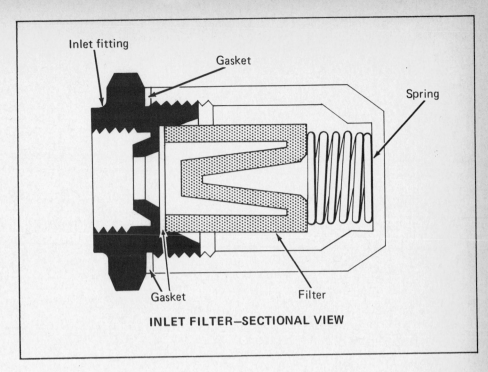

INLET FILTER—SECTIONAL VIEW

Various fuel-inlet nuts to house different filters, work with carb body designs and to accommodate 5/16 and 3/8-inch fuel lines. All of these are for the Q-jet.

when the fuel level is below the desired reference height—and shuts off the fuel when the desired height is reached. The float may have either one or two buoyant elements, sometimes referred to as "lungs" because their shape is sometimes similar to that of the human lung. "Pontoon" is another fairly common name. Over the past years floats were made from thin brass stampings soldered together into an airtight assembly. Today most floats are made of a closed-cellular material which is not affected by gasoline, alcohol or any other of the other commonly used fuels or fuel additives. The float is designed with adequate buoyancy so that fuel will be positively shut off by the inlet valve when the desired fuel level has been reached. The buoyant portion is usually mounted on a lever to multiply the buoyancy effects of the float itself and provide a surface for operating the inlet valve.

    A float-assist spring is sometimes added to the float-arm pivot pin to minimize vibration of the float. Float vibration, caused by engine or vehicle vibration or bouncing—or both—can cause wide fuel-level variations because a "jumping around" float allows the inlet valve to admit fuel when it is not needed. This type float assist can be most helpful in preventing pronounced and unneeded float drop (opening) during short-turn/acceleration maneuvers. Some carburetors may use a tiny spring inside the inlet valve itself instead of—or in addition to—a spring under the float. The Corvair H carburetor utilized both damping methods in some of its model years. Such springs can be especially helpful in dirt-track, off-road and marine applications.

    Float shape and mounting (pivot orientation) may be dictated by bowl size and the use of the carburetor. The float has to provide enough buoyancy to close the inlet valve and this sometimes requires making a very odd shape to get enough float volume to provide the required buoyancy.

    Float-lever length determines the mechanical advantage which the buoyancy of the float (and float spring, if used) can apply to close the inlet valve at a given fuel level.

    Every carburetor has a published float setting. This specification is established by the design engineers to take into account the fuel pressure, float buoy-

Arrow points to pull clip connecting float to inlet needle for positive opening when float drops.

Why a needle-to-float clip? If a vehicle or engine sits relatively long periods of time an accumulation of deposits may form in the fuel system. These deposits are quite heavy and sticky with a molasses-like consistency. In their first stages, these are known as gum deposits.

    Cases have been known where heavy gum concentration in certain fuels formed enough harmful deposits to hang up inlet valves so they could not open and close with the action of the float. Most pump fuels today are carefully guarded against incorrect blending because they are made according to standards set forth by organizations such as the American Society of Testing Materials. All fuels leave some deposits. Be sure the clip is reinstalled so it can do its job.

These Q-jet floats show some of the variety of shapes and sizes which are required to meet various needs. A closed-cellular plastic material is typically used in today's carburetors.

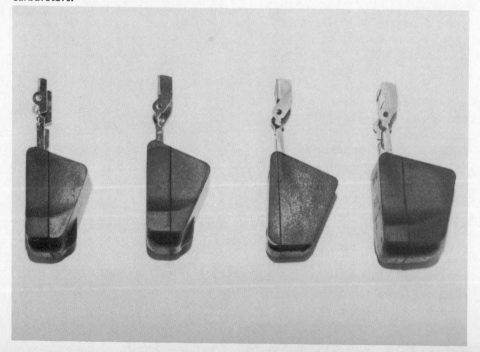

ancy, and float-lever effects of float- or needle-damper springs, seat size and the required fuel level which must be maintained in the fuel bowl. The float setting is the float location which closes the inlet valve when the required fuel level in the bowl is reached. This is referred to as the "mechanical setting" as a measuring instrument is used to determine its position.

A fuel level for a particular carburetor is established by the designer and test engineer so the carburetor will operate without problems in fast starts and stops and in maneuvers which would ordinarily be encountered in the particular vehicle for which the carburetor was made. And the level is set so that there will be no fuel spillage when a passenger car is parked or operated facing up, down, or sideways on a hill with approximately a 30% grade.

For best operation in high-speed cornering, bowls are equipped with center-pivoted floats with the pivot axis parallel with the axles of the car. For best resistance to the effects of acceleration and braking, the side-hung floats with pivots perpendicular to the axles have a slight advantage. Rochester Q-jets are center-pivot type, as are the 4GC's. Two-barrels and Monojets are side-hung.

**Inlet valve**—The rounded end of this valve rests against or is connected to the float-lever arm with a pull clip. The tapered end of the inlet needle closes against an inlet-valve seat as the float rises.

As mentioned in the float section, some inlet valves are hollow to include a tiny damper spring and pin to help cushion the needle valve so that it is protected from road shocks and vehicle vibrations.

The steel inlet valve needle in today's carburetors has a tapered seating surface tipped with Viton. Viton-tipped needles are extremely resistant to dirt and conform to the seat for good sealing with low closing forces.

Inlet valves are supplied with various seat openings. They are not marked . . . you have to measure the inlet orifice with plug gages or the shank of drill bits to discover which size you are looking at. Unless you have all number and fractional drills in the 0.060" to 0.140" range, you may not be able to determine the size with any degree of accuracy.

The seat diameter

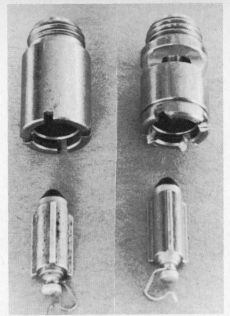

Two types of inlet valves. Engineers select the type which brings fuel into the bowl in the best relation to the vents and/ or metering orifices. Non-windowed type on left flows more fuel in a Q-jet than the other kind. Note pull clips on each needle. These are 0.135-inch Buick needle/seat assemblies.

and orifice length determine how much fuel flow occurs at a given fuel pressure. A smaller opening flows less fuel—a larger opening, more.

Typically, seat size is selected to allow reasonably quick filling of the bowl to handle quick accelerations after a hot-soak condition and to give tolerable restriction for high fuel demands such as occur at wide-open throttle at high RPM. Larger seats are also used to provide better purging of vapor from the fuel lines. A small needle seat provides the best control of hot fuel because vapor pressure acts against less area to force the needle off its seat. Any change in seat size must be accompanied by resetting the fuel level.

A fuel filter or screen may be found in the carburetor body or fuel-bowl cover as part of the fuel-inlet system. The filtering device is placed between the fuel pump and the inlet valve to trap dirt which could cause inlet valve seating problems. This filter is installed with a pressure-relief spring behind. If the filter becomes clogged from excessive amounts of dirt, the relief spring lets the filter move off its seat. This ensures fuel flow to the carburetor, even if the fil-

ter is clogged. Fuel filters for use in fuel lines are discussed in detail in another section of this book.

**Bowl Stuffers and Baffles**—Changes in the motion of the car, such as turning a corner or braking cause the fuel to rush to the side, front or rear of the bowl. Abrupt changes may move the fuel so that the main jets are uncovered and fuel may even splash out of the internal vents into the air horn. Three ways are used to help this condition. Bowl baffles are sometimes inserted into the bowl to reduce the fuel-slosh area. In the case of the 2G carburetors, a baffle is sometimes placed alongside the pump well. Semi-solid air-horn gaskets are another fix for the problem. These may be solid in the area below vent openings to serve as a baffle and prevent fuel from splashing out of the vents. Bowl stuffers (inserts) are also used to reduce bowl volume and to help control fuel splash in the bowl. Such stuffers are also helpful in reducing emissions.

**SUMMARY**

Before leaving the inlet system, we should list the items which can affect fuel level:

1. Inlet-pressure changes (fuel-pump output pressure)
2. Float-assist spring
3. Inlet-seat size
4. Float density (buoyancy)
5. Float size
6. Float-lever length and pivot-pin location relative to the inlet-needle.
7. Fuel density (density of gasoline remains fairly constant, so this is not usually a problem).

---

**Sour gas**—A common problem of yester-year, it seldom occurs with modern-day gasolines. We should mention what it is and what it does. Sour gas is caused by a chemical reaction of gasoline with brass parts. When it occurs, it is immediately obvious when the fuel bowl is taken off or uncovered. The gasoline stinks! Sour gas attacks Viton needle tips, gaskets and synthetic compounds used in pump cups. It also attacks the solder used in brass-float assembly and in some cases the cellular composition floats. Additives in gasoline sold nowadays usually prevent or control the problem.

---

# Main System

The main-metering system supplies air/fuel mixture to the engine for cruising speeds and above. During the time that engine speed or air flow is increasing to a point where the main-metering system begins to operate, fuel is fed by the idle and accelerator-pump systems, which are described separately. Under conditions of high load, when the engine must produce full power, added fuel comes from the power system, described in another section.

Many people believe that the throttle controls the volume of air/fuel mixture which is being pumped into the engine. This is not the case. Piston displacement never changes, so the volume of air pulled into the engine is constant for any given speed and the mixture volume which comes into the engine is always the same. The throttle controls the density or mass flow of the air pumped into the engine by the action of the pistons: least density of charge is available at idle, highest density is at wide-open throttle. A dense charge has more air mass, hence higher compression and burning pressures can be developed for higher power output. Simply stated, the piston displaces the same volume during each intake stroke. When the throttle is nearly closed, the piston pulls a thinner charge (less dense). In its attempt to inhale volume equal to its displacement, manifold vacuum is increased. Thus, the throttle controls engine speed and power output by varying the charge density supplied to the engine.

Now that we've discussed the throttle, let's proceed to the heart of the carburetor and the key to its simplicity: the venturi. It is the one part that really makes a carburetor function, so it is important to understand how the venturi operates before getting deeper into how the main-metering system works.

The venturi and its principle of operation are named after G. B. Venturi, an Italian physicist (1746-1822) who discovered that when air flows through a constricted tube, flow is fastest and the pressure lowest at the point of maximum constriction. In the internal-combustion engine, a partial vacuum is created in the cylinder by the downward strokes of the pistons. Because atmospheric pressure is nominally 14.7 pounds per square inch, air rushes through the carburetor and into the cylinder to fill the vacuum. On its way to the cylinders, the air passes through the venturi.

The venturi is a smooth-surfaced restriction in the path of the incoming air. It "necks down" the inrushing air column, then allows it to widen back to the throttle-bore diameter. Air is rushing in with a certain pressure. To get through the necked-down area (venturi), it must speed up, reducing the pressure inside the venturi. A gentle diverging section is used to recover as much of the pressure as possible.

The venturi is the controlling factor on the carburetor because fuel discharges into the venturi at the point of lowest pressure (greatest vacuum). This minimum-pressure point applies a "signal" to the main-metering system.

The pressure drop or vacuum "signal" is measured at the discharge nozzle in the venturi. Because the fuel bowl is maintained at near-atmospheric pressure by the vent system, fuel flows through the main jet and into the low-pressure or vacuum area in the venturi.

Pressure-drop (vacuum) at the venturi varies with engine speed and throttle position, increasing with engine RPM. Wide-open throttle and peak RPM give the highest flow and the highest pressure difference between the fuel bowl and a discharge nozzle in the venturi, thus the highest fuel flow into the engine. The pressure difference ($\triangle P$) as engine speed changes is approximately proportional to the difference in velocity squared ($\triangle V^2$), $\triangle P \sim V^2$.

The pressure drop in the venturi also depends on the size of the venturi itself. A small venturi provides a higher pressure drop at any given RPM and throttle opening than a large venturi will provide. The Design Features portion of this section tells you more about this important consideration and its relation to performance.

No fuel issues from the discharge nozzle until flow through the venturi and hence, pressure drop, is sufficient to offset the level or "head" difference between the spill-over

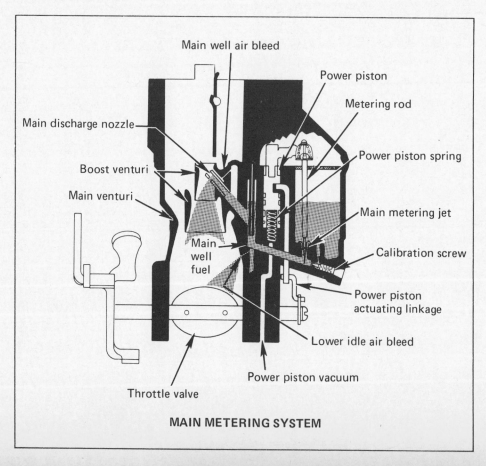

**MAIN METERING SYSTEM**

- Main well air bleed
- Power piston
- Metering rod
- Main discharge nozzle
- Power piston spring
- Boost venturi
- Main venturi
- Power piston
- Main metering jet
- Main well fuel
- Calibration screw
- Power piston actuating linkage
- Lower idle air bleed
- Throttle valve
- Power piston vacuum

**Simplifying the venturi idea**—The carburetor uses the venturi theory to maintain the correct air/fuel ratio. A venturi is merely a calibrated orifice or restriction which increases the speed and lowers the pressure of air flowing past it. The driver controls air flow by varying throttle position. The low pressure (vacuum) which exists at the venturi is determined by this flow of air— this "tells the carburetor" what air flow exists. At the same time, this vacuum acts on a nozzle connected to the fuel in the carburetor and "sucks" fuel from the nozzle. Thus varying air flows cause varying amounts of suction therefore varying fuel flow according to air flow. Over a given range the ratio or rate of air flow to the rate of fuel flow is a constant.

Because gasoline is much denser than air, more suction is required to draw gasoline through the carburetor than is required to draw the air. The venturi multiples the suction applied to the gasoline.

To multiply suction without sacrificing maximum air flow, a secondary or "booster" venturi is used. The nozzle is located in this secondary venturi with its high suction. When you look at a venturi you see a leading edge shaped similarly to the top-front edge of an aircraft wing. Air increases speed as it rushes over this curved surface creating the most depression at the highest point. The depression causes lift allowing the aircraft to fly. In the case of the venturi, this depression at the smallest diameter also causes lift or suction so fuel is pulled up and out of the bowl via the nozzle.

---

These drawings show typical vacuums inside a carburetor air section in throttled- and wide-open-throttle conditions. If we were to tap into the carburetor at critical points and connect our taps to vacuum gages, this is what we would see.

Note that the inlet shows a very slight vacuum representing only the drop across the air cleaner.

The gage downstream from the large venturi throat shows a higher vacuum because the air is still at a relatively high velocity. The vacuum returns almost to the inlet value just before the throttle plate. The throttled carburetor shows a very high manifold vacuum because there is a large pressure drop across the partially opened throttle. The wide-open-throttle carburetor has a low manifold vacuum indicating a heavy-load/dense-charge situation. In both cases, the highest carburetor vacuum is at the boost-venturi throat. This area supplies the signal to the main system.

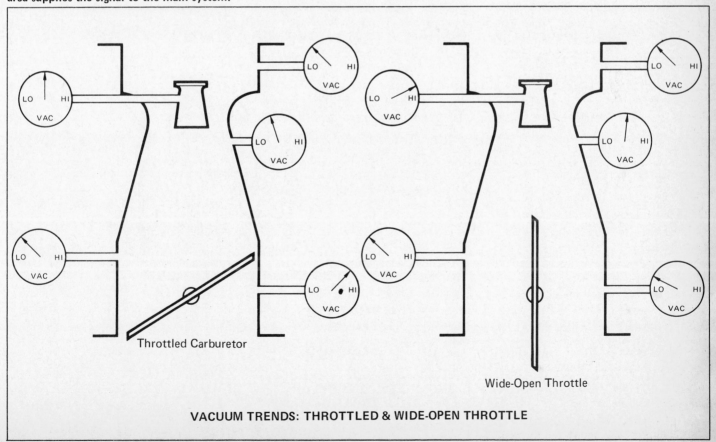

Throttled Carburetor

Wide-Open Throttle

**VACUUM TRENDS: THROTTLED & WIDE-OPEN THROTTLE**

point and the lower level of fuel in the bowl.

Once main-system flow is started, fuel is metered (measured) through a main jet in the fuel bowl. From the main jet, fuel passes into a main well. As fuel passes up through this main well, air from a main air or "high-speed" bleed is added to pre-atomize or emulsify the fuel into a light, frothy air/fuel mixture which issues from the discharge nozzle into the air stream flowing through the venturi. The discharge nozzle is often located in a small boost venturi centered in the main venturi (described with other Main System features).

There are two main reasons why the liquid fuel is converted into an air/fuel emulsion. First, it vaporizes much easier when it is discharged into the air flowing through the venturi. Second, the emulsion has a lighter viscosity than liquid fuel and responds faster to any change in the venturi vacuum (signal from the venturi applied through the discharge nozzle). It will start to flow sooner and quicker than purely liquid fuel.

The strong signal from the discharge nozzle is bled off or reduced by the main air bleed so that there is less effective pressure difference to cause fuel flow. The mixture will become leaner as the size of the bleed is increased. Decreasing the bleed size increases the pressure drop across the main jet to pull more fuel through the main system, giving a richer mixture. Main-air-bleed changes affect the entire range of main-metering-system operation. RPD establishes the main-air-bleed size for each carburetor to work correctly over that carburetor's air-flow range. Changes in main-air-bleed sizes are rarely necessary or advisable in field service. This is a valuable calibration aid in original calibration work. Calibration changes are easy to make by changing main metering jets.

The main air bleed also acts as an anti-siphon or siphon-breaker so that fuel does not continue to dribble into the venturi after air flow is reduced or stopped.

**DESIGN FEATURES/Main System**
**Throttle**—Because we have talked about capacity, air density and air through the venturi, let's say a few words about the simple control valves referred to as the throttle blades and shaft.

The throttle shaft is offset slightly—about 0.020-inch on pri-

maries—about 0.060 on secondaries—in the throttle bore so that one side of the throttle has a larger area which tends to cause self-closing. There are two reasons for this feature. First, idle-return consistency is greatly aided by the sizable closing force which is generated when the manifold vacuum is high—as at idle. Second, it is a safety measure to guard against overspeeding the engine if it is started without installing the linkage or throttle-return spring because there will be positive closing of the throttles.

Primary throttles are seldom closed against the throttle bore; instead, they are factory-set against a stop to provide a closed-throttle air flow which has been specified for that particular carburetor model.

**Venturi**—A few comments about venturi design. The most efficient (or ideal) venturi which creates the maximum pressure drop with the minimum flow loss requires a 20° entry angle and a diverging section with a 7° to 11° included angle on the "tail." The designers try to keep as close as possible to the "ideal" venturi entry and exit angles.

Although the theoretical low-pressure point and point of highest velocity would be expected at the minimum diameter of the venturi, friction causes the point to occur about 0.030-inch below the smallest diameter. This low-pressure, high-velocity point is called the "vena contracta." The center line of the discharge nozzle or the "tail" of the boost venturi is located at this point.

A venturi allows a much greater metering signal than a straight tube and has a minimum loss of air pressure due to friction losses. This is because the venturi's trailing edge conforms to the normal air stream and allows the mixture to recover the venturi pressure-drop while also slowing down to minimize friction losses. This pressure recovery would be much less if a flat-plate orifice were used. The venturi is a very efficient air-measuring device because of the high signal levels it provides in conjunction with minimum loss of pressure.

The venturi controls other things in addition to providing a way to supply fuel in the correct proportion to the mass of air rushing into the engine. Its size, as mentioned in the description of how it works, affects the pressure drop available to operate the main-metering system. The smaller the venturi, the greater

Top view of one primary Q-jet bore shows two booster venturis in a cluster or stack to increase signal for improved metering at low air flows.

**Throttle travel**—The usual throttle travel to get to the full-open 90° position is about 75° because the throttle is set at about a 15° angle at curb idle.

Cutaway of Q-jet primary bore showing how main fuel discharges into center of smallest booster venturi. Arrows indicate two openings in fuel nozzle.

the pressure drop, the sooner the main system will be brought into operation, and the better the mixing of fuel with air will be. The RPM at which main-system "spill-over" or "pull-over" starts is affected by demand of the engine which is pulling air through the carburetor.

To say this in a more practical way: if you keep engine size constant, a larger carburetor will require a higher RPM to bring the main system into operation; a smaller carburetor—a lower RPM. But, the size of the venturi also controls the quantity of air available at wide-open throttle for engine operation. If the venturi is too small, top-end HP will be reduced, even though the carburetor provides very good air/fuel mixing (vaporization) at all speeds. For this reason automobile manufacturers typically compromise—by using carburetors with smaller than optimum (maximum power) air-flow capacities. They do this for very good reasons, even if you—a performance enthusiast—may feel cheated. Good fuel vaporization promotes good distribution and improves smooth running and economy at around-town and cruising speeds.

Manufacturers offering more performance typically supply a carburetor with a secondary system which retains the advantages of a small primary venturi with the capability of higher air flow for top-end power.

**Boost Venturi**—Boost venturis act exactly the same as the larger venturi, but supply a stronger signal to the discharge nozzle because boost-venturi velocity is higher than that in the main venturi.

By increasing the available signal for main-system operation, the boost venturi allows the carburetor to work well at lower speeds (and therefore lower air flows) than would otherwise be the case. This is especially helpful in a performance-type carburetor because the boost venturi does not seriously affect the air-flow capacity of the carburetor.

The tail of the boost venturi discharges at the low-pressure point (vena contracta) in the main venturi. Thus, the boost-venturi air flow is accelerated to a higher velocity because it "sees" a greater pressure differential than the main venturi.

The stronger signal and higher velocity work hand in hand with the additional air-shearing at the boost-venturi tail to promote vaporization of the air/fuel emulsion.

Boost venturis also aid fuel distribution because the ring of air which flows between the two venturis directs the charge to the center of the air stream, helping keep some of the wet air/fuel mixture off the carburetor wall below the venturi. In so doing more of the air/fuel mixture reaches the manifold hot spot for improved vaporization. In addition, tabs, bars, wings and other devices are sometimes needed to provide correct directional effects for good cylinder-to-cylinder distribution with a particular manifold. All such development work is accomplished on the dynamometer during design and development of the carburetor/manifold combination.

Boost venturis allow using a much shorter main venturi so that the carburetor can be made short enough to fit under the hood of an automobile. Carburetor designers could achieve essentially the same results with a long (ideal) venturi as they can with one or more boost venturis "stacked" in the main venturi, but it is very tough to build a carburetor which allows the 20° entry to the venturi and a 7° to 11° "tail," and at the same time get venturi size down to the required diameter for adequate signal. Some RPD carburetors use as many as two boost venturis to get adequate signal for main-system operation.

**Main Jets**—These metering orifices are used to control fuel flow into the metering system. They are rated in flow capacity and are removable for carburetor-calibration purposes. It is commonly thought that main jets are selected solely on a trial-and-error basis, but this is not the case. The accompanying material clearly shows the relationship of main-jet size to venturi size. For a given venturi size, a small range of main jets will cover all conditions. However, the final selection of the correct jet for the application will have to be done by testing because design and operational variations (climate, altitude and temperature) affect jet-size requirements.

There is a basic misconception about jets—that size alone determines their flow characteristics. This is not the case because the shape of the jet entry and exit, length of orifice—as well as the finish—affect flow. RPD checks each jet on a flow tester and grades it according to flow. This is a production-line calibration tool used to assure the customer uniform air/fuel ratios. The tolerance range used during initial build for each size explains why a 69 jet found in a carburetor may not seem to give a richer mixture than a 68. If the 68 is on the "high" side of its allowable tolerance and the 69 is on its "low" side tolerance limit, the two jets will flow very close to the same amount of fuel. The tolerance range for a 68 is from 382 to 398

**Three types of main jets for Rochester carbs (except Q-jet, Monojet). With same orifice size, each jet flows a slightly different amount of fuel. Square entry (left) allows good flow at low end; acts as a larger orifice to allow easy nozzle start at low velocity. This design tends to reduce fuel at high flows and engineers use it to tailor fuel flow for specific applications. Circled digits in part number identify jet configuration.**

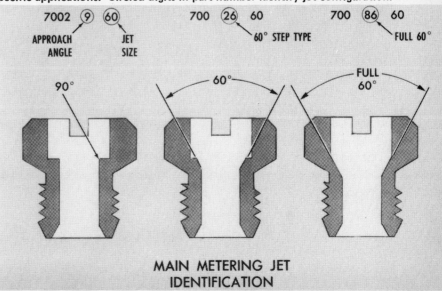

MAIN METERING JET
IDENTIFICATION

cubic centimeters per minute at a specified head with a given test fuel. The 69 ranges from 394 to 410cc.

Service jets will have a mean flow rate (average, in middle of tolerance range). A 69, for instance, will have approximately 402cc flow. For fine tuning it is best to have sets of jets of the same origin, preferably RPD service jets, as these will all have the mean flow characteristic rather than being toward a high or low tolerance. Don't try to tune with one jet from a carburetor, another from your dealer and still others from a specialty shop. There are plenty of variables in the tuning process without introducing unneeded ones!

A plus for tuners lies in the fact that Rochester supplies only the "mean" or average flow jets for service replacements. Therefore, when you select a 72 jet to replace a 71, you can be sure it will flow more fuel than a replacement jet. Jets used in RPD carburetors supplied to the various car makers may vary due to flow differences, but the carburetor has been checked on a flow stand to make sure that its mixture characteristics fall within the allowable tolerances. This means that a 71 jet in a new carburetor or one supplied on your automobile, truck or marine engine could flow about the same as a 70, or a 72. The only way to get around this problem is to use new service replacement jets so you will know what base you are working from.

Drilling out jets to change their size is never recommended because this always destroys the entry and exit features to a certain degree, and may introduce a swirl pattern, even if the drill is held in a pin vise and turned by hand. You cannot be sure of the flow characteristics of a jet that you modify by drilling—unless you can get that jet back on a flow machine to compare it with a standard jet.

**Main Air Bleed**—All Rochester carburetors are equipped with built-in fixed-dimension air bleeds.

# Power System

When the engine is called upon to produce power in excess of normal cruising requirements, the carburetor has to provide a richer mixture, as discussed in the Engine-Requirements chapter. Added fuel for power operation is supplied by the power system under control of manifold vacuum. Manifold vacuum accurately indicates the load on the engine. It is usually strongest at idle. As load on the engine increases, the throttle valve must be opened wider to maintain a given speed, thereby offering less restriction to the air entering the intake manifold and reducing the manifold vacuum.

A vacuum passage in the carburetor applies manifold vacuum to a power-valve piston. At idle or normal-cruising-load conditions, manifold vacuum acting against a spring holds the valve closed. As high power demands load the engine, and manifold vacuum drops below a pre-set point, usually at about 7 inches of mercury (Hg), the power-valve spring overcomes manifold vacuum and starts to open the power valve. Most

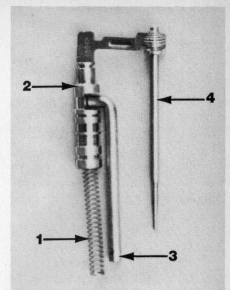

Monojet power metering-rod assembly shows power spring 1, power piston 2, actuating rod 3 is a mechanical override which works with a cam on the throttle lever to ensure power mixture at or near wide-open throttle, and metering rod 4 which centers in the main metering jet.

**Air velocity demand-enrichment system—** Vega Monojet, Vega 2G and some Corvair H carburetors have a power system which operates as follows. When demand for fuel at the boost venturi exceeds the main-metering jet's capacity (thereby creating additional depression in the main fuel channel to the nozzle), this depression lifts the power-enrichment valve so fuel can flow directly from the bowl to the main well and then to the boost venturi via the nozzle. This fuel by-passes the main jet. The point at which the extra fuel is added is controllable by the weight of the valve. A heavier valve unseats at higher air flows so that the extra fuel is supplied at higher RPM. So, if you would like to move the power fuel to a higher RPM, simply add weight to the needle. This is done in 1/4-gram increments by factory tuners.

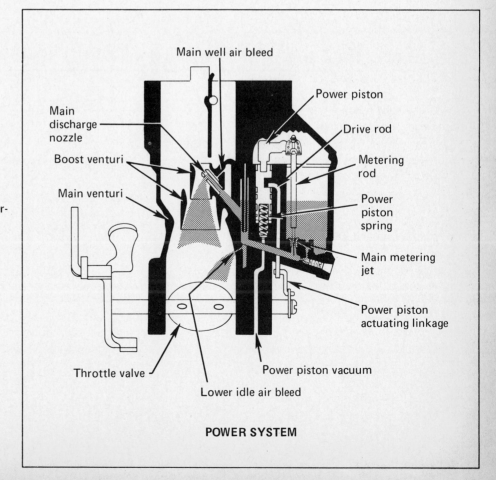

**POWER SYSTEM**

RPD power systems are staged to start opening at one vacuum, then progressively open to a full-flow position as manifold vacuum decreases to 3–5 inches Hg vacuum. In most 2G carburetors, the fuel flows through the power valve and through a power-valve restriction to join fuel from the main jet, thereby richening the mixture. The power-enrichment in a Monojet and Quadrajet is caused when the metering rod is lifted to decrease the diameter of the rod in the main jet. This occurs as the rod is lifted out of the main jet by power-piston/spring operation.

The power valve can be considered as a switch which turns on extra fuel to change from an economical cruising air/fuel mixture to a power mixture. The power valve itself is merely a gate operated by manifold vacuum and spring pressure. It is designed to operate at a given load.

When engine power demands are reduced, increasing manifold vacuum acts on the power piston to overcome the spring tension, closing the power valve and shutting off the added fuel supply.

The power-valve opening point is another variable which the carburetor designer uses to arrive at the best compromise between economy, exhaust emissions, drivability and performance.

Racing engines often have wild manifold-vacuum fluctuations at idle and low speeds. These vacuums can go low enough to cause the power valve to begin feeding additional fuel. It is important to adjust the power valve so that it will not open and close in response to these variations caused by valve timing instead of throttle position or engine load. The power valve must be altered so that it will still function, but at a lower vacuum than occurs at idle, as explained in chapter fourteen.

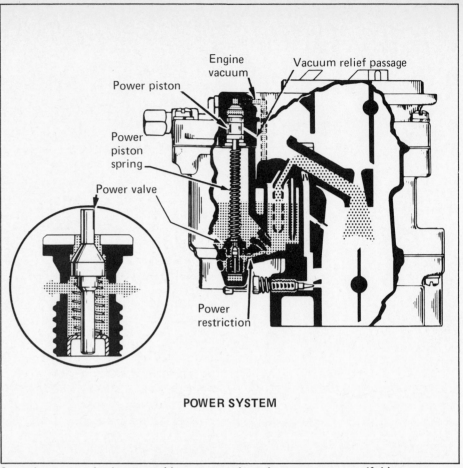

**POWER SYSTEM**

Screw-in power valve is actuated by a power piston in response to manifold vacuum. As the engine is loaded, manifold vacuum drops, allowing power-piston spring to open power valve. Power system fuel supplements that supplied by main jets.

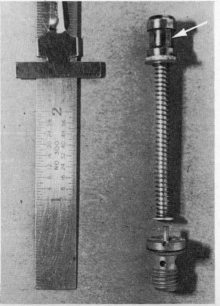

2G power system uses separate power valve (at bottom of photo) with vacuum-piston (arrow) actuated by manifold vacuum. When engine power demands more fuel, the spring on the piston stem overcomes the lowered manifold vacuum and the end of the piston pushes the center stem of the power valve to allow power fuel flow.

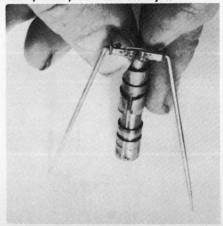

Monojet uses one power piston and one metering rod in the main jet. This Q-jet assembly has two metering rods, one for each primary bore and main jet.

# Accelerator Pump System

The accelerator pump has three functions:
1. To make up for the fuel that condenses onto the manifold surfaces when the throttle is opened suddenly.
2. To make up for the lag in fuel delivery when the throttle is opened suddenly, allowing more air to rush in without sufficient nozzle fuel.
3. To act as a mechanical injection system to supply fuel before main system starts.

As the throttle is opened quickly, the intake manifold vacuum instantly drops, moving the pressure towards atmospheric. A high manifold vacuum tends to keep the mixture well vaporized. As the pressure rises toward atmospheric, fuel drops out of the vapor, condensing into puddles and wet spots of liquid on the walls and floor of the intake manifold. Thus, the mixture which is available for the cylinders is instantly leaned out and the engine will hesitate or stumble unless more fuel is immediately added to replace that which has been lost to the manifold surfaces. This is especially important with big-port manifolds or manifolds with large plenum areas because these have more surface area onto which fuel can condense.

Making up for condensed fuel loss to the manifold and taking care of fuel-delivery lagging behind increasing air flow are both important, but the relative importance of the two has not been established.

A third function of the accelerator-pump system is supplying fuel when the throttle is quickly opened past the point where the idle-transfer system would have supplied fuel until the main system could begin its normal operation. In this case, the accelerator pump supplies the required fuel until the main system starts flowing. During the low-flow, low-vacuum period the accelerator pump injects fuel under pressure into the throttle bore. The duration of accelerator-pump operation must be carefully engineered to provide a "cover-up" of sufficient length to allow main-system flow to be established so that good vaporization will be ensured and correct air/fuel ratio reestablished.

The accelerator pump operates when the pump-operating lever is actuated by throttle movement. As the throttle opens, the pump linkage operates a pump plunger. Pressure in the pump forces the pump-inlet ball or floating cup onto its seat so that fuel will not escape from the pump into the fuel bowl. Pressure also raises the discharge needle or ball off its seat so that fuel is discharged through a "shooter" into the ven-

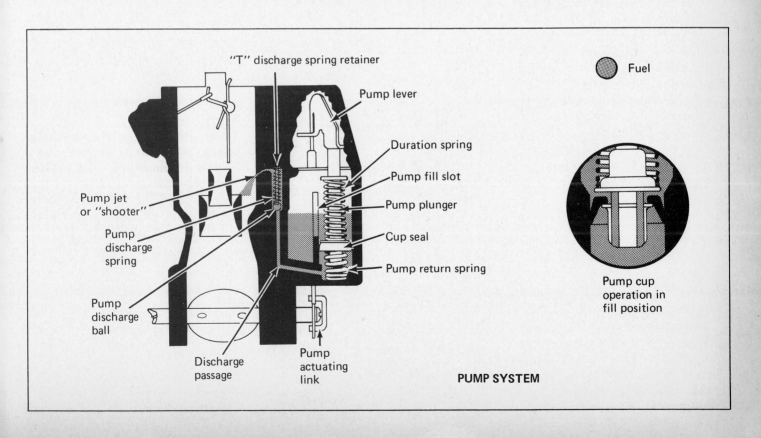

"T" discharge spring retainer

Pump lever

Duration spring

Pump fill slot

Pump plunger

Cup seal

Pump return spring

Pump jet or "shooter"

Pump discharge spring

Pump discharge ball

Discharge passage

Pump actuating link

Pump cup operation in fill position

● Fuel

**PUMP SYSTEM**

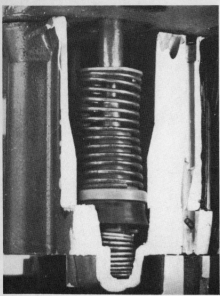

Q-jet pump cutaway showing pump cup at bottom of stroke with return spring compressed. Duration spring is on top of cup.

turi.

As the throttle is moved toward the closed position, the linkage returns to its original position and the pump-inlet ball or cup is moved off its seat to allow the pump to refill from the bowl. As the piston is positively pulled back to the at-rest position, a vacuum is created in the pump cavity so that quick refilling of the pump is ensured. As pump pressure is relieved, the discharge check needle or ball reseats. It remains seated while the pump cavity refills and so that the signal created by air passing by the pump shooters will not pull fuel out of the pump system. The weight of the valve is designed to keep it closed against this signal. In some carburetors, the discharge check is a lightweight ball at the bottom of the pump passage. In this instance, fuel is maintained (stored) in the passage between the check and the nozzle (shooter). Vapors can only be purged from the pump when the pump is operated or when pressure becomes sufficient to raise the discharge ball or pump cup off its seat.

Clearance around the pump-inlet ball and/or the vapor vent check ball in the pump plunger—or between the cup and the pump piston stem allows excess vapors to escape to the fuel bowl. Venting the pump ensures that a solid stream of fuel will be available from the pump system when it is needed. It also reduces hard-starting problems because

vapor pressure does not build up in the pump system to push fuel into the manifold.

NOTE: The late-style floating cup pump is illustrated in the accompanying drawing. An earlier version with an inlet check ball and a vapor-vent check ball in the plunger is shown in the two-barrel and 4GC chapters. All late-model RPD carburetors use the floating cup pump design.

## DESIGN FEATURES/Accelerator Pump System

**Pump Inlet Valves**—Up through the mid '60's RPD designs allowed fuel to enter the slotted pump well where it flowed past a check ball in the plunger head and around the pump plunger. Downward motion of the plunger seated the check ball so fuel could be forced through the pump system discharge channel. The check ball also serves as a vapor vent to relieve vapor pressure which might otherwise form in the pump well.

Some RPD models have an inlet check ball at the bottom of the pump well. This ball is in a pump-inlet passage supplied from the bowl. The passage may be protected by an inlet screen.

The second type to be discussed is known as the floating-cup design. Accelerator pumps used on most current RPD carburetors fill through center of pump cup. The cup is fitted onto the plastic pump-plunger body with some

vertical clearance. During the delivery stroke the cup is hydraulically forced up against the plastic pump-piston face sealing off the fill hole positioned there. During the return to an up position the cup drops a few thousandths away from the piston face, allowing fuel to enter through the center of the cup. Fuel fills the well through this hole so all is ready for the next shot.

The pump-shot and refill rate are designed to allow full-capacity shots at intervals of two to three seconds depending on duration springs, shooter size, etc.

Pump capacity depends on piston size and stroke. An accepted way to measure this capacity is as follows: Funnel the pumped fuel into a finely graduated cc beaker. With fuel level properly maintained, activate the throttle from closed to full-open position ten times consecutively. Move the throttle steadily and hold it in the full-open position approximately one second each time. This allows the fuel shot to be expelled to the burette.

At the end of ten strokes the cc's of collected fuel is your pump capacity. Small carburetor pumps like the Corvair H units will pump 3 to 5 cc's each. Larger carburetors will pump 15—20 cc's in ten strokes.

Ways to increase Q-jet carburetor pump capacities to nearly 50 cc are discussed in chapter fourteen.

# Idle System

Idling requires richer mixtures than part-throttle operation. Unless the idle mixture is richer, slow and irregular combustion will occur due to the high dilution of the charge by residual exhaust gases which exist at idle vacuums.

## DESCRIPTION OF OPERATION

The idle system supplies fuel at idle and low speeds, and should keep the engine running even when accessory loads are applied to the engine. These include the alternator, air conditioning and power-steering pump. The idle system also has to keep the engine running against the load imposed by placing an automatic transmission in one of the operating ranges (low, drive, reverse).

At idle and low speeds, not enough air is drawn through the venturi to cause the main-metering system to operate. Intake-manifold vacuum is high because of the great restriction to the air flow by the nearly closed throttle valve. This high vacuum provides the pressure differential for idle-system operation.

When the throttle is closed—or nearly so— the reduced pressure between throttle and intake manifold draws air/fuel mixture through the curb-idle port in the bore *below* the throttle plate. When the throttle is just opening, manifold vacuum is still high and additional mixture is drawn through the off-idle port as it is uncovered by the opening throttle. The amount of flow through the idle system depends on the channel restriction, the size of the idle-discharge ports, and the idle-mixture needle setting.

Backing the idle-mixture screw out, or opening the throttle to uncover the off-idle port or slot, allows more air to be drawn through the idle system. Pressure all along the idle passage becomes lower (1) due to the action of the throttle in exposing more port area to manifold vacuum and (2) due to increased pressure drop across the air bleed. This results in increased fuel flow from the idle well so the mixture stays at the desired air/fuel ratio.

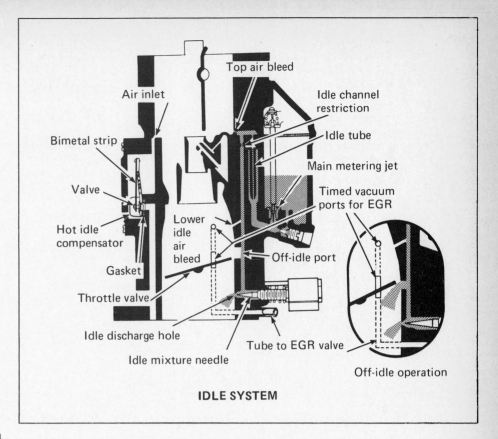

**IDLE SYSTEM**

Idle system operates like the main system, only on a smaller scale.

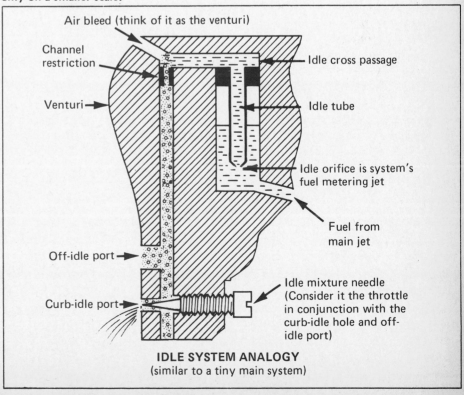

**IDLE SYSTEM ANALOGY**
(similar to a tiny main system)

29

The idle-fuel supply flows through the main jet into a vertical idle well. Idle-fuel metering is accomplished by the small orifice at the bottom of the idle tube. Metered fuel flows up the idle tube, along the idle cross-passage and into the idle down passage where it mixes with air entering through the idle air bleed.

As the throttle approaches wide open, there is low vacuum at the idle and off-idle ports and the idle system gradually ceases to deliver air/fuel mixture.

The idle ports are very small and are not intended to supply all of the fuel and air required by the engine even at a slow idle. The throttle is not closed completely at idle, or outside air comes around the throttle in some way. This air flow directly through the main bore of the carburetor is not enough to start up the main fuel system, so it is air alone, not air mixed with fuel. The air/fuel flow through the idle system must carry along enough fuel to make a proper idling ratio when added to the outside air flow past the throttle.

This highlights the use of the controls. A throttle stop or adjustment which regulates the amount of air flowing past the throttle is used to set idling speed. The idle-mixture needle, which regulates flow through the idle system, is used to set air/fuel ratio.

Most RPD carburetors have an intermediate idle system which discharges fuel through a hole below the venturi. This lower idle-air bleed hole is drilled from the lower venturi area into the vertical feed passage of the idle/off-idle system to serve two purposes. First, it bleeds air into the idle/off-idle fuel when these systems are in operation, providing better air/fuel mixing before the mixture enters the carburetor bore through the curb-idle and off-idle holes/slots. Second, because of its location between the venturi and the throttle blade, it is subjected to low pressure (vacuum) anytime the throttle blade is opened past the off-idle slot and to a point nearing the lower idle-air bleed hole. At this point the lower idle-air bleed becomes a feed to supply fuel to cover up any "hole" between the off-idle fuel and the start of the main system. When this bleed/feed has a strong signal (when throttle blade is near the bleed/feed hole), the idle/off-idle holes/slots will supply little—if any—

Curb-idle discharge is through hole indicated by black arrow. Off-idle slot just behind the throttle plate is barely visible (white arrow). Two arrows indicate fixed idle-air by-pass orifice.

fuel. The lower idle-air bleed/feed gets fuel from the idle system through idle-metering orifices supplied from the bowl.

The supplemental fuel from the lower idle-air bleed/feed diminishes as the throttle is opened past the bleed to a point where the main system becomes self-sustaining in its flow.

If the bleed/feed hole is made larger, it supplies more fuel and is more effective in covering up any "lean hole" or "sag" after the throttle passes the off-idle slots/holes and before the main system starts to flow. However, enlarging the bleed/feed hole leans out the idle/off-idle fuel and this should be remembered in making any changes to this supplementary transition fuel system. When the bleed/feed is shrouded with an eyebrow-shaped tube where the hole enters the lower venturi area, the bleed/feed feeds over a wider range of throttle opening. This is because high air flows create a depression at the shrouded "nozzle" so fuel continues to flow even though the throttle has moved up to a point farther open than the bleed hole.

As the throttle valve is opened and engine speed increases, air flow through the venturi is increased so that operation of the main-metering system begins discharging fuel through the discharge nozzle in the boost venturi. Flow from the idle system tapers off as the main system starts to discharge fuel. The two systems are designed to provide smooth, gradual transition from

idle to cruising speeds when the carburetor capacity is correctly matched to engine displacement.

In normal driving, flow swings quickly back and forth between idle and main operation as the vehicle is accelerated, slowed by closing the throttle, idled at stop, and then reaccelerated.

### DESIGN FEATURES/Idle System

**Throttle Stop**—Seating the throttle against a stop or idle-speed screw instead of closing it fully against the throttle bore ensures against the throttle sticking in the bore and makes the idle system less sensitive to mixture adjustments. RPD carburetors are factory-set to an idle air-flow specification and the primary throttles should never be readjusted to seat in the bore. Some models are set up so that the primary throttles are seated in the bore. This arrangement is usually done in a carburetor which has an idle air by-pass system.

**Idle-Air-Bleed Size**—Increasing the idle-air-bleed size reduces the pressure drop across the bleed, decreasing the amount of fuel which can be pulled over from the idle well. Thus, increasing idle-air-bleed size leans the idle mixture, even if the idle-feed restriction is left constant. Conversely, decreasing the size of the idle-air bleed increases the amount of pressure drop which can be obtained in the system and richens the idle mixture.

**Auxiliary Air Bleeds**—Auxiliary air bleeds are sometimes used in the idle system. Although these usually add air to the idle system downstream from the traditional idle-air bleed, they act in parallel with the idle-air bleed. The previously mentioned intermediate idle system is also an auxiliary air bleed.

**Idle-Speed Setting**—Before emission-control requirements became important, idle setting was typically the slowest speed at which the engine would keep running smoothly. Emission requirements have made higher idle speeds necessary in many cases. A higher idle speed reduces some of the exhaust-gas dilution which occurs so that leaner idle mixtures can be used without misfiring.

Cars with engines designed to pass emission requirements, i.e., production-type street machines, are typically set for lean best idle at specified RPM and a subsequent reduction in idle speed by leaning the mixture still further—as stated on a label in the engine compartment.

The manufacturers have carefully correlated this lean best idle and subsequent idle drop-off as that which provides the required carbon monoxide percentage (CO%) to pass the requirement. From this lean-best-idle point, and at this specified RPM figure, each of the idle-mixture-adjustment screws is turned in to provide a specified RPM drop-off. Where there are two screws, each is adjusted to provide 1/2 of the specified idle drop-off. Air/fuel ratios are usually stated for use when an analyzer is available.

Older non-emission-controlled cars and racing cars are typically idle-set for the desired idle RPM and best manifold vacuum. This is not a minimum-emission setting, however.

**Idle Limiter**—An idle-limiter cap limits idle-mixture-screw adjustment to less than 1/2 turn. This limiter is applied after the desired idle mixture has been factory-set. This is to prevent easy tampering with the idle-mixture adjustment.

The limiter is constructed so that removing it requires destroying the cap, thereby showing instantly that the carburetor has been readjusted and may not be providing specification emission performance.

**Idle-Air By-Pass System**—A number of RPD two- and four-barrel carburetor models had a system known as the "Idle-Air By-Pass System." A brief explanation of this follows.

The idle-air by-pass system used on some two- and four-bore carburetors allows the throttle valves to be completely closed during curb-idle operation. Gum and carbon formation around the throttle valves of conventional systems often disrupts engine idle. RPD's by-pass system ensures more consistent idle characteristics, even with a dirty carburetor.

Fuel flow in the system is basically the same as described for a conventional idle system. Idle air which normally by-passes the slightly open throttle valve is passed *around* the closed throttle valves through an idle-air by-pass channel.

This system takes idle air from the carburetor bore *above* the throttle valves, by-passes the closed throttle valves through a separate air channel and enters the carburetor bore just *below* the throttle valves. The amount of air supplied to the engine is regulated by an idle-air-adjustment screw

Idle-limiter cap is installed at factory after carb is set to correct idle fuel flow for emission performance (black arrow). White arrow indicates curb-idle speed-setting screw.

in the idle-air by-pass channel. The idle-air-adjustment screw is mounted in the float-bowl casting at the rear. Turning the screw inward (clockwise) decreases idle speed; turning it outward (counterclockwise) increases the engine speed.

To obtain sufficient idle air for adequate idle speed, a fixed idle-air supply is necessary in conjunction with the adjustable air or by-pass supply. The fixed idle-air path in the two-bore carburetor is provided by a hole in each throttle valve. These fixed idle-air holes maintain a constant idle-air flow for part of the idle-air requirements, while the idle-air-adjustment screw regulates the remainder of the idle air. Thus, idle speed is adjusted by the idle-air-adjusting screw. The throttle plates are not moved from their factory-set positions.

**Idle-Air Compensator**—An idle-air compensator is used on some RPD two-barrels to offset enriching effects caused by excessive fuel vapors created by fuel percolation during extreme hot-engine operation.

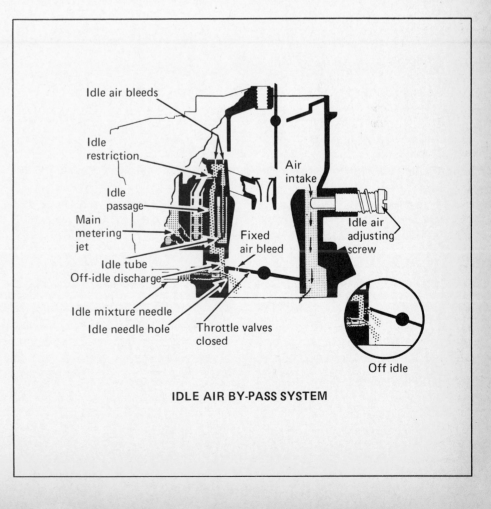

IDLE AIR BY-PASS SYSTEM

The compensator consists of a thermostatically controlled valve, usually mounted in the area above the main venturi or at the rear of the float bowl. The valve closes off an air channel leading from above the carburetor venturi to a point below the throttle valves.

The compensator valve is operated by a bi-metal strip which senses temperature. During extreme hot-engine operation, excessive fuel vapors entering the engine manifold cause richer than normally required mixtures. This causes rough engine idle and stalling. At a certain pre-determined temperature, when extra air is needed to offset the enrichening effects of fuel vapors, the bi-metal strip bends and unseats a valve to uncover the compensating air channel from the carburetor venturi to below the throttle valves. At this time, just enough air is added to the mixture to offset the richness and maintain a smooth idle. When the engine cools and the extra air is not needed, the bi-metal strip closes the valve and operation returns to normal mixtures.

To ensure correct idle adjustment, the valve should always be closed when setting idle speed and mixtures.

**Adjustable Off-Idle Air Bleed**—Some emission-control carburetors include an adjustable off-idle (AOI) air-bleed system. A separate air channel bleeds air past an adjustment screw (needle) into the idle system. Factory adjustment of the screw establishes very accurate off-idle air/fuel mixture ratios to meet emission-control requirements.

**Idle-Channel Restrictions**—Some idle systems have a calibrated restriction in the idle down channel. This secondary metering calibration for the idle system affects transition metering from 25 to 40 MPH, depending on the carburetor.

**Curb-Idle Port Size**—Emission-control carburetors may have smaller idle-mixture screw discharge holes than were used on pre-emission carburetors. This was done to eliminate over-enrichment by turning out the mixture-adjustment screws. As an example, 1970 and earlier Q-jets had 0.095-inch diameter curb-idle discharge holes.

1971 and later models have 0.080-inch or smaller holes. Let's go back to our analogy of the idle system as a small main metering system. You can see that reducing the opening limits the "throttle opening" and reduces the signal which could otherwise be obtained by screwing out the mixture/needle. No matter how far you back the idle mixture screws out, it is not possible to create a "rich-roll" adjustment.

**Off-Idle Discharge Ports**—These can be either one or more holes or a slot. Either method gives correct air/fuel mixtures and satisfactory operation. Slots are usually used because they are less expensive to manufacture.

**Fixed Idle-Air By-Pass**—A few Q-jet models have idle-air channels from the air horn to a point below the primary throttle valves. Extra idle air coming through these channels allows the throttle valves to be more closed at idle. It reduces the signal applied to the main fuel nozzles by the very efficient venturi cluster. This eliminates *nozzle drip* at idle, an especially annoying problem on large-displacement engines operating at the high idle speeds required for emission control.

**Systems interrelated**—As you consider the role of the idle system, keep in mind that it accomplishes correct metering up to a certain throttle opening (higher flow) which brings the main system into operation. As a matter of fact, it flows fuel any time there is sufficient vacuum signal. The amount of fuel contributed by the idle system to engine demands may be an insignificant part of the total volume when the main system is working. The same is true when demands get too high for the main system and the power system enters the picture. Then three systems are operating: idle, main and power.

Idle flow may be diminished when the power system comes in (depending on manifold vacuum and throttle position) but the power system depends on the main system continuing to feed fuel at up to its maximum capacity. This is true with almost all carburetors. Variances are found in the manner in which each system blends with and supports the other systems. This is worked out in the original design and development of the particular carburetor.

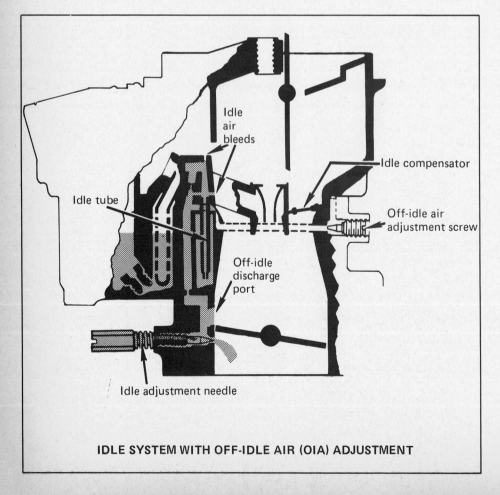

Idle air bleeds

Idle compensator

Idle tube

Off-idle air adjustment screw

Off-idle discharge port

Idle adjustment needle

**IDLE SYSTEM WITH OFF-IDLE AIR (OIA) ADJUSTMENT**

# Choke System

The choke system provides the richer mixture required to start and operate a cold engine, as discussed in the Engine Requirements section. Cranking speeds for a cold engine are often around 50 to 75 RPM. These speeds are low compared to engine operating speeds, hence very little manifold vacuum is created to operate the idle system. A closed choke valve, which conforms closely to the inlet air horn, causes a vacuum below it so that fuel is pulled out of both the idle and main-metering systems during cranking. At times fuel is even pulled from air-bleed holes.

As you might imagine, this surplus fuel creates an extremely rich mixture of approximately equal fuel and air. The super-rich mixture is needed because there is not much manifold vacuum to help vaporize the fuel and the manifold is cold, so most of the fuel puddles onto the manifold surfaces as it immediately recondenses. Also, the fuel is cold and not volatile or high with vapors. Liquid fuel cannot be evenly distributed to the cylinders and when it arrives there, it will not burn correctly. Only a small portion of the fuel ever reaches the cylinders as vapor during starting.

Once the engine starts, off-center mounting of the choke-plate shaft causes air flow to open the choke partially against torque of the bimetal spring so that the mixture is leaned out somewhat.

In the case of the automatic choke, there is a vacuum "break" diaphragm which pulls the choke valve to a pre-set opening once the engine starts. This opening provides a partial lean-out of the starting mixture. When the choke assumes this position, it is still providing a 20 to 50 percent richer-than-normal mixture during the warm-up period. This mixture is further leaned out towards a normal operating mixture as the engine warms up and the bimetal choke thermostat weakens, allowing the choke to come off (open fully). Up through the '60s chokes were calibrated to be off in 1.5 to 3.0 miles city driving. Today's chokes are for the most

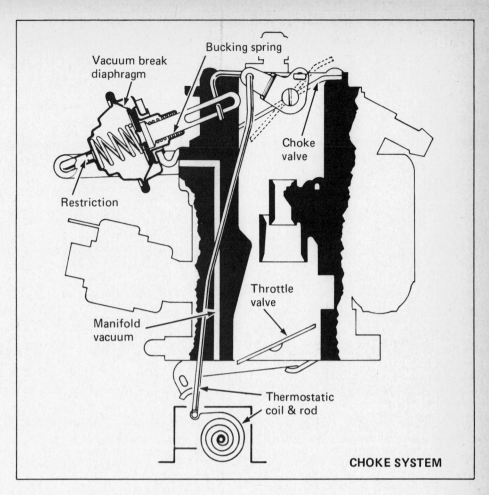

**CHOKE SYSTEM**

Thermostatic-coil assembly for choke actuation mounts on intake manifold, cylinder head, or exhaust manifold. This arrangement is called a "divorced" choke.

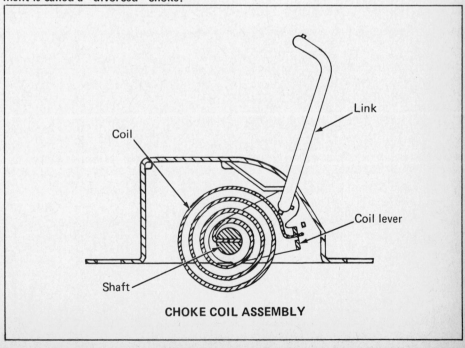

**CHOKE COIL ASSEMBLY**

Cadillac installation showing stat coil mounted on intake manifold over exhaust-heat crossover passage.

part effectively off at 0.8 to 1.5 miles as part of the current emission-reduction efforts.

Although it appears to be a simple valve atop the air horn, the choke system is often one of the most complex in the carburetor. There are schemes to take away part of the choking effect after the engine starts and in the event the driver accelerates before the engine is warm. You'll find these discussed in the specific model chapters in this book.

**Divorced Choke**—The divorced choke is operated by a thermostatic coil on the intake manifold. This arrangement provides the most accurate reflection of engine starting mixture requirements. The divorced thermostatic coil closes the choke only when the engine gets cold.

Adjustment of the divorced choke is accomplished by changing the length of the rod connecting the thermostatic coil to the choke linkage on the carburetor. Shortening the rod applies more closing force to the choke so it closes at a higher ambient temperature and stays closed for a longer period.

**Integral Choke**—This arrangement places the thermostatic coil on the carburetor. Heat applied to the thermostatic coil is usually hot air from a "stove" around the exhaust manifold. When the heat source is engine water, the choke is called a *hot-water* choke.

An integral choke which is hot-air heated can close when the engine is still hot, even though a choke-supplied rich mixture and fast idle are not needed.

Adjustment of the integral choke is accomplished by changing the setting of the *choke index*. The thermostatic coil is factory-set so the choke will just close (on a new carburetor) when the choke bimetal is at 75°F. Tapping the carburetor lightly helps overcome any shaft friction so the choke will seek the position being set by the bimetal. If less choke is desired at this temperature, then the choke index can be moved one mark to unwrap bimetal tension. If more choke is required, the index can be moved one mark in the other direction. An arrow on the housing shows the direction the index must be moved to change the choke operating characteristic. A choke mixture change rarely requires more than one index-mark change from the factory setting.

**Unloader**—This device works to counter the problem caused when the engine is hot but the choke thermostatic coil has cooled down.

If you park your vehicle on a cool, breezy day for 20 minutes to an hour, the effect of blowing air can cool the choke coil and housing (external engine components). The coil signals for full choke (maximum enrichment) when the engine is re-started, even though the internal components of the engine (heads, combustion chambers, etc.) do not cool totally during the period. The choke sends a full cold start supply of fuel which the warm engine vaporizes readily as it ingests this mixture into the combustion chambers. Sometimes it is too rich to fire. To get the engine started, push the throttle to the floor as you crank. This causes the unloader linkage to open the choke blade to admit fresh air to the induction system where it mixes with the excess fuel, making it combustible.

The unloader is a tang on the throttle lever which contacts the fast-idle cam to open the choke plate sufficiently to allow clearing (unloading) the excess fuel from the manifold.

The unloader aids in starting a flooded engine, regardless of the cause for flooding.

**What's a bimetal?** The bimetal referred to in the accompanying automatic-choke discussion is two different metals bonded into a strip and formed into a coil. Because the metals have unequal thermal-expansion characteristics, this coil unwraps when warmed and wraps up again when cooled. The outer or "free" end of the coil attached to the choke linkage holds the choke plate closed—or loads it to close the plate when the throttle is opened—until the bimetal is warmed. Warming can be by exhaust-warmed air, jacket water, heating of the manifold from engine operation, or an electric heating element—depending on the type of choke actuation. Bimetal temperature-response characteristics are built in according to the metals selected to make the strip. Most automotive-choke bimetals wrap up (choke just closed) at 75°F.

As mentioned before, the choke mixtures are required because fuel will not vaporize correctly until the exhaust hot spot in the manifold has warmed up sufficiently to ensure good vaporization.

While the choke is "on," the engine idles quite fast by design. Usually this is approximately 1,000 RPM with a cold engine. This happens because the choke linkage includes a fast-idle cam to keep the revs up once the engine starts to aid in vaporizing fuel and in overcoming cold-engine friction loads.

Oldsmobile carbs typically use an integral choke. Black arrow indicates connection for exhaust-heat tube. White arrow points to choke index. This is a 4MC version of the Q-jet. "C" in the model designation indicates choke-coil assembly attached to carburetor.

Olds choke with stat cover removed to show bimetal in cover. Tab or loop in end of bimetal attaches to choke tang (arrow).

## DESIGN FEATURES/Choke System

**Automatic Chokes**—There are two main types of automatic chokes: integral and divorced. The *integral* type uses a tube to connect heated air from the exhaust manifold or exhaust-heat crossover to the bimetal spring inside a housing on the carburetor. The *divorced* type uses a bimetal spring which is mounted directly on the intake manifold or in a pocket in the exhaust-heat passage of the intake manifold. A mechanical linkage from the bimetal spring operates the choke lever on the carburetor.

Also, if the driver should open the throttle fully, reduced manifold vacuum on the choke vacuum break piston or diaphragm will tend to allow the choke to close, giving an excessively rich mixture when it is not required. The unloader ensures that the choke valve will be opened so air can flow through the carburetor without creating an excessively rich mixture.

**Fast Idle**—During the warm-up period the engine has higher frictional forces to overcome. To prevent stalling the engine must run at a higher idle speed than would be required if it were warm. A fast-idle screw on the throttle contacts a fast-idle cam linked to the choke. Rotation of this cam in relation to the choke valve holds the throttle valve farther open than a curb idle to increase idle RPM during warm-up. When the choke valve moves to its fully open position the fast-idle cam moves out of the way of the fast-idle screw so idle can return to its normal curb-idle setting.

**Secondary Lockout**—During the early stages of engine warm-up it is not a good idea to allow the secondaries to function. The sudden burst of air to the cylinders via cold manifold runners creates a certain hesitation or backfire. To prevent this, lock-out devices are interconnected with the choke. When the choke is on any appreciable amount the secondaries are locked out so they cannot be operated. After engine warm-up and the choke comes off, the secondaries are free to operate.

**Vacuum Break**—As you read more about chokes you will see constant reference to the term *vacuum break*. Carburetors with a choke thermostat housing on the carburetor body have a manifold-vacuum passage to draw exhaust-heated air into the choke and across the bimetal strip which controls choke opening according to tem-

Secondary throttle-blade lockout (white arrow) prevents secondary operation until choke link (some of choke mechanism removed for clarity) is pulled away by choke opening.

Some Q-jets use an air valve lockout instead of locking out the secondary throttles. In the left picture the choke is partially on. Black arrow indicates lockout lever; white arrow points to connecting link to fast-idle cam and thermostatic coil. In right picture the choke is fully off and the air-valve lockout has retracted to permit secondary operation.

perature. This vacuum also operates a piston or diaphragm to force the choke slightly open once the engine starts—just far enough to allow the engine to run without loading or stalling. The choke valve position relates to the torque of the thermostatic coil balanced against vacuum piston or diaphragm pull and air velocity acting on the offset choke valve. Because the piston or diaphragm operates from manifold vacuum, any heavy acceleration removes the vacuum so the choke can close slightly

to give a richer mixture to handle the acceleration.

On models with the piston arrangement, once the choke opened, the piston covered the vacuum port, closing off or *breaking* the vacuum when the choke opened due to air velocity and heating of the thermostatic coil. The term *vacuum break* is still applied to any vacuum-operated choke mechanism, including the diaphragm types, even though these do not actually include a vacuum break as such.

# Secondary Throttle Operation

For many years U.S. carburetors were single-staged. V-8 engines were equipped with a two-barrel: merely a single casting containing two one-barrel carburetors side by side. One was used for each level of the traditional two-level cross-H manifold. All past and current Rochester two-barrels are single-stage carburetors: The throttles operate together on one shaft.

The late 1940's saw increased emphasis on vehicle performance. Because more air flow means more power, single-stage carburetors became bigger and bigger—and so did the drivability problems. The large venturis caused the main systems to start flowing at higher RPM or greater air flows. Under some conditions the idle system could be made to cover up the late entry of the main system. But, because the idle system is controlled by manifold vacuum, a great deficiency was felt at low manifold vacuum and low air flows, such as in a high-load, low-RPM condition. In addition, the low venturi velocities at low RPM resulted in poor fuel vaporization and poor air/fuel mixture quality. These caused cylinder-to-cylinder distribution problems and erratic operation.

Simply stated, the single- or two-barrel carburetor's metering range was too narrow to satisfy all requirements. The answer was obvious: Use a staged carburetor to stretch the metering range and get back to venturis small enough to get the main systems flowing at low RPM and provide good vaporization—while having the required capacity for high-RPM operation when needed.

Six-cylinder engines of medium displacement remained the economy workhorses, so no staged two-barrels were ever used on a U.S. car until the 1970 Pinto and the 1973 Vega, even though such systems had long been used on European and Japanese cars. There the costs could be outweighed by the need to get maximum economy and maximum performance from small-displacement engines.

However, in the early 1950's, a whole host of four-barrel carburetors were introduced for V-8's. The primary side of these carburetors was smaller than it had been on the single-stagers, returning all of the benefits provided by small venturis: early start of the main system, good vaporization and air/fuel quality—and good distribution. The primary carburetor barrels were used for the cruising loads and light accelerations encountered in normal traffic. Flexibility of operation and economy were regained.

Coupled to the primary carburetor barrels were two secondary carburetor barrels—designed to operate when maximum air flow was required for more power. Essentially, the metering range of the carburetor was doubled, combining good part-throttle operation with relatively unrestricted flow for maximum-power conditions.

Secondary car-

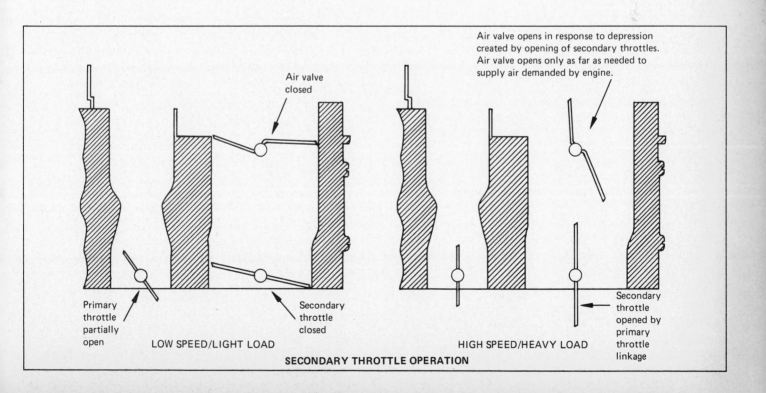

Air valve closed

Air valve opens in response to depression created by opening of secondary throttles. Air valve opens only as far as needed to supply air demanded by engine.

Primary throttle partially open

Secondary throttle closed

Secondary throttle opened by primary throttle linkage

LOW SPEED/LIGHT LOAD

HIGH SPEED/HEAVY LOAD

**SECONDARY THROTTLE OPERATION**

buretors are simply another carburetor in parallel with the first one. They always have their own main metering system. Some have their own idle system to one degree or another because this gives better distribution and idle or off-idle stability, thereby improving emissions by allowing the use of leaner idle settings. A secondary idle is sometimes used to purge the secondary carburetor so that varnish does not form because the driver fails to use the secondary side. Rochester Products does not always include an idle system on the secondary side or in secondary carburetors.

Some of the carburetors have power systems in the secondary side.

The secondary side of a carburetor has no choke. Lockout devices prevent secondary throttle (or air valve, depending on the application) operation when the choke is on (cold engine). There would be extreme engine stumbling due to lean mixtures if the secondary throttles were allowed to operate while the primary side choke was on.

Operation of the secondaries can be accomplished in several ways:
1. Mechanical only
2. Diaphragm
3. Mechanical with velocity valve
4. Mechanical with air valve.
RPD carburetors use the third and fourth methods in the 4GC and Q-jet, respectively. RPD used the second method in a number of applications to open the end carburetors on 3 x 2 setups for Olds, Pontiac and Chevrolet. And, the mechanical-only method was used on a few of the 3 x 2 setups, plus the 4 x 1 setup on the 140 HP Corvairs.

## DESCRIPTION OF OPERATION

**Mechanical**—Mechanically operated secondaries are very simple in operation. The secondary throttles are opened by a direct link from the primaries, usually on a progressive basis so that the secondary opening is delayed until the primaries have reached approximately 40° of opening. Closing of the secondary throttles is very positive because of the return spring, because the throttles are offset on their shaft so that air flow aids their closing and because a return link from the primary throttle pulls them closed.

**Diaphragm**—Secondary throttles are opened by a vacuum-operated diaphragm. Vacuum for the diaphragm is obtained from one of the primary venturis.

At maximum speed, the secondary throttles are opened fully. However, if the carburetor is too large for the engine, the diaphragm automatically "sizes" the carburetor so that it flows the needed amount of mixture by only partially opening the secondary throttles. The secondaries remain closed when the primary throttles are opened wide at low RPM. This eliminates bogs or sags and allows use of the carburetor with a wide range of engine displacements, gear ratios, car weights, etc.

**Mechanical with velocity valve**—This system as used in the 4GC carburetor has velocity valves above the secondary throttles and below the rudimentary venturi structure. The secondary throttles are progressively linked to the primaries. When the primaries are open about 45° from the closed position, the secondary throttles are opened mechanically by the linkage. However, little air flow occurs through the secondaries until the depression under the offset velocity valves acts on the long side of the offset blade to start the opening of the velocity valves. The velocity valves delay the secondary air flow for smooth transition during secondary throttle valve opening. If the engine RPM is high enough at this time, the velocity of the inrushing air forces the spring-loaded velocity valves farther open until they reach a full-open position.

This entire sequence occurs in about one-half second from the time the primary throttles are fully opened until the velocity valves are full open *if the engine RPM is high enough.*

When full throttle is applied at low engine speeds the velocity valves may start to open but they will close almost immediately because not enough air is coming past them to hold them open. The spring tension is set so the velocity valves will only open when the engine demands more air/fuel for power operation.

In the case of the 4GC, the secondary side was sometimes equipped with an off-idle system: (1) to offset a lean condition which would otherwise occur when the secondaries were just slightly cracked open during high-speed cruising; and (2) to help eliminate slight bogging which tends to occur when the secondary throttles start to open.

A power system was used only on the primary side of the 4GC.

**Mechanical with Air Valve**—The Quadrajet utilizes this system. The secondary throttles are progressively linked to the primaries. When the primaries are open about 35° from the closed position, secondary throttles are opened mechanically by the linkage. No air flow occurs through the secondaries until the depression under the offset air valve causes it to start opening.

The air valve opening continues until its open area (plus that of the primaries) is sufficient to handle the air flow required by the engine at the throttle position being used. If the primaries are at full throttle, which also fully opens the secondaries, the air valve will open as far as needed to provide the engine's air-flow requirements (up to the 750–800 CFM flow capacity of the Q-jet at 90° air-valve opening). The air-valve opening rate is controlled by a damping device so that bogging does not occur.

A transition system provides fuel from ports just above or just below the air valve as the air valve starts to open. This can be considered similar to a secondary accelerator pump because it supplies a brief shot of fuel until the secondary main system begins feeding fuel.

The main system has a variable metering system which uses tapered metering rods in fixed metering orifices (or jets).

A cam on the air valve raises the metering-rod hanger attached to the metering rods. Increasing the air-valve opening lifts the tapered metering rods so a smaller diameter is in the jets, allowing more fuel flow. When the air valve is from 60° to 80° (80° is fully open), the straight non-tapered portion of the metering rod provides a richer fuel-power mixture.

There is no idle system in the secondary side of the Q-jet.

# Other Functions
## (Not Systems)

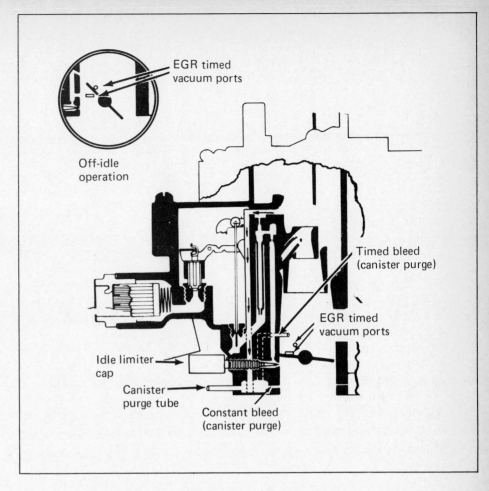

EGR timed vacuum ports

Off-idle operation

Timed bleed (canister purge)

EGR timed vacuum ports

Idle limiter cap

Canister purge tube

Constant bleed (canister purge)

The carburetor also provides control for other important functions, especially those related to spark advance/retard and emission controls. Signals created by various pressure areas in the carburetor, especially as related to throttle position, are used to good advantage by automobile designers.

**Vacuum Advance Ports**—Some carburetors have a slot or holes drilled just above and alongside (not connected!) the off-idle discharge slot or holes. These feed vacuum to operate the distributor vacuum advance unit when the throttle is opened past idle. You will sometimes hear this carburetor-controlled distributor vacuum advance referred to as *timed spark*.

**Canister purging**—Certain 1970 and all 1971 and later GM model cars have completely closed fuel tank venting to control evaporative emissions. The vent from the fuel tank leads into a vapor collection canister. Starting in 1972, all Q-jets also have a bowl vent connected to the canister.

Because the fuel tank is not vented to atmosphere and the carburetor is vented only to the canister when the engine is stopped, fuel vapors are collected in the vapor canister. Purge ports for the canister are provided in the carburetor throttle body on certain models. The purge ports lead through passages to a common chamber in the throttle body to a purge tube hose connected to the vapor canister. The purge ports may consist of a constant-bleed purge and a separate timed canister purge, or a separate timed canister purge only.

The constant-bleed purge operates during idle, to purge the canister continuously when the engine is running. Along with the constant-bleed purge for the canister, a timed purge may also be used. The timed purge port is located in each bore adjacent to the off-idle discharge ports. The timed purge operates during off-idle, part-throttle and wide-open throttle. This provides a large purge capacity for the vapor canister and prevents over-rich mixtures from being added to the carburetor at any time.

**Exhaust gas recirculation**—Some 1972 models and all 1973 models use an exhaust-gas-recirculation system to reduce oxides of nitrogen emissions. The EGR valve is operated by a vacuum supply signal taken from the carburetor. Two punched ports, one just above the throttle valve and one mid-way between the throttle valve and upper surface of the throttle body are located in the primary bore.

As the primary throttle valve is opened beyond the idle position, the first vacuum port for the EGR system is exposed to manifold vacuum to supply a vacuum signal to the EGR valve. To control the vacuum signal at the lower port the upper port bleeds air into the vacuum channel and modulates the amount of vacuum signal supplied by the lower EGR port. In this manner, the EGR valve can be timed for precise metering of exhaust gases to the intake manifold depen-dent upon location of the ports in the carburetor bore and by the degree of throttle valve opening. As the primary throttle valve is opened further in the part-throttle range, the upper port ceases to function as an air bleed and is gradually exposed to manifold vacuum to supplement the vacuum signal at the lower port and helps maintain correct EGR valve position.

The upper and lower vacuum ports connect to a cavity in the throttle body which, in turn, through a passage supply the vacuum signal to an EGR tube pressed into the front of the throttle body. The tube in the throttle body is hose-connected to the EGR valve on the intake manifold. The EGR valve remains closed during periods of engine idle and deceleration to prevent rough idle from excessive exhaust gas contamination in the idle air/fuel mixtures. A different throttle-body-to-float-bowl gasket is required on all 1973 models with the addition of the EGR port system in the carburetor throttle body.

# Monojet Carburetors

**R**ochester has been a high-volume supplier of single-barrel carburetors since 1950 when they introduced the B series carburetor. A need for metering sophistication came about in the early 60's and concurrent with this need came the Monojet design. It has been the mainstay for General Motors single-barrel carburetor requirements since 1968. The many varied metering controls designed into this carburetor are a big plus for emission calibrations.

Monojet carburetors are designated by the letter M. Manual-choke models use only the M designation. Automatic-choke units use a second letter V to indicate the addition of a vacuum break diaphragm on the carburetor and a thermostatic coil mounted on the engine. Automatic-choke models are designated MV.

## GENERAL DESCRIPTION

The Monojet is a single-bore downdraft with a tube nozzle used in conjunction with a multiple venturi. Fuel flow through the main-metering system is controlled by a mechanically and vacuum-operated variable-area jet. A specially tapered rod which operates in the fixed-orifice main-metering jet is connected directly by linkage to the main throttle shaft. The same tapered rod is also vacuum-operated by a power piston to form a power-enrichment system in conjunction with the main-metering system. This arrangement provides good performance during moderate to heavy accelerations.

An exception to the above type metering was made on the Monojet developed for the Vega "base" (lowest HP) engine. In place of the metering rod and vacuum-operated power system it has a conventional orificed metering jet with a velocity-actuated power-enrichment valve. System choice was heavily influenced by economy goals for this small-displacement engine. A wise choice in my opinion because the acceleration and cruise vacuum is nil which would cause a vacuum-controlled system to call prematurely for fuel enrichment.

This 1974 Monojet for the Vega uses an idle-stop solenoid as an anti-dieseling device. Many other Monojets also include this feature. Curb idle setting is made by turning the energized solenoid in its threaded boss in the float bowl to obtain the specified idle RPM. The ignition-off setting which closes the throttle blade farther to prevent dieseling is set by turning a small setscrew in the outer end of the solenoid with the solenoid de-energized. Monojets may also be equipped with a CEC valve to provide a high idle speed during deceleration as a means of controlling hydrocarbon emissions on the over-run. All 1974 models have bushing for the extra throttle closing spring. Six-cylinder models have the closing spring bracket attached to the air horn.

A—Conventional Monojet. B—Vega Monojet with thin throttle body required for hood clearance. Note difference between external (A) and internal (B) vacuum-break assemblies.

System drawings—Monojet system drawings are used as examples in the previous chapter on how your carburetor works. Turn back to those pages and refer to the drawings as you read the paragraphs about the various Monojet systems.

Other Monojet features include an aluminum throttle body for decreased weight. A thick throttle-body-to-bowl insulator gasket keeps excessive engine heat from the float bowl. The carburetor has internally-balanced venting through a vent hole in the air horn. An external idle-vent valve is used on some pre-1971 models.

Simplicity in design of the Monojet carburetor leads to ease in servicing.

## FLOAT SYSTEM

The float system is similar to other RPD carburetors. The exception is that the Monojet designers brought the fuel in at the bottom of the bowl to reduce the effect of vapors entering with the liquid fuel. By introducing the incoming fuel (and vapors) through liquid fuel in the bowl, the vapor tends to condense.

An integral fuel filter mounts in the fuel bowl behind the fuel-inlet nut.

## IDLE SYSTEM

The Monojet has a conventional idle-off-idle system using a vertical slot for the off-idle feed between curb-idle and main-system operation.

At least three bowl vent designs have been used on the Monojet carburetors through 1974 models. Although only two of these are truly *idle* vent valves, we will discuss all three types here. Type A—1968 and 1969 models have an idle vent valve atop the air horn. It is pushed open by a tang on the accelerator-pump shaft when the throttle is at idle. Any fuel vapors formed during hot engine operation or a hot soak are vented to the atmosphere. The valve is closed at off-idle and larger throttle openings.

Type B—Some 1968-69 models have a vent valve similar to type A except that no closing spring is used on the vent valve. The valve operates at idle like Type A but it is also a pressure-relief valve for venting to atmosphere anytime excessive vapor pressures build in the float bowl. This prevents richness during hot-engine operation because the vapor pressures escape instead of forcing excessive fuel into the carburetor bore to cause uneven running or hard starting.

Type C—1970 and later models use a wafer-type pressure-relief valve in an air horn vent connected to the vapor canister. Pressure build-up in the float bowl pushes the valve off its seat so vapor escapes to the canister.

Because fuel is not forced into the carburetor bore, rough running and hard starting during hot operation or after a hot soak are greatly reduced.

An exhaust gas recirculation system (EGR) is part of all 1973 Monojets. The separately mounted EGR valve is operated by a vacuum signal obtained from timed ports (or a timed slot) in the throttle body. This signal is available at off-idle and part throttle.

## MAIN METERING SYSTEM

The main system is augmented by an adjustable-flow feature set during manufacture to control the fuel mixtures more accurately than could be done with a fixed orifice. This adjustable-part-throttle (APT) fuel is channeled from the fuel bowl to the nozzle feed well, independent of the metering jet but in parallel with it.

## POWER SYSTEM

The power system is an integral part of the main metering system because both operate on the same tapered metering rod in the main metering jet. During part-throttle and cruising, high manifold vacuum holds the power piston down against spring tension. The upper side of the power-piston groove is held against the top of the drive rod so the metering rod is kept low in the jet for maximum economy.

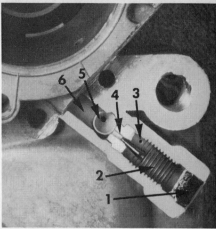

Some Monojets are equipped with an Adjustable Part-Throttle (APT) feature. Cutaway bowl casting shows plug 1, adjusting needle 2, orifice from bowl 3, APT orifice 4, screw-in metering jet 5, and main-well channel 6. Needle is factory-set to provide correct air/fuel ratio for main-system operation during part-throttle operation.

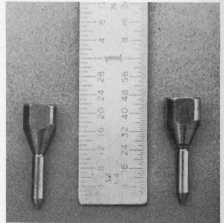

Vega-style power valves are weights. Tuning information is on page 234.

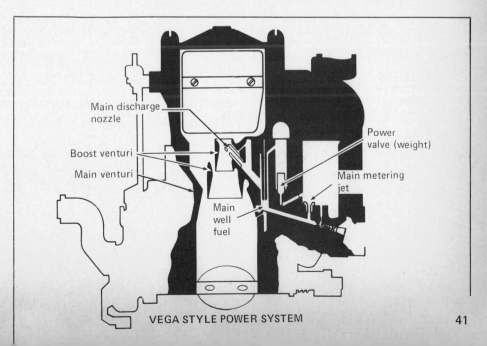

VEGA STYLE POWER SYSTEM

41

Acceleration or opening the throttle to get more power reduces manifold vacuum. The power piston spring pushes the power piston up until the lower edge of the power-piston groove is against the bottom of the drive rod. This raises the metering rod to allow more fuel to flow through the main jet.

The amount of power enrichment is controlled by clearance between the power-piston groove and the drive-rod diameter. 1969 Oldsmobiles with manual transmissions used a compression spring on the bottom of the float bowl to hold the drive rod down, thereby removing all play from the power-piston linkage. This aided emission control by providing better control of fuel metering.

## VEGA POWER SYSTEM

The Monojet used on 1971-74 base engines does not use the vacuum-controlled metering rod for power enrichment. An air-velocity demand power-enrichment system in these carburetors operates as follows.

When the demand for fuel at the boost venturi exceeds the main metering jet's capacity, the power-enrichment valve unseats so fuel can flow directly from the bowl to the main well and then to the boost venturi via the nozzle. This fuel by-passes the main jet. This system is built into a special Monojet bowl casting which is used only in this application.

## CHOKE SYSTEM

The Monojet is designed to accommodate either manual or automatic choke systems. A conventional choke valve is in the air-horn bore. On automatic-choke models the vacuum-diaphragm unit is either an integral part of the air horn or installed on a bracket. The automatic-choke coil is manifold-mounted and link-connected to the choke-valve shaft.

The choke system on 1968 units had a feature to give added enrichment during cold starting. It greatly reduced starting time, yet allowed the use of low-torque thermostatic coils for increased economy. The choke blade actuated a rod which was the upper end of a valve. When the choke blade contacted the rod, the valve opened a starting fuel passage to feed extra fuel into the air horn above the venturi. This feature was discontinued in 1969.

Starting in 1973 an integral

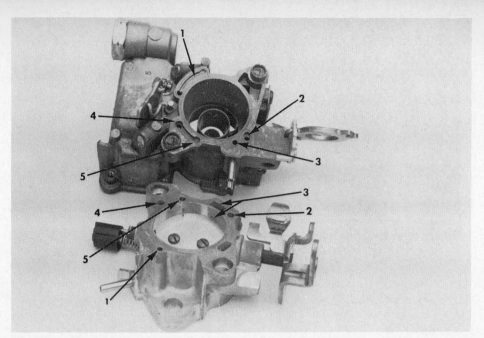

Passages in float bowl and throttle body are identified in this photo. 1—manifold vacuum to power valve, 2—timed spark port, 3—manifold vacuum to purge canister, 4—idle down channel, 5—off-idle down channel.

bucking spring was added to the diaphragm plunger on Vega MV models. This allows the thermostatic coil to modulate choke-valve position according to temperature, maintaining leaner mixtures during warm temperatures and richer mixtures when the weather is cold.

During extreme cold operation the thermostatic coil exerts more closing pressure than during warmer temperatures. The thermostatic coil compresses the

plunger bucking spring, offsetting the vacuum diaphragm pull so it cannot open the choke valve as far as it would otherwise. This gives richer mixtures for cold-weather operation. During warmer temperatures the thermostatic coil's reduced tension during the starting period does not compress the plunger bucking spring quite so far. Thus the vacuum diaphragm opens the choke farther to supply a leaner mixture.

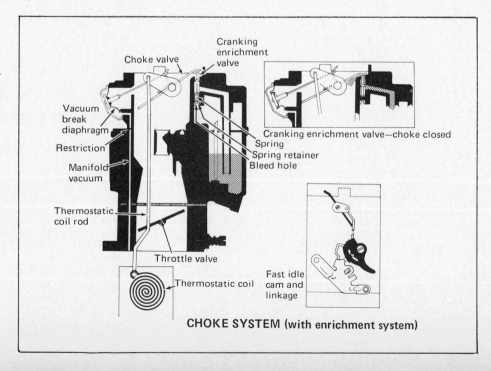

CHOKE SYSTEM (with enrichment system)

## Model MV Monojet—Exploded View

1 Air horn assembly
2 Air horn—long screw
3 Air horn—short screw
4 Air cleaner stud bracket
5 Bracket attaching screw
6 Idle vent valve kit
7 Air horn gasket
8 Choke shaft & lever assembly
9 Choke valve
10 Choke valve screw
11 Vacuum break link lever
12 Vacuum break link assembly
13 Vacuum break diaphragm
14 Vacuum break cover
15 Cover screw
16 Choke rod
17 Choke lever
18 Choke lever screw
19 Fast idle cam
20 Cam attaching screw
21 Float bowl assembly
22 Idle tube assembly
23 Main metering jet
24 Pump discharge ball
25 Pump discharge spring
26 Pump discharge guide
27 Needle and seat assembly
28 Needle seat gasket
29 Idle compensator assembly
30 Idle compensator gasket
31 Idle compensator cover
32 Cover screw
33 Float assembly
34 Float hinge pin
35 Power piston assembly
36 Power piston spring
37 Power piston rod
38 Metering rod & spring assembly
39 Fuel inlet filter nut
40 Filter nut gasket
41 Fuel inlet filter
42 Fuel filter spring
43 Slow idle screw
44 Pump assembly
45 Pump actuating lever
46 Pump return spring
47 Throttle body assembly
48 Throttle body gasket
49 Idle needle
50 Idle needle spring
51 Throttle body screw
52 Pump and power rods lever
53 Lever attaching screw
54 Power piston rod link
55 Pump lever link

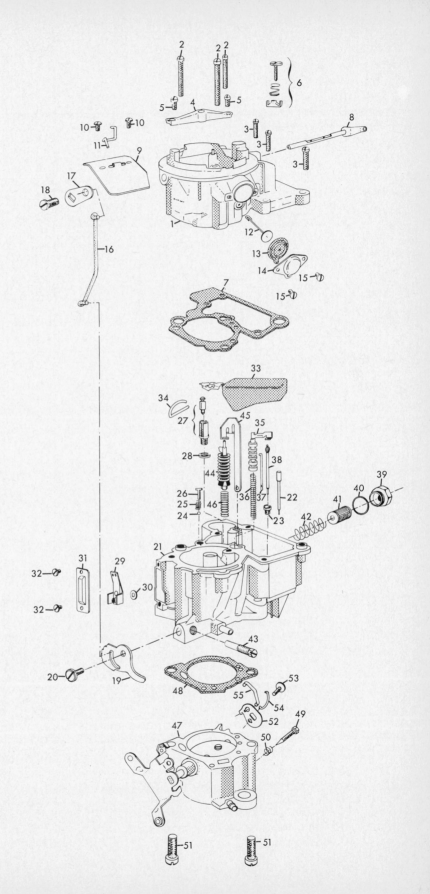

## Model MV Monojet
## with external vacuum break
## and emission-control equipment

1 Air horn assembly
2 Air horn—long screw
3 Air horn—short screw
4 Air cleaner stud bracket
5 Bracket attaching screw
6 Air horn gasket
7 Choke shaft & lever assembly
8 Choke valve
9 Choke valve screw
10 Choke vacuum break unit
11 Vacuum break hose
12 Vacuum break lever
13 Vacuum break link
14 Vacuum break lever screw
15 Choke lever
16 Choke rod
17 Fast idle cam
18 Cam attaching screw
19 Float bowl assembly
20 Idle tube assembly
21 Main metering jet
22 Pump discharge ball
23 Pump discharge spring
24 Pump discharge guide
25 Needle and seat assembly
26 Needle seat gasket
27 Idle compensator assembly
28 Idle compensator gasket
29 Idle compensator cover
30 Cover screw
31 Float assembly
32 Float hinge pin
33 Power piston assembly
34 Power piston spring
35 Power piston rod
36 Metering rod & spring assembly
37 Fuel inlet filter nut
38 Filter nut gasket
39 Fuel inlet filter
40 Fuel inlet spring
41 Idle stop solenoid
42 Pump assembly
43 Pump actuating lever
44 Pump return spring
45 CEC valve
46 CEC vacuum tube
47 CEC valve nut
48 CEC valve bracket
49 CEC valve bracket screw
50 Throttle body assembly
51 Throttle body gasket
52 Idle needle limiter cap
53 Idle needle
54 Idle needle spring
55 Throttle body screw
56 Pump and power rods lever
57 Lever attaching screw
58 Power piston rod link
59 Pump lever link

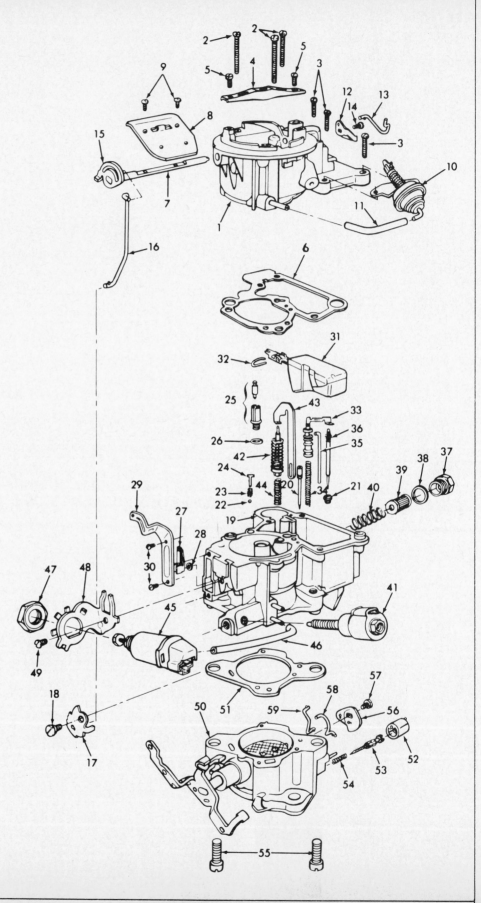

## MAJOR SERVICE OPERATIONS—DIS-ASSEMBLY, CLEANING, INSPECTION & ASSEMBLY PROCEDURES

### DISASSEMBLY

### AIR-HORN REMOVAL

1. Carefully look over the fast-idle cam, choke rod and choke lever to ensure correct reassembly. Make drawings if possible.
2. Remove fast-idle-cam pivot screw and disengage fast-idle cam and choke rod from the choke lever.
3. Remove six air-horn-to-float-bowl attaching screws. There are three long and three short screws. Be sure to note the location of any brackets, etc. also fastened by these screws.
4. Remove air horn by lifting straight up, then twist the float side of the air horn up to disengage the choke lever from the fast-idle linkage. Invert air horn and place on clean bench. Air-horn-to-bowl gasket can remain on bowl for later removal.

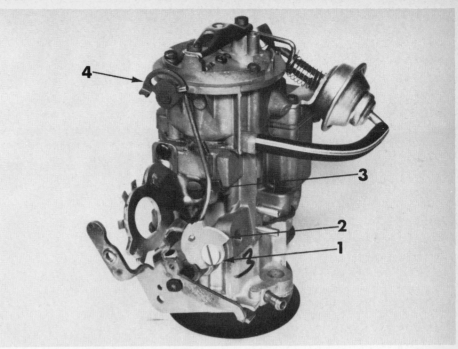

Monojet with external vacuum break. 1—fast-idle cam pivot screw, 2—fast-idle cam, 3—choke rod, 4—choke lever.

4/Once screws clear the bowl, tip air horn to slide link out of choke-lever slot. These levers and links vary by year and model so it is wise to make a sketch of any assembly before you take it apart.

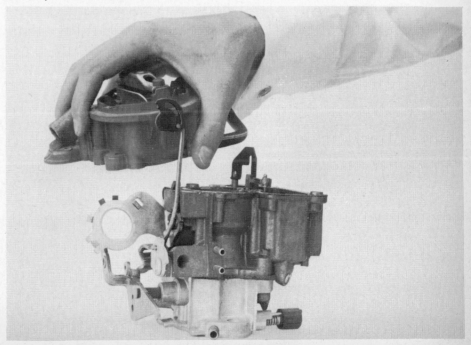

3/Be sure to note the location of any brackets attached to the carburetor by the air-horn-to-float-bowl attaching screws.

## AIR-HORN DISASSEMBLY

1. Remove two vacuum-break-diaphragm-cover screws. Then carefully remove diaphragm cover. On some models the vacuum break is mounted remotely on a bracket with choke-assist spring.

2. To remove a vacuum-break diaphragm and plunger rod, hold choke valve open. Then push upward on diaphragm rod until the rod's looped end slides out off vacuum-break lever attached to choke valve. Remove diaphragm plunger rod through hole in air horn. To remove rubber diaphragm, carefully slide off plunger stem.

3. If necessary the choke valve, vacuum-break lever and choke shaft can be removed from air horn by removing two choke-valve screws. File off staking on choke-valve screws before removing so as not to ruin threads or distort choke shaft. The air horn should not be disassembled for normal cleaning.

4. No further disassembly of the air horn is necessary.

NOTE: On early-model Monojets the idle-vent valve can be removed by turning screw head out of plastic guide. A repair kit is available if replacement parts are needed. The cranking enrichment valve is not removable. After cleaning make sure cleaning solution is completely removed from valve cavity and bleed hole in valve retainer is open.

1/Disassembled internal vacuum-break assembly. Note rod connecting diaphragm to choke valve. Rod should hook over choke as shown. Be careful not to rupture diaphragm on reassembly.

Underside of air horn shows valve for venting fuel vapors from bowl to canister (arrow). Models with the choke enrichment feature have a tube which protrudes below the gasket surface of the air horn.

## FLOAT BOWL DISASSEMBLY

1. Remove air-horn-to-float-bowl gasket. Slit in gasket next to metering-rod lever allows sliding gasket over lever for ease in removal.

NOTE: Vega Monojet does not have a metering-rod lever.

2. Remove float assembly from float bowl by lifting upward on float hinge pin. Remove hinge pin from float arm.

3. Remove needle, seat and gasket. To prevent damage to needle seat, use seat-removal tool BT-3006 or a screw driver wide enough to engage both slots fully.

4. Remove fuel-inlet nut and gasket, then remove filter element and pressure-relief spring.

5. Using long-nosed pliers, remove T-shaped pump-discharge guide. Pump-discharge spring and ball are removed by inverting the bowl.

6. The idle tube will fall out at same time when you invert the bowl.

NOTE: On the Vega Monojet, the power valve (used instead of a power piston-metering rod assembly) also falls out when the bowl is inverted.

7. To remove accelerator-pump plunger and power-piston/metering-rod assemblies, remove actuating lever on throttle shaft by removing attaching screw in shaft end.

8. Hold the power-piston assembly down in float bowl, then remove power-piston-drive link by sliding out of hole in power-piston-drive rod. The power-piston/metering-rod assembly can now be removed from float bowl.

NOTE: The metering rod can be removed from power-piston holder by pushing downward on end of rod against spring tension. Then slide narrow neck of rod out of rod-holder slot.

9. Remove power-piston spring from power-piston cavity.

10. Remove power-piston-drive link from throttle-actuating lever by aligning protrusion on rod and notch in lever.

11. Remove actuating lever from accelerator-pump-drive link in same manner. Note position of actuating lever for ease in reassembly.

12. Hold the pump plunger down in bowl cavity and remove drive link from pump-plunger shaft by rotating link until protrusion on link aligns with notch in plunger shaft.

13. Remove pump-plunger assembly from float bowl.

14. Remove pump-return spring from pump well.

15. Remove main-metering jet from bottom of fuel bowl.

16. Remove two screws from idle-compensator cover. Then remove cover, hot-idle compensator and seal from recess in bowl beneath compensator.

NOTE: The Vega carburetor does not have a hot-idle compensator.

1—fuel inlet seat, 2—T-shaped discharge valve-spring guide, 3—idle-tube, 4—power-piston metering-rod assembly, 5—accelerator-pump assembly.

3/Use wide-bladed screwdriver to avoid damaging inlet seat.

6/Vega Monojet float bowl. Power valve 1 replaces the power-piston metering-rod assembly. 2 is idle tube.

7/Black arrow indicates screw holding pump and power-piston lever. Link connects to power piston and pump (white arrow).

15/Removing main jet with screwdriver. Arrow indicates hole which feeds APT fuel system.

## THROTTLE BODY REMOVAL & DISASSEMBLY

1. Remove idle-mixture needle and spring. 1970 and later carburetors have idle-limiting caps over the mixture needles that must be removed carefully to avoid bending the mixture needles. The limiting caps are destroyed when they are removed.

2. Invert carburetor bowl on bench and remove two throttle-body-to-bowl attaching screws. Throttle body and insulator gasket may now be removed.

NOTE: Due to the close-tolerance fit of the throttle valve in the bore of the throttle body, do not remove the throttle valve or shaft.

1/One method of removing idle-limiting caps on late-model carburetors. Be careful not to damage or bend the needle.

2/Two throttle-body attaching screws are shown by arrows.

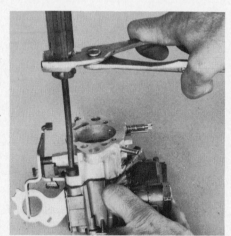

2A/Use a pair of pliers for extra leverage on a large phillips-head screwdriver if the screws are seized in the aluminum float bowl. Better yet, use an impact driver. To avoid damage to the thin metal (domed) main well plug (bleed), it is best to take the throttle body screws out before removing the air horn.

Throttle body, gasket and retaining screws after disassembly from bowl.

# ASSEMBLY

## THROTTLE BODY ASSEMBLY & INSTALLATION

1. Install idle-mixture needle and spring until *lightly* seated. Back out 1-1/2—2 turns as a preliminary idle adjustment.
2. Invert float bowl and install new throttle-body-to-bowl insulator gasket, making sure all gasket holes align with float-bowl holes.
3. Install throttle body on bowl gasket so all throttle-body holes align with gasket holes.
4. Install two throttle-body-to-bowl attaching screws. Tighten evenly and securely (12—15 ft-lbs of torque).

## FLOAT BOWL ASSEMBLY

1. Install slow-idle adjustment screw, if removed.
2. Install seal into recess in idle-compensator cavity in float bowl, then install idle-compensator assembly (except Vega).
3. Install idle-compensator cover, retaining with two attaching screws. Tighten securely.
4. Install main-metering jet into fuel-bowl bottom. Tighten securely.
5. Install pump-return spring into pump well. Make sure spring is correctly seated in bottom of well.
6. Install pump-actuating lever to lower end of pump-drive link by aligning rod protrusion with notch in lever. Projection on actuating lever points downward. Install power-piston-actuating link into opposite end of actuating lever. Lower end of link has retaining protrusion and faces outward away from throttle bore.
7. Install pump-plunger assembly into pump well with actuating shaft protruding through bottom of bowl casting. Push downward on pump plunger and install pump-drive link into hole in lower end of plunger shaft. Ends of drive link point towards carburetor bore. Bend in link faces toward fuel inlet. Protrusion on end of link retains link to pump shaft.
8. Install power-piston spring into power-piston cavity.
9. Install end of power-piston drive rod into groove on side of power piston. Then install power-piston/metering-rod assembly and drive rod into float bowl. Metering-rod end must enter jet orifice.
10. Hold complete assembly downward in bowl, then install power-piston-drive link into hole in lower end of power-piston-drive rod beneath bowl. Align D hole in actuating lever with flats on throttle shaft

1/Idle-mixture needle installed. Black arrow indicates curb-idle discharge hole (nearest gasket surface). White arrow indicates bottom of off-idle slot.

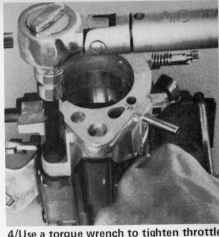

4/Use a torque wrench to tighten throttle body screws to 12—15 ft-lbs torque.

7/Arrow indicates pump actuating connecting arm.

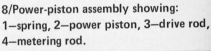

7A/Push down on pump arm to allow attaching link (arrow). Note that lever is not yet attached to throttle shaft.

8/Power-piston assembly showing:
1—spring, 2—power piston, 3—drive rod, 4—metering rod.

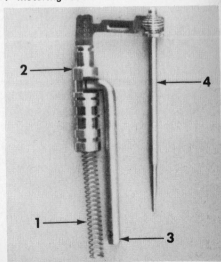

9/Install power-piston rod end into groove in power piston (arrow).

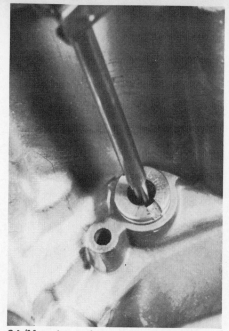

9A/Metering rod must enter jet orifice. Guide it carefully or you may damage the rod.

and install lever on end of throttle shaft. Install retaining screw in end of throttle shaft and tighten securely.

11. Install idle tube into float-bowl cavity.

12. Install pump-discharge ball, spring and spring retainer. Make sure spring retainer is in flush with top of bowl casting.

13. Install fuel-filter-relief spring, fuel-inlet filter, filter nut and gasket. Tighten securely.

NOTE: Open end of filter should face hole in fuel-inlet nut.

14. Install float-needle seat and gasket. Tighten securely using needle-seat installation tool BT-3006 or a wide-bladed screw driver.

15. Install float needle valve into needle seat.

16. Insert float hinge pin into float arm. Then install float and hinge pin into float bowl.

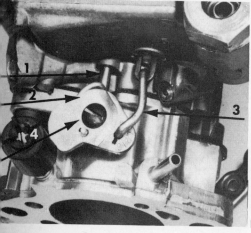

10/ 1 is power-piston actuating rod, 2 is link from rod 1 to lever, 3 is accelerator-pump link, 4 is pump lever.

16/Insert hinge pin into float arm and lower assembly into recess provided in bowl.

11 & 12/ 1—idle tube, 2—main-well plug with bleed hole, 3—T-shaped pump-discharge-spring retainer, 4—fuel-inlet needle, 5—velocity-actuated power valve weight (Vega only). Unit in left photo is Monojet with metering rod. Carburetor in right photo is Vega type (no metering rod).

## FLOAT-LEVEL ADJUSTMENT

1. Adjust float level per details in Figure 1 of adjustment drawings at end of this chapter.

## METERING-ROD ADJUSTMENT

1. Measure metering rod setting and make necessary adjustments by following details in Figure 2 of adjustment drawings at end of this chapter.

2. After adjustment, install metering rod. Refer back to disassembly information for attachment details.

## AIR HORN ASSEMBLY & INSTALLATION

1. Install air-horn gasket on float bowl by carefully sliding slit portion of gasket over metering-rod holder. Then align gasket with dowels provided on top of bowl casting and press gasket firmly in place.

2. Install idle-vent valve assembly, if removed.

3. Install choke shaft, choke valve and vacuum-break lever, if removed. Align choke valve, tighten two retaining screws and stake securely.

4. Install vacuum-break rubber diaphragm on plunger stem. The diaphragm fits around the plunger head. Then install plunger assembly into cavity at side of air horn. With choke valve in the open position, slide plunger-rod eyelet over end of vacuum-break lever on choke valve.

5. Seat vacuum-break diaphragm over sealing bead on air-horn casting. With diaphragm held in place, carefully install diaphragm cover and two retaining screws. Tighten screws securely.

6. Install air horn to float bowl by lowering gently onto float bowl until seated. Install three long and three short air-horn-to-float-bowl attaching screws. Tighten securely.

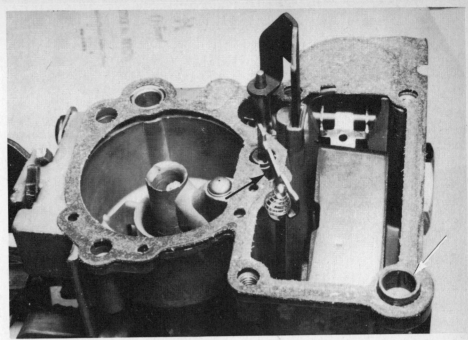

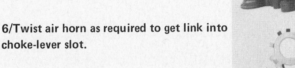

1/Black arrow indicates gasket slit which slides over metering rod holder. White arrow indicates gasket and air-horn-to-bowl alignment dowel.

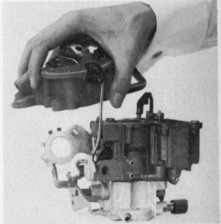

6/Twist air horn as required to get link into choke-lever slot.

Installation of external vacuum-break diaphragm with choke-assist spring.

Do not forget any brackets that are also fastened to the carburetor by the air-horn-to-float-bowl attaching screws.

7. Install lower end of choke rod into keyed hole in fast-idle cam. Numbers on the cam should be on the side opposite the rod.

8. Hook the end of the choke rod through the long slot in the choke lever. The rod should be on the carburetor side of the lever.

9. Bring the fast-idle cam down to its mounting boss and secure it with the pivot screw. Letters on the cam should face outwards and the steps should face the throttle-lever idle tang.

## OTHER ADJUSTMENTS

See adjustment drawings on following pages for:
1. Off-car idle-vent adjustment.
2. Fast-idle adjustment on bench.
3. Choke-rod adjustment.
4. Vacuum-break adjustment.
5. Unloader adjustment.
6. Choke-coil adjustment.
7. Fast-idle adjustment on car.

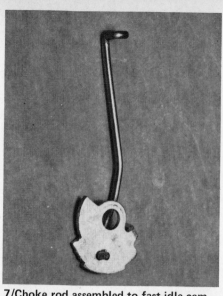

7/Choke rod assembled to fast-idle cam.

9/Assembled fast-idle mechanism.

## MONOJET MAIN METERING JETS

The main metering jet used in the Monojet differs from and should not be interchanged with other models. It has five radial lines stamped opposite the jet identification number. The stamped number indicates orifice size. The number can be two or three digits, depending on orifice size. Jet size can be determined by subtracting 100 from the last three digits of the part number. For example:

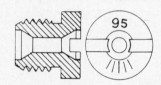

| | Stamped On Jet | Diameter Of Orifice |
|---|---|---|
| 7034195 | 95 | .095'' |
| 7034205 | 105 | .105'' |

### MONOJET MAIN METERING JETS

| Part | Stamped On Jet | Part | Stamped On Jet |
|---|---|---|---|
| 7034154 | 54 | 7034203 | 103 |
| 7034192 | 92 | 7034204 | 104 |
| 7034195 | 95 | 7034205 | 105 |
| 7034198 | 98 | 7034206 | 106 |
| 7034199 | 99 | 7034215 | 115 |
| 7034200 | 100 | 7034218 | 118 |
| 7034201 | 101 | 7034225 | 125 |
| 7034202 | 102 | 7034228 | 128 |

## MONOJET MAIN METERING RODS

The Monojet main metering rod can be identified as follows. A three-digit number stamped on the shank of the metering rod is the diameter in thousandths at point A.

Example:

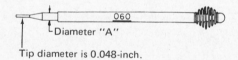

Diameter "A"

Tip diameter is 0.048-inch.

| Stamped | | Diameter A |
|---|---|---|
| 060 | | .060'' |

| Part | Stamped On Rod | Part | Stamped On Rod |
|---|---|---|---|
| 7035929 | 070 | 7037662 | 066 |
| 7035930 | 072 | 7037974 | 076 |
| 7035931 | 074 | 7037979 | 094 |
| 7035932 | 080 | 7040630 | 078 |
| 7035934 | 088 | 7040770 | 060 |
| 7035935 | 096 | 7044791 | 081 |
| 7036226 | 086 | 7045843 | 098 |

**Don't mix jets**—Jets for the Q-jet and Monojet have square shoulders under the head of the jet. Jets for other Rochester carburetors are *tapered* under the head of the jet. Refer to page 214 so you can tell the difference.

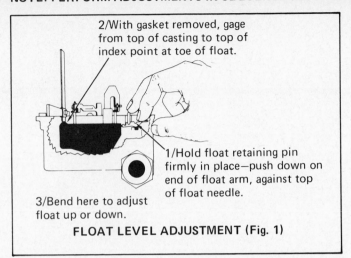

2/With gasket removed, gage from top of casting to top of index point at toe of float.

1/Hold float retaining pin firmly in place—push down on end of float arm, against top of float needle.

3/Bend here to adjust float up or down.

**FLOAT LEVEL ADJUSTMENT (Fig. 1)**

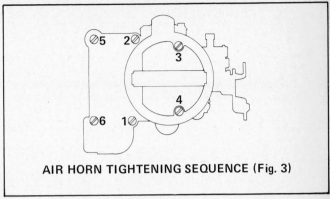

**AIR HORN TIGHTENING SEQUENCE (Fig. 3)**

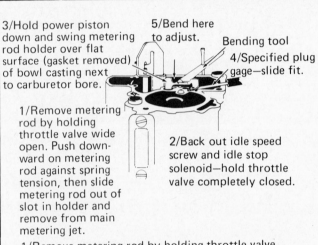

3/Hold power piston down and swing metering rod holder over flat surface (gasket removed) of bowl casting next to carburetor bore.

5/Bend here to adjust.

Bending tool

4/Specified plug gage—slide fit.

1/Remove metering rod by holding throttle valve wide open. Push downward on metering rod against spring tension, then slide metering rod out of slot in holder and remove from main metering jet.

2/Back out idle speed screw and idle stop solenoid—hold throttle valve completely closed.

1/Remove metering rod by holding throttle valve wide open. Push downward on metering rod against spring tension, then slide metering rod out of slot in holder and remove from main metering jet.

2/To check adjustment back out slow idle screw and rotate fast idle cam so that fast idle cam follower is not contacting steps on cam. With throttle valve completely closed, apply pressure to top of power piston and hold piston down against its stop.

3/Holding downward pressure on power piston, swing metering rod holder over flat surface of bowl casting next to carburetor bore.

4/Use specified plug gage and insert between bowl casting sealing bead and lower surface of metering rod holder. Gage should have a slide fit between both surfaces.

5/To adjust, carefully bend metering rod holder up or down at point shown.

After adjustment install metering rod.

Install air horn gasket on float bowl by carefully sliding slit portion of gasket over metering rod holder. Then align gasket with dowels provided on top of bowl casting and press gasket firmly in place.

**METERING ROD ADJUSTMENT (Fig. 2)**

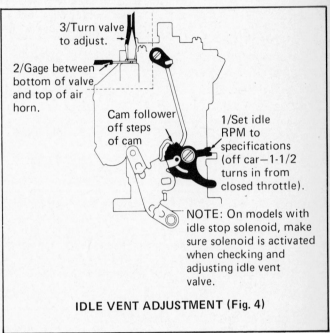

3/Turn valve to adjust.

2/Gage between bottom of valve and top of air horn.

Cam follower off steps of cam

1/Set idle RPM to specifications (off car—1-1/2 turns in from closed throttle).

NOTE: On models with idle stop solenoid, make sure solenoid is activated when checking and adjusting idle vent valve.

**IDLE VENT ADJUSTMENT (Fig. 4)**

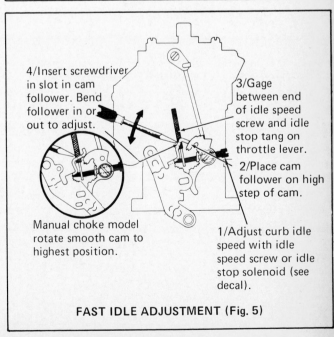

4/Insert screwdriver in slot in cam follower. Bend follower in or out to adjust.

3/Gage between end of idle speed screw and idle stop tang on throttle lever.

2/Place cam follower on high step of cam.

Manual choke model rotate smooth cam to highest position.

1/Adjust curb idle speed with idle speed screw or idle stop solenoid (see decal).

**FAST IDLE ADJUSTMENT (Fig. 5)**

Continued on following page.

## FAST IDLE ADJUSTMENT (Fig. 5, continued)

Automatic choke models with steps on fast idle cam.

1/Set normal engine idle speed. (Initial idle speed setting off car is 1-1/2 turns in on idle speed screw from closed throttle valve position.)

2/Place fast idle cam follower tang on highest step of cam.

3/With tang held against cam check clearance between end of slow idle screw and idle stop tang on throttle lever. It should be as specified.

4/To adjust insert end of screwdriver in slot provided in fast idle cam follower tang and bend inwards (toward cam) or outward (away from cam) to obtain specified dimension.

Manual choke models with smooth contour cam surface.

1/Use same procedure as above except in step 2 rotate the fast idle cam clockwise to its farthest up position. Always recheck fast idle setting on car.

Fast idle adjustment (on car)

1/Warm up engine.

2/Place fast idle cam follower on specified step of fast idle cam.

3/With cam follower held against specified step, insert screwdriver in adjustment slot and bend tang towards or away from cam to obtain specified RPM.

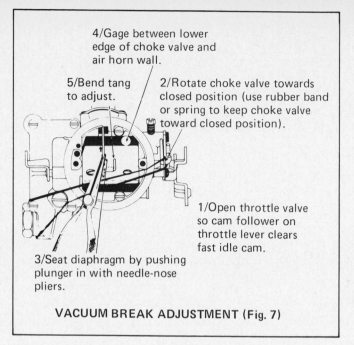

4/Gage between lower edge of choke valve and air horn wall.

5/Bend tang to adjust.

2/Rotate choke valve towards closed position (use rubber band or spring to keep choke valve toward closed position).

1/Open throttle valve so cam follower on throttle lever clears fast idle cam.

3/Seat diaphragm by pushing plunger in with needle-nose pliers.

### VACUUM BREAK ADJUSTMENT (Fig. 7)

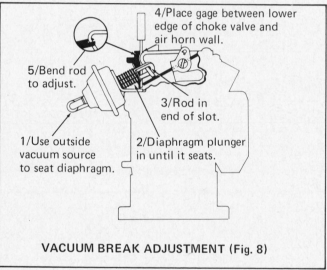

4/Place gage between lower edge of choke valve and air horn wall.

5/Bend rod to adjust.

3/Rod in end of slot.

1/Use outside vacuum source to seat diaphragm.

2/Diaphragm plunger in until it seats.

### VACUUM BREAK ADJUSTMENT (Fig. 8)

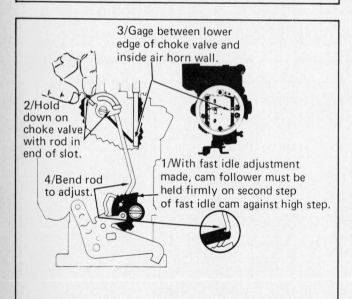

3/Gage between lower edge of choke valve and inside air horn wall.

2/Hold down on choke valve with rod in end of slot.

4/Bend rod to adjust.

1/With fast idle adjustment made, cam follower must be held firmly on second step of fast idle cam against high step.

NOTE: Manual choke models with smooth contour cam:

Use the same procedure as above except for step 1. As there are no steps on manual choke cam, the index line on side of cam should be lined up with contact point of the fast idle cam follower tang.

### CHOKE ROD (FAST IDLE CAM) ADJUSTMENT (Fig. 6)

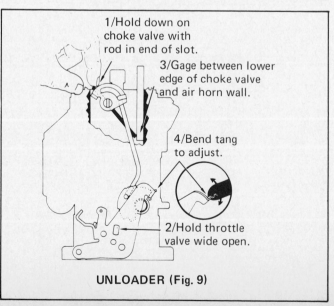

1/Hold down on choke valve with rod in end of slot.

3/Gage between lower edge of choke valve and air horn wall.

4/Bend tang to adjust.

2/Hold throttle valve wide open.

### UNLOADER (Fig. 9)

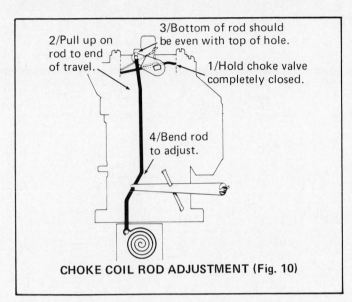

2/Pull up on rod to end of travel.

3/Bottom of rod should be even with top of hole.

1/Hold choke valve completely closed.

4/Bend rod to adjust.

**CHOKE COIL ROD ADJUSTMENT (Fig. 10)**

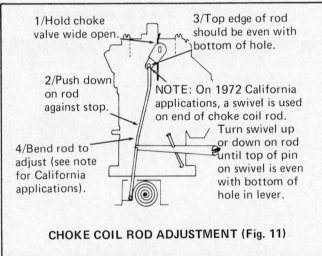

1/Hold choke valve wide open.

3/Top edge of rod should be even with bottom of hole.

2/Push down on rod against stop.

NOTE: On 1972 California applications, a swivel is used on end of choke coil rod.

4/Bend rod to adjust (see note for California applications).

Turn swivel up or down on rod until top of pin on swivel is even with bottom of hole in lever.

**CHOKE COIL ROD ADJUSTMENT (Fig. 11)**

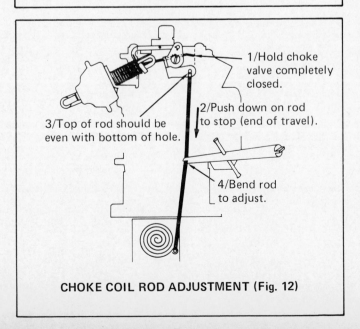

1/Hold choke valve completely closed.

2/Push down on rod to stop (end of travel).

3/Top of rod should be even with bottom of hole.

4/Bend rod to adjust.

**CHOKE COIL ROD ADJUSTMENT (Fig. 12)**

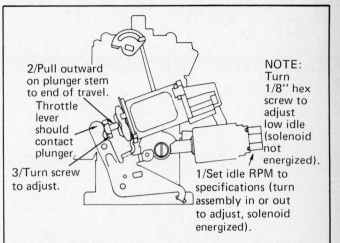

2/Pull outward on plunger stem to end of travel. Throttle lever should contact plunger.

3/Turn screw to adjust.

NOTE: Turn 1/8'' hex screw to adjust low idle (solenoid not energized).

1/Set idle RPM to specifications (turn assembly in or out to adjust, solenoid energized).

This adjustment is to be made only after: (1) replacement of solenoid, (2) major overhaul of carburetor, or (3) throttle body removed and replaced.

The following procedure is used to adjust the CEC valve on a running engine (in "neutral" for manual or in "drive" for automatic transmissions), with air conditioning off, distributor vacuum hose removed and plugged, and fuel tank hose from vapor canister disconnected.

Before proceeding follow instructions on vehicle tune-up sticker.

1/Some applications us an idle stop solenoid. Adjust curb idle speed to specifications by turning the idle stop solenoid in or out with solenoid energized. At this point with the solenoid de-energized (electrically disconnected), check the low idle speed setting and, if necessary, turn the 1/8'' hex screw at the rear of the idle stop solenoid to adjust.

CAUTION: Do not turn the hex screw in too far or damage to the idle stop solenoid will result.

On models not using a carburetor mounted idle stop solenoid, turn low idle speed screw for low idle setting.

2/Manually extend the CEC valve plunger to contact the throttle lever and pull outward on plunger stem to end of travel.

3/Turn plunger screw to adjust engine speed to car manufacturer's specifications.

|  | Passenger | Truck |
|---|---|---|
| A/T | 650-D | 750-D |
| M/T | 850-N | 1000-N |

CAUTION: Do not use the CEC valve to set curb idle speed.

**CEC VALVE ADJUSTMENT
MODELS M AND MV
(Fig. 13)**

# Single Barrels of the Past

## THE MODEL B CARBURETOR

The B series single-barrel carburetor was found on most of the Chevrolet six-cylinder engines produced for a 15-year period from 1949–63. Various letter designations denote choke variations: B had a manual choke and was used exclusively on trucks, BC had a thermostat choke mounted on the air horn and was used on cars and trucks, BV operated with a remote thermostatic coil on the exhaust manifold.

Major design features of this carburetor were the circular float bowl and the metering passages in the air horn. The float bowl completely encircled the venturi which, combined with the centrally located nozzle, prevented fuel spill-over during violent maneuvers such as quick turns or stops. All metering passages (except for the idle passages) are in the air horn. Because these passages are kept cooler by the air-horn gasket and the gasoline in the float bowl, the B models offered good metering consistency and less chance of vapor lock than previously used carburetors.

The B series carburetor suffered a partially deserved notoriety for many years after its 1949 introduction. The first year's production had a very thin air-horn mounting surface which warped badly. For years afterward, even though the problem had been cured, mechanics still replaced B series carburetors at the first hint of problems. Shop-talk myths linger on far beyond the manufacturer's fixes and thusly cost the motorist many unnecessary replacement costs.

The only really unusual feature is the use of a "snatch-idle" system. The idle tube is located across the nozzle from the main fuel system. Idle fuel is then "snatched" across the air gap into the idle down passage. One side benefit of this method of obtaining idle fuel is that the main-nozzle passage is always primed and ready to provide fuel on nozzle demand. This allows a small-capacity accelerator pump with a short stroke because there is less lag in main system fuel start-up.

Systems operations are covered in chapter two, "How Your Carburetor Works."

Model BV choke operates from remote thermostatic choke coil on exhaust manifold. B models were the main air/fuel mixers for 1949-63 six-cylinder GM cars. Circular fuel-bowl design keeps fuel in throttle bowl during violent maneuvers. All metering components attach to air horn to keep these relatively cool. This helps to ensure good metering after a hot-soak or hot-idle period.

"Snatch-idle" system pulls fuel across top of throttle bore. Idle fuel is "snatched" across main-discharge nozzle. Idle pickup tube is in main well. System allows main system to start quickly because idle system keeps fuel at the main discharge nozzle. Note power piston held up against spring by manifold vacuum. Power valve is closed.

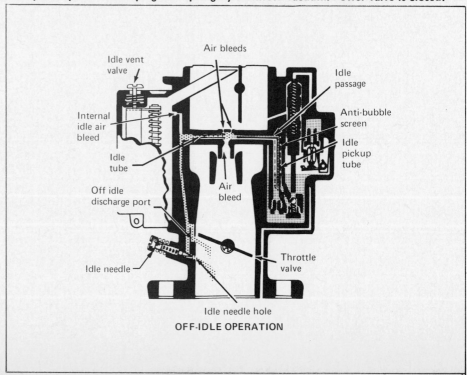

OFF-IDLE OPERATION

Air horn of B model shows "snatch-idle" connection to idle down channel 1, main discharge nozzle 2, power piston and stem 3, power valve 4, main jet 5, and float hinge pin 6. Cutaway gasket serves as a baffle to reduce fuel slosh out of vents.

"Circular" fuel bowl of B model surrounds throttle bore.

With engine under load, manifold vacuum drops, allowing power piston spring to push power-piston stem against power valve. Fuel flow through power valve supplements fuel supplied to main well through main metering jet.

**Economy tuning**—If you decide to tune your B model for utmost economy, see the hints provided in the economy and drivability section of chapter fourteen. Power valve assembly 7004475 cuts in at 5 in. Hg and is all in by 3 in. Hg manifold vacuum. Other power valve assemblies for the B models start at 8 and 11 inches, respectively. This information will be helpful if you are tuning for economy or altitude or if you are running a camshaft giving low manifold vacuum at idle.

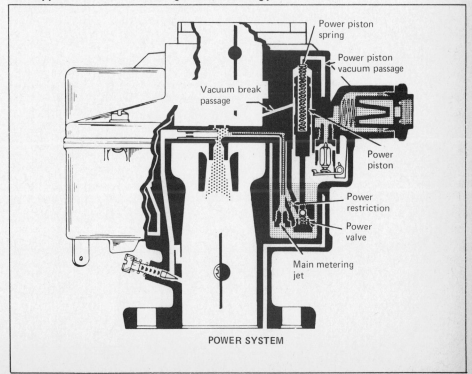

POWER SYSTEM

## Models B, BC, BV—Typical Assembly

1  Stat-cover screw
2  Plain retainer
3  Toothed retainer
4  Stat cover and coil assembly
5  Stat-cover gasket
6  Baffle plate
7  Choke-housing screw
8  Choke-piston pin
9  Choke piston
10  Choke housing
11  Fuel-inlet fitting
12  Fuel-inlet gasket
13  Fuel-inlet-filter gasket
14  Fuel filter
15  Fuel-filter spring
16  Choke-lever screw
17  Piston lever and link assembly
18  Choke shaft assembly
19  Choke valve
20  Choke-valve screw
21  Air-cleaner support
22  Air-horn screw
23  Air horn assembly
24  Needle and seat assembly
25  Power-piston spring
26  Power piston
27  Main-well support
28  Main-well-support screw
29  Main metering jet
30  Power valve assembly
31  Float-hinge pin
32  Float assembly
33  Air-horn gasket
34  Pump-assembly retainer
35  Pump-duration spring
36  Pump plunger assembly
37  Retainer pin
38  Pump link
39  Pump-discharge guide
40  Pump-discharge spring
41  Pump-discharge ball
42  Pump-return spring
43  Choke-rod clip
44  Choke rod
45  Choke-rod pin
46  Cam attaching screw
47  Fast idle cam
48  Float bowl assembly
49  Throttle-body gasket
50  Idle needle
51  Idle-needle spring
52  Choke-tube packing
53  Choke-tube nut
54  Throttle body assembly
55  Throttle-body screw
56  Idle-stop-screw spring
57  Idle-stop screw
58  Throttle-body gasket
59  Vacuum break diaphragm assembly (Model BV)
60  Screw
61  Link
62  Choke-shaft lever
63  Retainer pin
64  Vacuum break assembly
65  Vacuum hose
66  Idle vent valve assembly

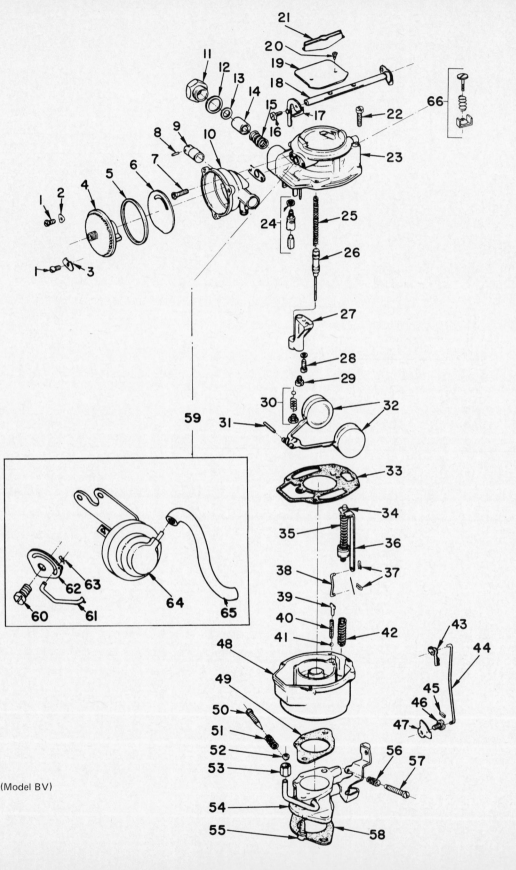

NOTE: PERFORM ADJUSTMENTS IN PROPER SEQUENCE

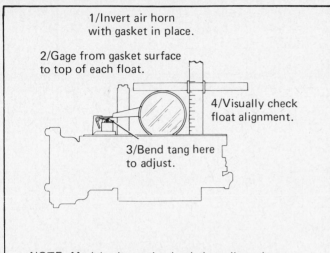

1/Invert air horn with gasket in place.

2/Gage from gasket surface to top of each float.

4/Visually check float alignment.

3/Bend tang here to adjust.

NOTE: Model using spring-loaded needle and seat assembly only: Place 0.030" shim between head of float-needle pin and float arm. With float arm resting freely on shim, check float height with gage. Bend float arms until each pontoon is set to specified dimension; remove shim from between float needle and float arm after adjustment.

**FLOAT LEVEL ADJUSTMENT (Fig. 1)**

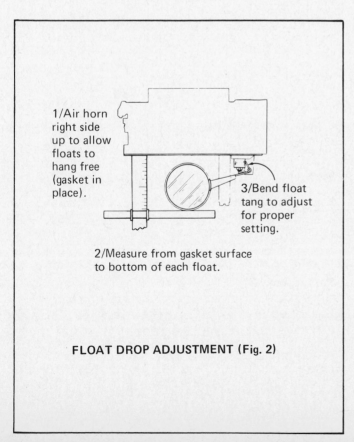

1/Air horn right side up to allow floats to hang free (gasket in place).

3/Bend float tang to adjust for proper setting.

2/Measure from gasket surface to bottom of each float.

**FLOAT DROP ADJUSTMENT (Fig. 2)**

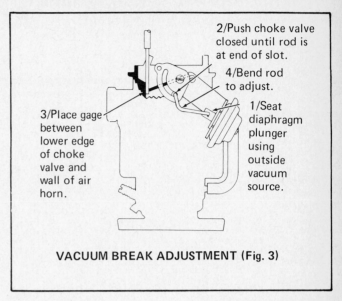

2/Push choke valve closed until rod is at end of slot.

4/Bend rod to adjust.

1/Seat diaphragm plunger using outside vacuum source.

3/Place gage between lower edge of choke valve and wall of air horn.

**VACUUM BREAK ADJUSTMENT (Fig. 3)**

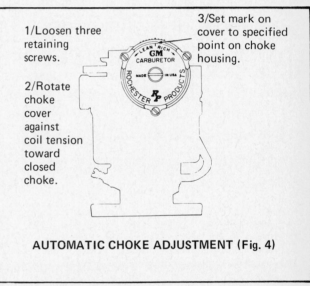

1/Loosen three retaining screws.

2/Rotate choke cover against coil tension toward closed choke.

3/Set mark on cover to specified point on choke housing.

**AUTOMATIC CHOKE ADJUSTMENT (Fig. 4)**

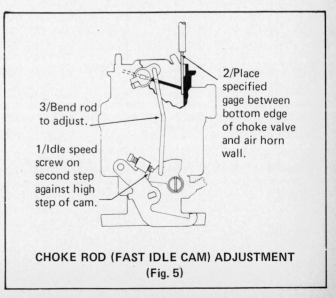

2/Place specified gage between bottom edge of choke valve and air horn wall.

3/Bend rod to adjust.

1/Idle speed screw on second step against high step of cam.

**CHOKE ROD (FAST IDLE CAM) ADJUSTMENT (Fig. 5)**

NOTE: PERFORM ADJUSTMENTS IN PROPER SEQUENCE

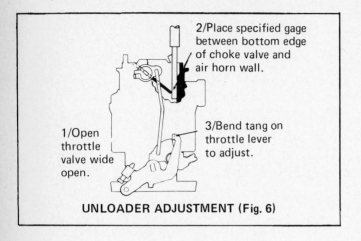

2/Place specified gage between bottom edge of choke valve and air horn wall.

1/Open throttle valve wide open.

3/Bend tang on throttle lever to adjust.

**UNLOADER ADJUSTMENT (Fig. 6)**

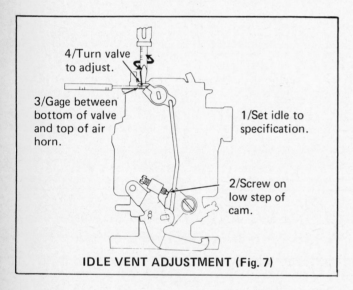

4/Turn valve to adjust.

3/Gage between bottom of valve and top of air horn.

1/Set idle to specification.

2/Screw on low step of cam.

**IDLE VENT ADJUSTMENT (Fig. 7)**

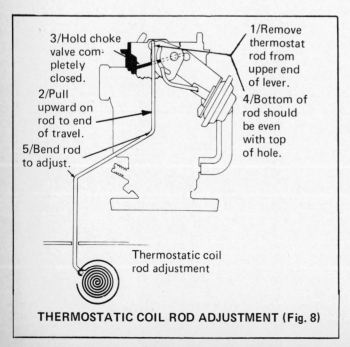

3/Hold choke valve completely closed.

2/Pull upward on rod to end of travel.

5/Bend rod to adjust.

1/Remove thermostat rod from upper end of lever.

4/Bottom of rod should be even with top of hole.

Thermostatic coil rod adjustment

**THERMOSTATIC COIL ROD ADJUSTMENT (Fig. 8)**

NOTE: PERFORM ADJUSTMENTS IN PROPER SEQUENCE

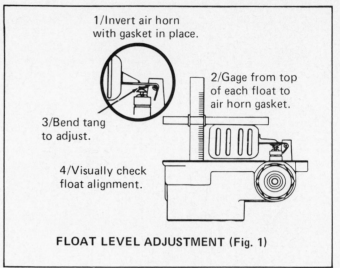

1/Invert air horn with gasket in place.

2/Gage from top of each float to air horn gasket.

3/Bend tang to adjust.

4/Visually check float alignment.

**FLOAT LEVEL ADJUSTMENT (Fig. 1)**

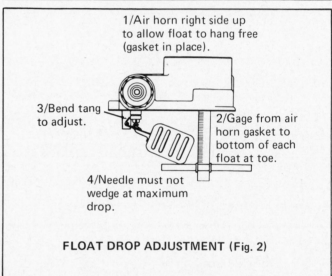

1/Air horn right side up to allow float to hang free (gasket in place).

3/Bend tang to adjust.

2/Gage from air horn gasket to bottom of each float at toe.

4/Needle must not wedge at maximum drop.

**FLOAT DROP ADJUSTMENT (Fig. 2)**

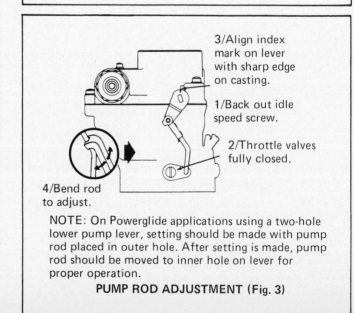

3/Align index mark on lever with sharp edge on casting.

1/Back out idle speed screw.

2/Throttle valves fully closed.

4/Bend rod to adjust.

NOTE: On Powerglide applications using a two-hole lower pump lever, setting should be made with pump rod placed in outer hole. After setting is made, pump rod should be moved to inner hole on lever for proper operation.

**PUMP ROD ADJUSTMENT (Fig. 3)**

**NOTE: PERFORM ADJUSTMENTS IN PROPER SEQUENCE**

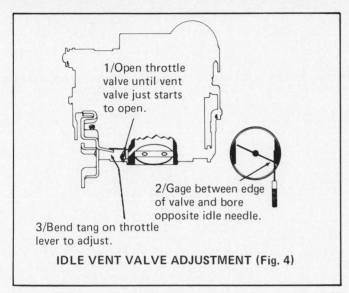

1/Open throttle valve until vent valve just starts to open.

2/Gage between edge of valve and bore opposite idle needle.

3/Bend tang on throttle lever to adjust.

**IDLE VENT VALVE ADJUSTMENT (Fig. 4)**

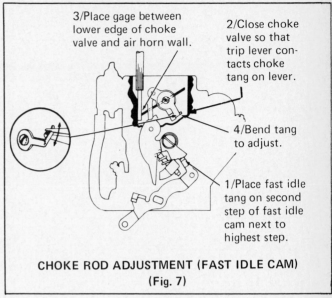

3/Place gage between lower edge of choke valve and air horn wall.

2/Close choke valve so that trip lever contacts choke tang on lever.

4/Bend tang to adjust.

1/Place fast idle tang on second step of fast idle cam next to highest step.

**CHOKE ROD ADJUSTMENT (FAST IDLE CAM) (Fig. 7)**

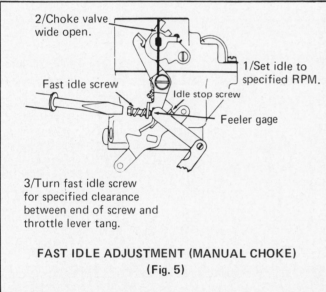

2/Choke valve wide open.

Fast idle screw

1/Set idle to specified RPM.

Idle stop screw

Feeler gage

3/Turn fast idle screw for specified clearance between end of screw and throttle lever tang.

**FAST IDLE ADJUSTMENT (MANUAL CHOKE) (Fig. 5)**

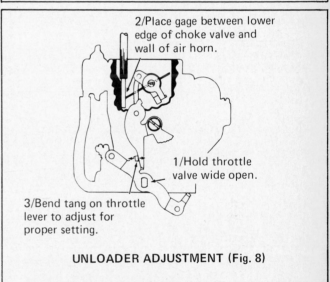

2/Place gage between lower edge of choke valve and wall of air horn.

1/Hold throttle valve wide open.

3/Bend tang on throttle lever to adjust for proper setting.

**UNLOADER ADJUSTMENT (Fig. 8)**

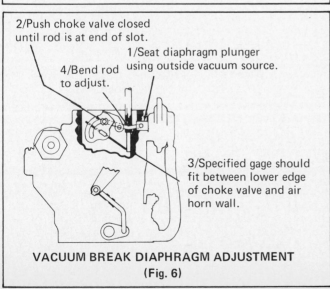

2/Push choke valve closed until rod is at end of slot.

1/Seat diaphragm plunger using outside vacuum source.

4/Bend rod to adjust.

3/Specified gage should fit between lower edge of choke valve and air horn wall.

**VACUUM BREAK DIAPHRAGM ADJUSTMENT (Fig. 6)**

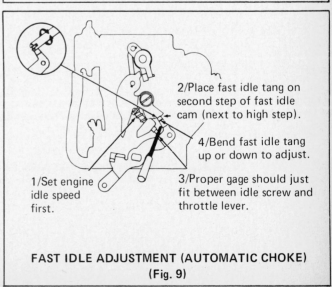

2/Place fast idle tang on second step of fast idle cam (next to high step).

4/Bend fast idle tang up or down to adjust.

1/Set engine idle speed first.

3/Proper gage should just fit between idle screw and throttle lever.

**FAST IDLE ADJUSTMENT (AUTOMATIC CHOKE) (Fig. 9)**

H models are used in dual (2 x 1) or four-carburetor (4 x 1) installations on Corvairs.

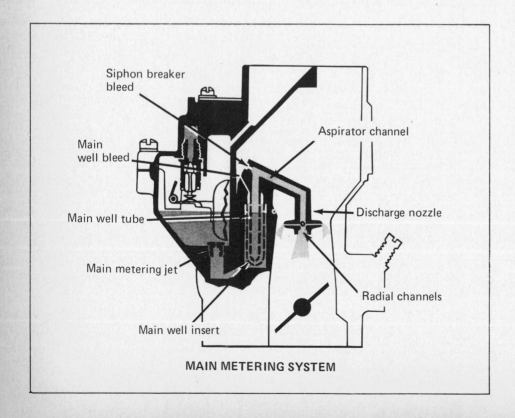

Siphon breaker bleed

Main well bleed

Aspirator channel

Main well tube

Discharge nozzle

Main metering jet

Radial channels

Main well insert

**MAIN METERING SYSTEM**

## THE MODEL H CARBURETOR

The Model H carburetor had a unique history in that it was designed, developed and produced exclusively for one vehicle: the Corvair. Two of the carbs were used in a dual installation. These were connected by a throttle-rod and air-cleaner arrangement. Some applications used four H carburetors (4 x 1) with a progressive linkage adapted to the cross-the-engine throttle-rod and air-cleaner arrangement.

The H carburetor had many new and distinct features and we have described the foremost of these in the following paragraphs.

The throttle body and float bowl were united in a single aluminum casting. This greatly simplified maintenance in the field and reduced initial building costs. Repair kits cost less and rebuild time is less than that required on other carburetors.

The fuel-discharge nozzle assembly was a completely new design. Instead of the conventional type of boost venturi, a radial-type discharge nozzle was used. It has four "spokes" or arms which discharge near the surface of the venturi. It's unusual and it works.

Main-well tube inserts are used in conjunction with the main-well tubes for improved hot-engine idle stability and to help prevent fuel percolation and general hot weather problems.

Early H carburetors did not have a power system as such. The Corvair relied on pulse enrichment created by sharply-defined pulsing caused by the carburetor placement, manifold design and the firing order. As exhaust-emission regulations increased the demand for leaner light-throttle metering, a power system was added in 1965 models. It provided enrichment only during high inlet-air velocities as opposed to more commonly used vacuum-operated systems. A similar system is used in the Monojet for Chevrolet Vega base (low-HP) engines.

The fuel filter, float, and metering systems were very conventional. Systems operations are covered in chapter two, "How Your Carburetor Works."

## Models H, HV—Typical Exploded View

1 Choke-valve screw
2 Choke valve
3 Choke shaft and lever assembly
4 Fuel-inlet nut
5 Fuel-inlet-filter gasket
6 Fuel-inlet-nut gasket
7 Fuel-inlet filter
8 Fuel-inlet-filter spring
9 Air horn assembly
10 Air-horn screw (short)
11 Lockwasher-air horn screws
12 Choke lever and collar assembly
13 Trip lever
14 Trip-lever screw
15 Pump shaft and lever assembly
16 Needle-seat gasket
17 Float needle seat
18 Pump-lever inside springclip
19 Pump inside lever
20 Vacuum break control assembly
21 Air-horn screw (long)
22 Control-rod clip
23 Vacuum control rod
24 Upper pump-rod retainer
25 Float-hinge pin
26 Torsion spring
27 Float needle
28 Float assembly
29 Vacuum-control hose
30 Pump-plunger clip
31 Pump assembly
32 Pump-return spring
33 Air-horn gasket
34 Venturi-cluster screw (short)
35 Venturi-cluster screw (long)
36 Cluster-screw lockwasher
37 Venturi-cluster assembly
38 Venturi-cluster gasket
39 Main-well-tube insert
40 Power valve
41 Pump-discharge valve
42 Main metering jet
43 Body and bowl assembly
44 Pump rod
45 Lower pump-rod retainer
46 Pump-lever attaching screw
47 Pump actuating lever
48 Choke rod
49 Fast idle cam
50 Cam attaching screw
51 Slow idle screw spring
52 Slow idle adjustment screw
53 Idle needle spring
54 Idle needle
55 Idle-vent valve
56 Idle-vent attaching screw

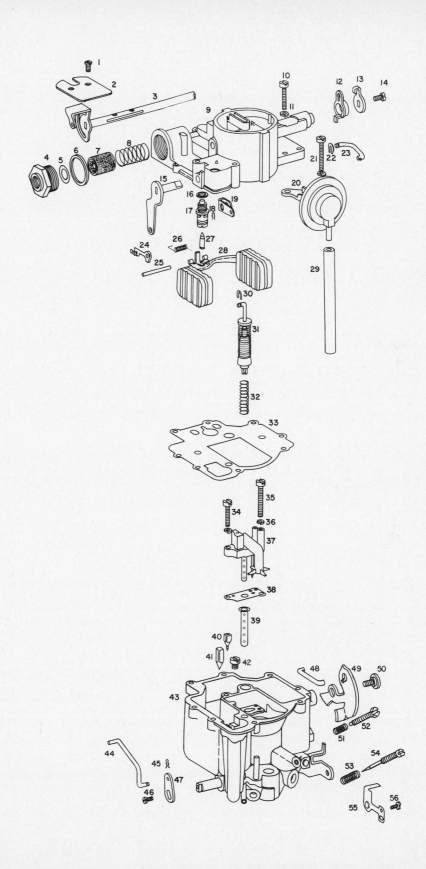

# 2G Carburetors

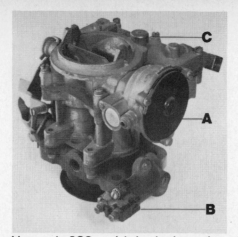

There are three basic G Rochester two-bore carburetor models: 2G, 2GC and 2GV. The basic model designation is G. 2 ahead of the G means the carburetor has two bores, two venturis and two separate identical metering systems: one of each per carburetor bore. The two-bore carburetor is normally used on V-8 engines with each bore supplying air and fuel to four cylinders through a divided intake manifold.

Two SAE throttle-body flange sizes are used: 1¼ and 1½ inch. The flange size used for a particular application is usually determined by the carburetor-bore size. Larger engines normally require more air capacity, hence, larger carburetor bores and flanges. Model 2G is equipped with a manually operated choke valve. It is usually used on truck and marine engines where an automatic choke is not necessary.

Model 2GC is the basic 2G unit with an automatic choke (designated by C) as an integral part of the carburetor. The automatic choke housing and thermostatic coil may be located on the air horn or throttle body, depending on the application.

Model 2GV also has an automatic choke. A vacuum-break diaphragm unit is used in place of the conventional choke housing and choke piston. The choke thermostatic coil ("stat coil") is located on the engine exhaust manifold and link-connected to the choke valve.

The Rochester two-bore carburetor has been kept basically simple for simple and easy servicing. Most of the calibrated metering parts are contained in the venturi-cluster assembly. This planned simplicity has been a big factor in bringing about the reputation of outstanding reliability enjoyed by Rochester Products' two-barrel carburetors since the current design was introduced in 1955.

Even though millions of happy customers have used them, this carburetor has seldom been touted as an

Very early 2GC model clearly shows the integral choke assembly A, throttle-actuated engine starting switch B (Buick feature) and the now-obsolete capped external bowl vent C.

2GV illustrates the vacuum-break diaphragm design.

These large 1-1/2-inch-bore two-barrel carburetors have been popular on many large V8's in both 2GC (as shown) and 2GV models. Note that the choke thermostat housing is located on the throttle body on this model.

This low-profile 1-inch-bore aluminum throttle body 2GC was designed specifically for small aluminum V8's sold by Buick and Oldsmobile in the early 60's. Choke thermostat housing is mounted on air horn in this example.

1974 Chevrolet 2GV features teflon-coated throttle shaft for low friction. A spun-in plastic bushing in the throttle lever is the bearing surface for the dual throttle-return spring required by 1974 safety standards. A longer outside pump lever and pump rod mates with the revised throttle lever. A heavier pump-return spring gives increased throttle closing tension. An electrically operated idle-stop solenoid maintains curb idle speed. The speed screw is set at a lower RPM so when ignition switch is turned off, de-energized solenoid allows throttle valves to close further so engine will not diesel (run on).

**Small vs. large bore**—Small-bore **2G, 2GC, 2GV** carburetors have a relatively small throttle body and are commonly referred to as the 1-1/4" size whereas large-bore **2G, 2GC, 2GV** carburetors have a relatively large throttle body and are commonly referred to as the 1-1/2" size. The 1-1/4 or 1-1/2" refers to the SAE flange size in relation to stud spacing.

ideal high-performance unit. This writer has helped a number of individuals tailor RPD two-barrels for many and varied race applications. After having helped install a pair of 1½-inch-bore Rochesters on a sprint car in Michigan, this writer watched gleefully as the car easily took first in the trophy dash, heat and main at the once-popular Jackson Raceway (no longer standing). A number of you older readers will recall the 348 Chevys with 3 x 2 carburetion. In 1958 those stood among the best in factory-performance machinery. Vic Hickey of Ventura, California selected the common 1-3/16-inch venturi RPD two-barrel for his Mini Boot which ran several off-road races including the Granddaddy Baja events. Vic was a front-runner in off-road race car building and the promotion of that type of racing. When a man chooses a unit costing less than $50 to put on $10,000–$20,000 vehicles, it is not with-

out thought. In short the two-barrel Rochester carburetors are good for general passenger-car use, circle-track installations and drag racing. For slaloms, road courses and off-road racing they require some race blueprinting. The high-performance section provides tips to help you accomplish this.

A number of aftermarket manifold suppliers have designed their unit for use with the Rochester two-barrel. All in all, it is a good serviceable unit that lends itself to many applications.

This writer cannot think of one carburetor that has stood the test of time as well as the RPD two-barrel units have. For nearly two decades this unit has "stood off" most competitive two-barrel units so that few others were used on General Motors Cars as original equipment. The following paragraph briefly summarizes RPD two-barrel features.

Vic Hickey, famous off-road race car builder, at right, watches his experimental "Mini Boot" going through its paces. It used a Rochester two-barrel carburetor on a four-cylinder Chevrolet engine.

The venturi cluster fits on a flat portion of the carburetor float bowl at the side of the main venturi. Idle tubes and main-discharge nozzles are permanently installed in the cluster body with a precision pressed fit. Main nozzles and idle tubes are suspended in the fuel in the main wells of the float bowl. This assembly and design method insulates the main metering parts from engine heat, preventing fuel vapors from disrupting efficient metering during hot-engine operation. The main metering jets are a fixed-orifice type. Metering calibration is accomplished through a system of calibrated orifices and air bleeds which supply the correct air/fuel mixtures to the engine throughout all operational ranges. The float bowl is located so each system will give instantaneous response for maximum efficiency and performance.

## OPERATING SYSTEMS

Six basic systems are used in the Rochester Model G two-bore carburetor: float, idle, main-metering, power, accelerator-pump and choke. We are not going to detail the operation of each system because a thorough study of this is provided in chapter two. It would be repetitious and of no particular value to

you, so we will touch only on important features or differences.

## FLOAT SYSTEM

Like most carburetors, the fuel level of the two-barrel RPD carburetor is controlled to design levels by a needle/seat assembly and float mechanism. The float is hinged from one side and consequently levels will vary to some extent during hard cornering from left to right. This is easily compensated for when the vehicle corners in the same direction—as in circle-track racing. Should you choose to use it on a race car for road racing it gets a little more difficult.

## IDLE SYSTEMS

Idle-system design and function are basically the same as most domestic and foreign units. One circumstance can exist with these carburetors which may cause annoying drivability problems. The idle tubes get their fuel from the main-nozzle well. This simply means that fuel has to pass through the main jets to get into that area. The main jets are at the back of the fuel bowl. During medium-hard long brake stops (such as a stop from speeds of 40 to 60 MPH for an unexpected red light) fuel stacks up in the front of the bowl, low-

ering the fuel level over the main jets and in the nozzle fuel wells. This can cause the 2G idle system to stop feeding. In such cases the engine stalls or falters badly. The tune-up cure is simply to richen the idle-mixture-screw settings. A small amount of enrichment is generally sufficient. This enrichment provides more signal (suction) on the idle fuel circuit, which will lift fuel from a lower level.

Current emission laws put a crimp in such procedures for late-model cars. Because of the "Clean Air Act, Amended December, 1970" idle needles are "set" at the factory to meet a specific requirement and black idle-limiter caps are installed. When the carburetor is removed for overhaul the caps are generally removed. After reassembly, the carburetor is adjusted according to the factory tune-up label on the vehicle. Any time the car is inspected, it will be obvious that the carburetor has been overhauled and perhaps readjusted.

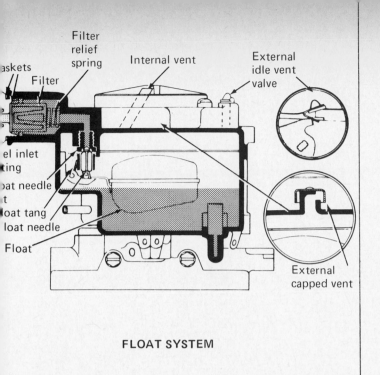

**FLOAT SYSTEM**

Labels: Gaskets, Filter, Filter relief spring, Internal vent, External idle vent valve, Fuel inlet fitting, Float needle seat, Float tang, Float needle, Float, External capped vent

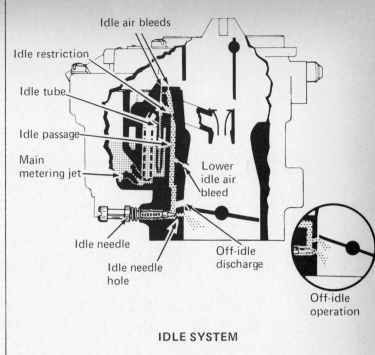

**IDLE SYSTEM**

Labels: Idle air bleeds, Idle restriction, Idle tube, Idle passage, Main metering jet, Lower idle air bleed, Idle needle, Idle needle hole, Off-idle discharge, Off-idle operation

**Vacuum-operated vent system**—Some California **2G** models have a vacuum-vent switch in the air horn. It ensures against any pressure buildup in the float chamber during hot soaks when the engine is not running by venting any vapor buildup to the canister. Once the engine starts, manifold vacuum switches the vent to vent the float bowl to air-horn pressure. A tang on the pump arm closes the vent mechanically at low manifold vacuums to ensure maintaining internal pressure balance in the carburetor. This system is not illustrated here.

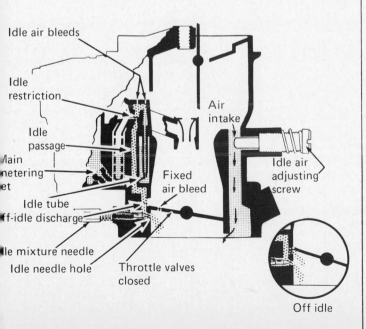

**IDLE AIR BY-PASS SYSTEM**

Labels: Idle air bleeds, Idle restriction, Idle passage, Main metering jet, Idle tube, Off-idle discharge, Idle mixture needle, Idle needle hole, Air intake, Fixed air bleed, Idle air adjusting screw, Throttle valves closed, Off idle

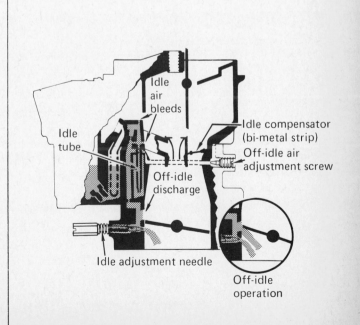

**IDLE SYSTEM WITH OFF-IDLE AIR ADJUSTMENT**

Labels: Idle air bleeds, Idle tube, Idle compensator (bi-metal strip), Off-idle air adjustment screw, Off-idle discharge, Idle adjustment needle, Off-idle operation

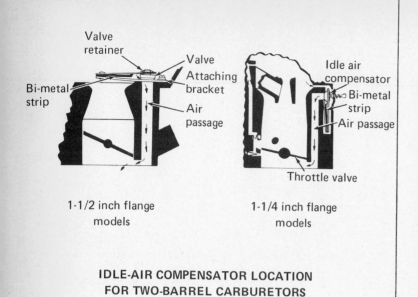

IDLE-AIR COMPENSATOR LOCATION
FOR TWO-BARREL CARBURETORS

Off-idle slots (arrows) are current practice for most carburetors as they are easier to machine and work similarly to a series of drilled holes. White arrows indicate holes in throttle plates, a feature often found on 2G carburetors. This allowed setting the throttle plates nearly closed, even on large-displacement engines, to ensure good operation of the off-idle system.

## THROTTLE-BODY VENTING

An important design feature in pre-1971 Rochester two-bore carburetors was throttle-body venting. It gave quicker hot-engine starting after short shut-downs.

During extreme hot-engine operation fuel in the carburetor tends to boil and vaporize. Some of the fuel vapor reaches the carburetor bores, condenses on the throttle valves and seeps into the engine manifold. By venting the area just above the throttle valves, hot-engine-starting time can be reduced to a minimum.

Two methods are used to vent the throttle-bore area.
1. A special throttle-body-to-bowl gasket has cut-out areas which vent fuel vapors from the carburetor bores just above the throttle valves to atmosphere. The vapors are emitted to the under-hood area.
2. Holes through the throttle-body casting just above the throttle valves serve the same purpose as the vented gasket. The vent holes are located so they will not disrupt idle or off-idle operation. They are above the throttle valves on the side opposite the mixture screws, in an area where the transfer from idle to main metering will not be

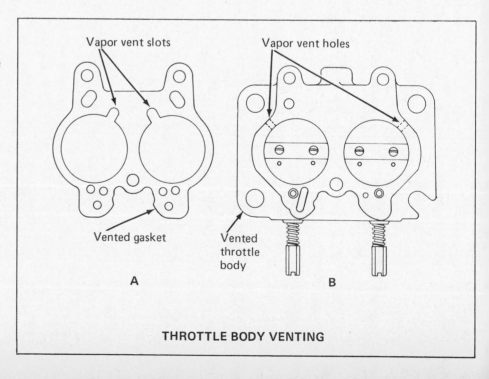

THROTTLE BODY VENTING

Here a drill has been inserted through the vent above the throttle plate. Arrow indicates similar vent hole in other barrel. This type venting was discontinued about 1970.

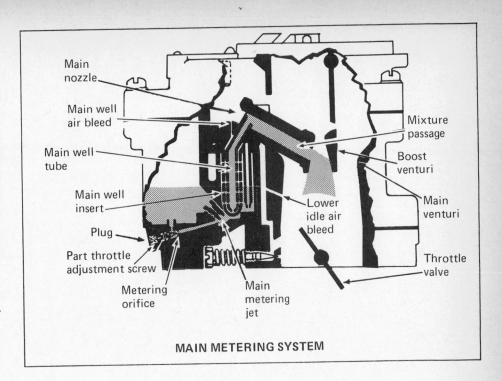

**MAIN METERING SYSTEM**

affected. Starting in the early 70's, these vents (and cut-out gaskets, too) were eliminated to reduce evaporative losses.

## MAIN METERING SYSTEM

Except for shape and location, this system is quite conventional. See the main-system section in chapter three for details of operation.

On some two barrels (especially those on late-model emission-controlled engines) the main system is supported with an adjustable-flow feature which enables production to control the fuel mixtures more accurately than could be done with a fixed orifice. This adjustable-part-throttle (APT) fuel is channeled from the fuel bowl to the nozzle feed well, independent of the metering jet but in parallel with it.

Most main well tubes (emulsion tubes) in 2G carburetors are crushed (flattened) and bent slightly away from the main jet passage. This reduces the possibility of lean mixtures and engine surge due to fuel hitting the end of the main well tube and interfering with fuel entering the tube. Some tubes contain a flat brass insert called a *splitter* to help reduce vapor effects by breaking up any bub-

bles entering the main well tube. Main well inserts may be used around the main-well tube to break up heat-caused vapor bubbles formed in the tube during hot-engine operation. The inserts prevent vapor bubbles from disrupting carburetor metering and help to ensure even fuel flow through the main well tubes and out of the discharge nozzles. The inserts may be either brass or plastic. Slashed plastic inserts are installed with the longer side of the insert next to the idle tube well to help keep vapor bubbles out of the idle circuit for consistent idle quality.

Some large-bore 1-1/2-inch 2G carburetors use a pull-over enrichment circuit at high speeds to allow leaner cruising or part-throttle mixtures for emission control. The system's two added holes in the air horn just above the choke valve are connected to fuel in the bowl by channels and tubes extending into the bowl just above the main metering jets.

Occasionally you will note boost venturis with metal inserts at the bottom on one or both sides. These help mixture distribution to the various cylinders. Some venturis are recessed for the inserts but do not use them. The use

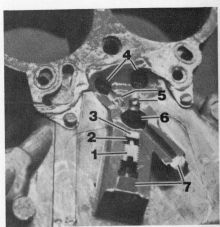

Adjustable part-throttle fuel system includes screw 1, fuel supply from bowl 2, metering orifice 3, and main wells 4. Power-valve channel restrictions 5 and power-valve cavity 6 do not affect the operation of the system. Plugs at 7 close the bowl passages after the carburetor has been machined.

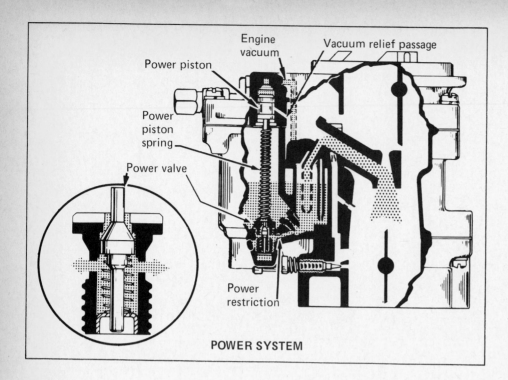

POWER SYSTEM

of inserts such as these is to solve a mani-folding problem—not a carburetor problem.

## POWER SYSTEM

To obtain the mixtures required for more power or sustained high-speed driving, most Rochester two-bore carburetors use a vacuum-operated power system. The 2GV carburetor used on the performance option 1971-72 Vega is an exception. Its air-velocity demand power-enrichment system operates as follows.

When the demand for fuel at the boost venturi exceeds the main-metering jets' capacity, the power-enrichment valve unseats so fuel can flow directly from the bowl to the main well and then to the boost venturi via the nozzle. This fuel bypasses the main jets.

A hole drilled from the carburetor air horn to the power-piston cavity relieves any vacuum that might leak around the power piston to the top of fuel in the bowl. Any vacuum acting on the fuel would affect carburetor calibration.

The power valve (see drawing inset) is a self-contained assembly consisting of plunger and a closing spring. The power piston forces the power valve plunger off

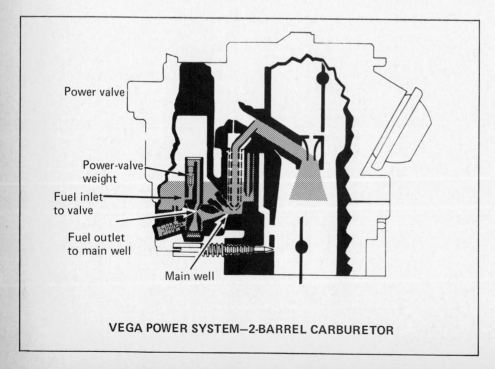

VEGA POWER SYSTEM—2-BARREL CARBURETOR

**Vega-type power valve contains a weight which raises off its seat at high air flows. The signal to the check valve is applied via the main fuel well which responds to the signal from the venturi. The inlet to the valve from bowl at 1 is calibrated to provide the desired amount of fuel.**

# MODELS 2G, 2GC, 2GV—Exploded View

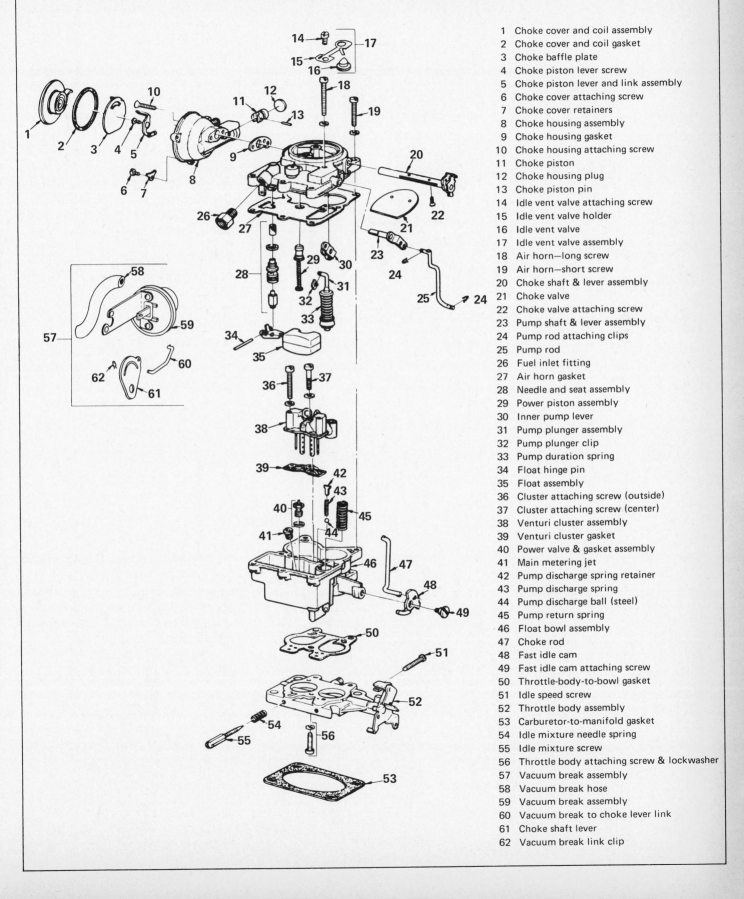

1. Choke cover and coil assembly
2. Choke cover and coil gasket
3. Choke baffle plate
4. Choke piston lever screw
5. Choke piston lever and link assembly
6. Choke cover attaching screw
7. Choke cover retainers
8. Choke housing assembly
9. Choke housing gasket
10. Choke housing attaching screw
11. Choke piston
12. Choke housing plug
13. Choke piston pin
14. Idle vent valve attaching screw
15. Idle vent valve holder
16. Idle vent valve
17. Idle vent valve assembly
18. Air horn—long screw
19. Air horn—short screw
20. Choke shaft & lever assembly
21. Choke valve
22. Choke valve attaching screw
23. Pump shaft & lever assembly
24. Pump rod attaching clips
25. Pump rod
26. Fuel inlet fitting
27. Air horn gasket
28. Needle and seat assembly
29. Power piston assembly
30. Inner pump lever
31. Pump plunger assembly
32. Pump plunger clip
33. Pump duration spring
34. Float hinge pin
35. Float assembly
36. Cluster attaching screw (outside)
37. Cluster attaching screw (center)
38. Venturi cluster assembly
39. Venturi cluster gasket
40. Power valve & gasket assembly
41. Main metering jet
42. Pump discharge spring retainer
43. Pump discharge spring
44. Pump discharge ball (steel)
45. Pump return spring
46. Float bowl assembly
47. Choke rod
48. Fast idle cam
49. Fast idle cam attaching screw
50. Throttle-body-to-bowl gasket
51. Idle speed screw
52. Throttle body assembly
53. Carburetor-to-manifold gasket
54. Idle mixture needle spring
55. Idle mixture screw
56. Throttle body attaching screw & lockwasher
57. Vacuum break assembly
58. Vacuum break hose
59. Vacuum break assembly
60. Vacuum break to choke lever link
61. Choke shaft lever
62. Vacuum break link clip

its seat so fuel can flow through the valve, past calibrated restrictions and on into the main well. Some carburetors have a two-step or two-stage power valve. The first stage unseats the plunger and fuel is metered between the plunger and the valve body for part-throttle (light power) operation. When the plunger opens the valve to the second stage (at still lower manifold vacuum and higher load) the fuel is metered solely by the power restrictions for full-power operation.

## PUMP SYSTEM

Accelerator pump purpose and operation is covered thoroughly in chapter two.

During high-speed operation a vacuum exists at the pump jets (shooters) because of their location in the venturi area. A flat ledge just below the jets breaks this vacuum so fuel will not be pulled out of the pump jets into the venturi area when the pump is not in operation. This is termed pump pull-over protection.

Underside of bowl and topside of throttle body passages are identified by numbers in this photo. Choke vacuum 1, idle compensator 2, idle down channel 3, timed spark vacuum 4, power system manifold vacuum 5 and manifold vacuum port 6.

**Accelerator-pump size**—The 1-1/2-inch-flange units have a 3/4-inch diameter pump cup; 1-1/4-inch units are 5/8-in.

Most Model 2G carburetors with 1-1/2-inch flange have an inlet check ball in addition to the fill slot for the floating pump cup. This is because the fill slot cannot always provide complete filling of the pump well during some maneuvers. Some models use an expander or garter spring to maintain constant pump-cup-to-pump-wall contact. You will find some 2G carburetors which do not use the floating-pump-cup design. These have a vapor-vent check ball in the pump plunger head.

**1972 Vega 2G is only 2G with pump-shooter nozzles (arrows). Pump shooters are usually drilled holes in the main body or cluster castings on other RPD carburetors.**

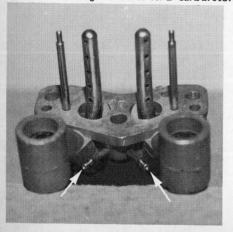

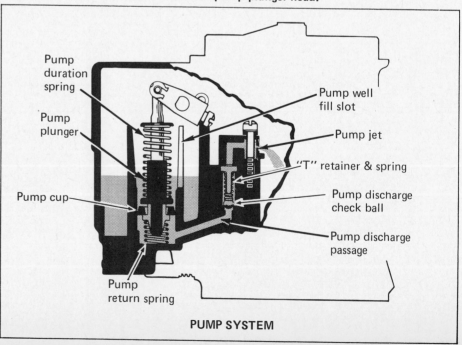

PUMP SYSTEM

## CHOKE SYSTEM

The two-barrel RPD carburetors have had manual 2G, integrated choke coil 2GC, and manifold-mounted 2GV chokes during their many years of existence. The basic operation of these types is quite similar to the descriptions, operation and service details shown in chapter two.

There is also a hot water choke system and a split-linkage system.

In the case of the 2GC choke system, when the engine is cold the thermostatic coil is calibrated to hold the choke valve closed. As the engine is started, air velocity against the offset choke valve causes it to open slightly against the torque of the thermostatic coil. Additionally, intake manifold vacuum applied to the choke piston also tends to open the choke valve. The choke valve assumes a position where the thermostatic-coil torque is balanced against vacuum pull on the choke piston and air velocity against the offset choke valve. Here the choke piston is in the vacuum-break position as shown in the drawing inset. This results in a regulated air flow into the carburetor which provides the richer mixture needed during the warm-up period.

As the engine warms up, hot air from the exhaust manifold is drawn into the thermostatic-coil housing by manifold vacuum. The hot air

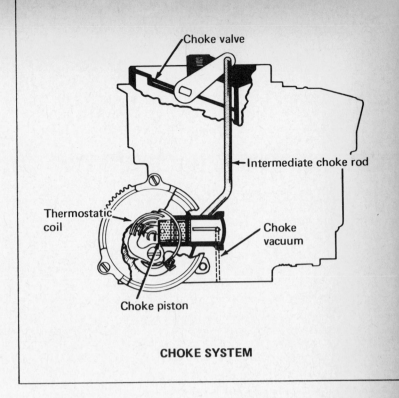

**CHOKE SYSTEM**

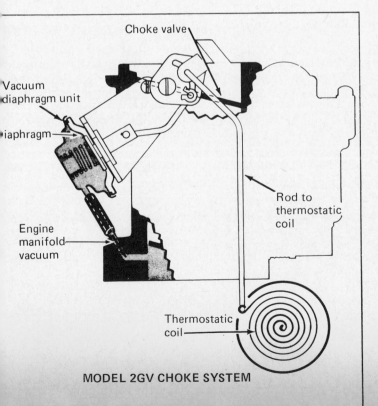

**MODEL 2GV CHOKE SYSTEM**

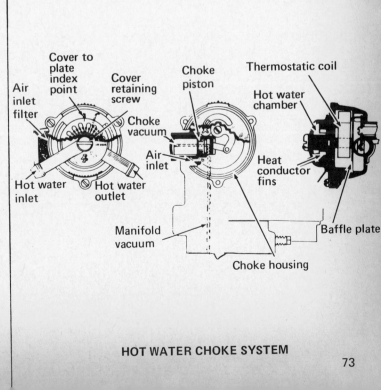

**HOT WATER CHOKE SYSTEM**

causes the coil to relax its tension slowly so the choke valve moves gradually to its full open position.

During the warm-up period the choke piston also modifies the choke action to compensate for varying engine loads on acceleration. Any acceleration or increased load decreases the vacuum pull on the choke piston. This allows the thermostatic coil to increase choke-valve closure momentarily to provide the engine with a richer mixture for acceleration.

**Choke Modifier**—Some 2GC models have a choke modifier to lessen the thermostatic coil's tension acting on the choke valve during heavy acceleration on cold driveaway. During heavy acceleration the manifold vacuum drops so the vacuum pull on the choke piston is less. If the thermostatic coil tends to close the valve too much, excessively rich mixtures will cause engine loading. The choke modifier overcomes this problem as follows.

The thermostatic coil is attached to a shaft protruding through the choke cover. The other end of the shaft has a lever attached by a rod to a lever on the throttle shaft. During cold driveaway when the throttle valves are opened, the choke modifier linkage rotates the choke-coil shaft to lessen the thermostatic-coil tension. This compensates for the decrease in manifold-vacuum pull on the choke piston, preventing over-choking and engine loading during cold-engine operation.

**Split Linkage Choke**—On some model 2GC carburetors a split linkage allows the choke valve and fast-idle cam to work independently. The choke coil and piston operate the same as a conventional system. The split linkage has an intermediate choke rod attached to a slot near the outer end of the intermediate choke lever and the choke rod is attached to a slot approximately halfway out on the lever. As the thermostatic coil warms up and the intermediate choke lever rotates clockwise, the intermediate choke rod moves farther than the choke rod. This allows the choke valve to open fully while maintaining a fast idle. This design feature is used to provide a short choking period with adequate fast idle for a cold engine.

**Vacuum Break**—Two methods of vacuum break are used in the 2GC carburetors with a piston controlling choke position according to manifold vacuum during starting and initial running: (1) Slotted-cylinder type

and (2) By-pass type.

The slotted-cylinder type has two grooves in the piston 180° apart extending from the vacuum end part way down the bore. When the choke is closed, the piston is in the solid part of the bore. The choke opens as the engine starts and the piston moves into the bore. A groove in the piston uncovers the slots in the bore to *break* the vacuum on the end of the piston.

A small hole may be drilled lengthwise in the piston to increase the flow of heated air onto the thermostatic coil and make the choke open faster. Air flow into the thermostatic coil housing is controlled by a calibrated hole in the vacuum passage or in the heat inlet.

In a by-pass type manifold vacuum is applied to the end of the choke piston, causing the piston to move toward the vacuum source until the vacuum port is

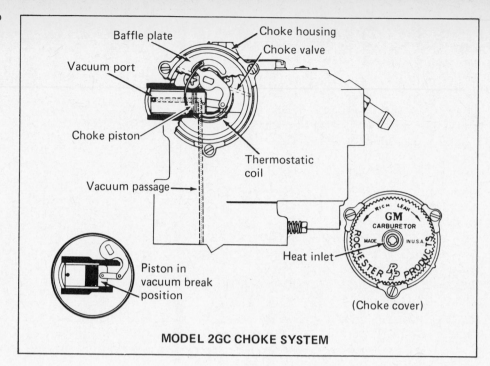

**MODEL 2GC CHOKE SYSTEM**

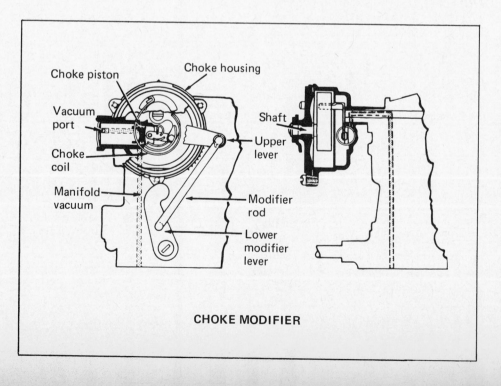

**CHOKE MODIFIER**

covered, thereby *breaking* the vacuum. Some of these types have a small by-pass hole drilled from the vacuum passage into the rear of the choke housing to increase the flow of heated air across the thermostatic coil. As the choke valve continues to open, the piston moves deeper into the bore until a hole drilled in a groove around the piston aligns with the vacuum port, causing a large amount of hot air to flow past the thermostatic coil so it fully relaxes. This keeps the choke off longer when the engine is stopped and helps to reduce hard starting with a warm engine.

**Delayed Vacuum-Break System**—Some 2GC and 2GV carburetors use a delayed vacuum break system with an internal check valve or a calibrated restriction to delay the choke valve from opening too fast during cold-engine starting. Once the engine starts, vacuum applied through the check valve causes the vacuum-break diaphragm plunger to move slowly inward. The few seconds required to seat the vacuum break diaphragm allow time for distribution to be established in the manifold and for engine friction to be somewhat reduced so lean stall will not occur. The system then operates like any other vacuum break except that check-valve models are designed to "pop" the check valve off the seat when spring force inside the diaphragm unit is greater than the vacuum pull. This gives immediate enrichment as required during heavy acceleration with a relatively cold engine.

**Bucking Spring**—Some 2GC and 2GV models incorporate a vacuum break diaphragm assembly with an integral bucking spring on the diaphragm plunger. This allows the thermostatic coil to modulate choke-valve position according to temperature, maintain leaner mixtures during warm temperatures and richer mixtures when the weather is cold.

During extreme cold operation the thermostatic coil exerts more closing pressure than during warmer temperatures. The thermostatic coil compresses the plunger bucking spring, offsetting the vacuum diaphragm pull so it cannot open the choke valve as far as it would otherwise. This gives richer mixtures for cold-weather operation. During warmer temperatures the thermostatic coil's reduced tension during the starting period does not compress the plunger bucking spring quite so far. Thus the vacuum diaphragm opens the choke further to supply a leaner mixture.

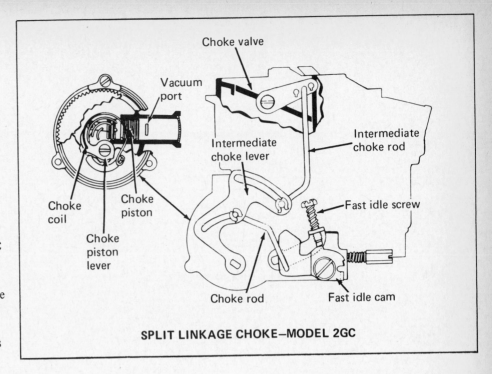

**SPLIT LINKAGE CHOKE—MODEL 2GC**

**Auxiliary Vacuum Break**—A second vacuum-break diaphragm is used on some 2GV models (notably 1972 and later Buicks) to improve cold-start and driveaway operation. The auxiliary vacuum-break feature is a refinement on the delayed vacuum break system previously described. The main vacuum-break diaphragm opens the choke valve to a point where the engine can run without loading or stalling. As fuel distribution is improved by engine warm-up and engine friction decreases, the second vacuum unit's delayed action gradually opens the choke valve still further to prevent loading. Several seconds delay in the auxiliary unit is provided by an internal check valve with a small bleed orifice in the tube fitting to which the hose attaches. Bleed air may enter the orifice through a filter element placed over the hose fitting. The auxiliary unit also includes a bucking spring on the plunger to offset choke thermostatic coil tension as described in the preceding paragraph.

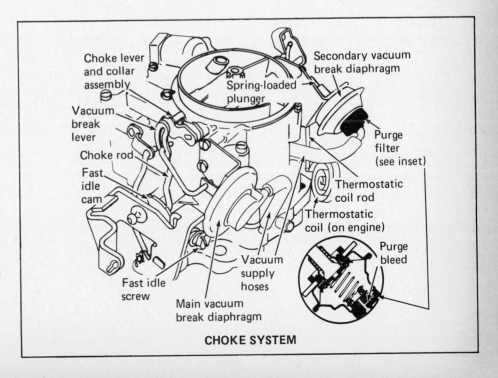

**CHOKE SYSTEM**

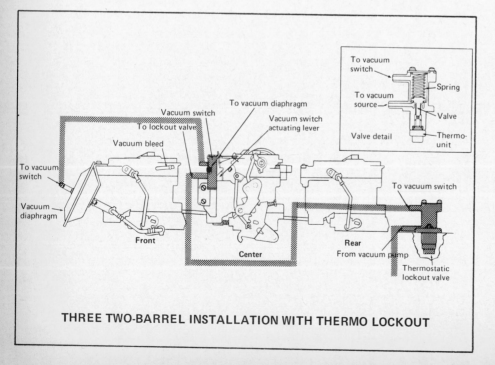

**TYPICAL THREE TWO-BARREL CARBURETOR INSTALLATION**

Front
Center
Rear

Vacuum switch
To vacuum diaphragm
Vacuum switch actuating lever
Choke lockout lever rod
Vacuum bleed
Engine vacuum
Choke lockout lever

**THREE TWO-BARREL INSTALLATION WITH THERMO LOCKOUT**

Vacuum switch
To vacuum diaphragm
To lockout valve
Vacuum switch actuating lever
Vacuum bleed
To vacuum switch
Vacuum diaphragm
Front
Center
Rear
From vacuum pump
To vacuum switch
Thermostatic lockout valve

To vacuum switch
To vacuum source
Spring
Valve
Thermo-unit
Valve detail

Some high-performance applications have used three two-barrel Rochester carburetors in tandem. These are referred to as a 3 x 2 setup. The center or primary carburetor contains all conventional systems: inlet, idle, main metering, power, accelerator pump and choke. The front and rear (secondary) carburetors contain only inlet, pump and main metering systems.

Only the primary carburetor is used for idle, warm-up and part-throttle operation. During these operational phases the secondary carburetors are kept out of operation. Closing springs attached to the throttle shafts hold the throttles closed. Throttle valves and accelerator pumps on the secondaries are operated by a vacuum diaphragm on one of the end carburetors. It is controlled by a vacuum switch on the center carburetor. The vacuum switch is connected to a vacuum canister (tank). Throttle shafts on the end carburetors are connected by a common rod so they operate simultaneously under control of the vacuum diaphragm.

On normal acceleration the center carburetor is the engine's only source of air and fuel until the throttle valves are opened about 60 degrees. At this point the tang on the accelerator pump lever opens the vacuum switch to apply vacuum to the diaphragm. Vacuum moves the diaphragm so the linkage opens the end carburetors simultaneously. When the center carburetor throttle is closed, the vacuum switch is shut off so vacuum is no longer applied to the diaphragm. Air is bled from one of the carburetor air horns through another line between the vacuum switch and the vacuum diaphragm so the diaphragm can return to its normal position under spring tension. This closes the throttles in both end carburetors.

On some installations the carburetor with the diaphragm has a choke lockout lever rod-connected to the choke valve on the center carburetor. Other applications use a temperature-controlled thermo-lockout valve instead of a choke lockout linkage. This valve shuts off vacuum to the switch on the center carburetor so the end carburetors cannot operate until engine temperature reaches 145°F.

# Major Service Operations

Vega 2GV carburetor. Large tube extending from air horn (arrow) vents vapors from the fuel bowl to the charcoal canister.

## DISASSEMBLY, CLEANING, INSPECTION AND ASSEMBLY PROCEDURES

The following disassembly and assembly procedures may vary somewhat between applications due to specific design features. However, the following will basically pertain to all Rochester two-bore carburetors of the side-bowl G design.

Many vehicles use the 2G, 2GC and 2GV carburetors. When removing the carburetor it is a good plan to identify where the different hoses, etc. are to be reinstalled. This is easily accomplished by attaching masking-tape tags onto the hoses and/or by making a simple drawing with labels of the various components and connections.

1. Remove fuel-inlet fitting, gasket and filter screen (if used). If an integral fuel filter is used, remove inlet nut, fuel filter, filter-relief spring and gaskets.

2. On 2GV only, remove vacuum-break diaphragm and link assembly.

3. On 2GC carburetors with a vacuum break, remove clips from intermediate choke rod and remove rod from upper choke lever and intermediate choke lever on choke housing.

NOTE: Omit steps 4, 5, and 6 if the automatic choke is not on the air horn.

4. Remove choke cover and coil assembly and gasket by removing three choke-cover screws and retainers.

5. Remove baffle plate inside choke housing.

6. On later models the choke piston and housing can be removed without removing the choke valve and shaft. On these units proceed as follows:

a. Remove choke piston, lever and link assembly from choke housing by removing attaching screw in choke-shaft end. Remove choke piston from lever and link by shaking piston pin into palm of hand.

b. Remove two choke-housing attaching screws, then remove choke housing and gasket.

7. Remove fast-idle-cam attaching screw, then remove fast-idle cam. Choke rod can be removed after air horn is removed.

NOTE: On early models, remove trip-lever

Some 2GC models use an integral choke mounted on the carburetor. This large 1-1/2-inch-bore 1972 Oldsmobile has a vacuum-break diaphragm (arrow). Early model 2GC's had a piston vacuum-break system.

screw in end of choke shaft, then remove trip lever, upper choke lever and choke rod.

8. Remove idle-vent valve and shield (if used) by removing small attaching screw.

9. Remove retaining clips from pump rod, then remove rod from pump lever and throttle lever.

10. Remove eight air-horn attaching screws and remove air horn from float bowl by carefully lifting upward. Invert air horn and place on a clean bench.

11. Remove float hinge pin and float assembly.

12. Remove float needle; then remove float-needle seat and gasket, using a wide-bladed screwdriver.

13. Remove air-horn-to-float-bowl gasket.

14. Remove power-piston assembly by depressing shaft and allowing spring to snap upward, thus forcing piston retainer from casting. Six to ten of these striking blows are generally required.

NOTE: If heavy staking is encountered, remove from around power-piston retaining washer. The staking is metal from the lip of the piston well that has been peened to overlay the retaining washer in two or three places. Vega 2GV carburetors have a power-valve assembly in the bottom of the float bowl and no power-piston assembly. This screws out after loosening with a wide-bladed screwdriver.

15. Remove retainer from end of pump-plunger shaft, then remove pump assembly from inner pump arm. The pump lever and shaft assembly may be removed by loosening setscrew on inner arm and removing outer lever and shaft assembly from air horn.

## FLOAT BOWL DISASSEMBLY

1. Remove accelerator-pump-return spring from pump well.

NOTE: On carburetors with pump-inlet channel in bottom of pump well, remove small aluminum inlet-check ball from bottom of pump well and pump-inlet screen from fuel-bowl bottom.

2. Remove power valve and gasket using tool BT-3007 or wide-bladed screwdriver.

3. Remove two main metering jets.

4. Remove three venturi-cluster attaching screws, then remove cluster and gasket. Center screw has smooth shank for accelerator pump fuel by-pass and fiber sealing washer under screw head (instead of lock-washer).

3/1972 Oldsmobile 2GC illustrating choke components. A—intermediate choke rod, B—choke lever, C—intermediate choke lever, D—choke rod, E—fast-idle cam, F—vacuum break, G—stat coil housing.

7/Use a good screw driver when loosening the fast-idle screw as it is generally quite tight. A—fast-idle-cam attaching screw, B—fast-idle cam, C—choke rod.

9/A—pump rod, B—pump lever, C—throttle lever.

10/One of the eight air-horn attaching screws on this Vega 2GV is extra long and has a phillips head. Expect screw-type variances.

11/A—float hinge pin, B—float assembly, C—power-piston assembly.

15/A—pump assembly, B—pump shaft and lever assembly. Note the Vega air-horn assembly has no power-valve actuating assembly at C.

1/ 1—accelerator-pump-return spring,
2—power valve, 3—two metering jets,
4—three cluster-attaching screws.

1A/Vega 2GV float bowl showing unique
power-valve assembly (arrow).

4/Lifting out the cluster after three screws
were loosened. Arrows indicate extended
idle bleeds which are often used on 2GV
carburetors to prevent the possibility of
ice formation in these orifices during certain operating conditions. Extended pump
shooters identify this cluster as a Vega part.

5. Using needle-nose pliers, remove pump-discharge ball spring T-shaped retainer. Then remove pump-discharge spring and steel check ball.

6. Remove main-well insert tubes from main fuel wells (if used). Consult parts list for application on particular carburetor model.

7. Remove idle-air by-pass adjusting screw and spring at rear of float bowl (where used).

8. Remove distributor vacuum fitting from float bowl (where used).

9. Invert carburetor and remove three throttle-body-to-bowl attaching screws. Remove throttle body and throttle-body-to-bowl gasket.

## CHOKE DISASSEMBLY

1. Where the choke housing is mounted on the throttle body, disassemble as follows:

a. Remove choke cover and coil assembly by removing three choke-cover attaching screws and retainers.

b. Remove baffle plate inside choke housing.

c. Remove choke piston, lever and link assembly from intermediate choke shaft by removing attaching screw in end of shaft. Choke piston can be removed from lever and link by shaking piston pin into hand.

d. Remove two choke-housing attaching screws, remove choke housing and gasket. Remove intermediate choke shaft and lever from choke housing.

NOTE: On units with split choke linkage, remove the fast-idle cam screw, then remove cam and choke rod as an assembly. The fast-idle cam and intermediate choke lever can be disassembled further by removing clips on choke-rod ends.

2. Remove idle-mixture adjusting needles and springs.

3. Remove slow- and fast-idle adjustment screws (where used) from throttle lever if replacement is necessary.

---

**Do not completely disassemble the throttle body.** Never remove the throttle valves because the idle and spark holes are drilled in direct relation to the throttle-valve locations. Removal of the valves will upset this location. The throttle body is serviced as a complete unit with valves intact.

---

9/Photo showing bottom of Oldsmobile F85 1-inch-bore aluminum-throttle-body 2GC carburetor. This is the smallest Rochester 2-barrel. Retaining-screw locations (arrows) are positioned as shown on all RPD two-barrel units. An impact driver may be needed to loosen these screws. Note idle-mixture needles at front of bowl.

## CLEANING PARTS

Refer to chapter eight for suggestions on cleaning carburetors.

## PARTS INSPECTION

1. Inspect carburetor castings on upper and lower surfaces to see that the small sealing beads are not damaged. Damaged beading may result in air or fuel leaks at the point of damage.

2. Inspect holes in pump lever, fast-idle cam, and throttle-shaft lever. If holes are worn excessively or out of round to affect carburetor operation, replace the parts.

3. Inspect the steps on the fast-idle cam for excessive wear. If worn excessively, replace the cam to ensure correct fast-idle operation during the warm-up and choking periods.

4. Inspect the pump plunger. If pump-plunger cup is worn excessively or damaged, replace the plunger. Shake pump plunger to make sure vapor-vent check ball is free in plunger head. It should rattle after cleaning.

5. Inspect throttle-body assembly. Make sure all passages and vacuum channels are clean.

6. Inspect throttle shaft axial clearance in housing. Movement of the shaft, measured with a dial indicator on one side of the shaft, should be less than 0.005 inch. If you do not have access to a dial indicator, good judgment will be sufficient: if the shaft cannot be wobbled freely in an obviously worn hole, go ahead and put it together. Too much clearance will be evidenced by an erratic idle. Small-displacement engines may idle too fast with the idle-speed screw backed fully off because of the air entering the engine through the excessive clearance around the throttle shaft. In conclusion, if the clearance seems excessive and either of the two problems exists, you should order a new throttle body and replace it when it arrives. This part will typically take four to six weeks to arrive. Dealers will not usually carry such items in their stocks. Some may prefer not to order them unless you pay for the parts in advance.

7. Check throttle-valve screws for tightness. If loose, tighten and stake properly. Problems seldom exist in this area.

## ASSEMBLY

1. Install idle-mixture needles and springs finger-tight. Back out the needles 1½ turns as a preliminary idle adjustment. 1971 and later models have plastic caps that should be put on after final settings are made. Leave them off for now.

2. If removed, install the slow and fast-idle screws in the throttle levers.

3. Install throttle-body gasket on bottom of float bowl, making sure gasket holes line up with holes in float bowl. Then install throttle-body assembly on bowl using three attaching screws and lockwashers. Tighten evenly and securely.

4. Install distributor vacuum fitting into float bowl, if removed.

5. Install idle-air by-pass adjustment screw (where used) into rear of float bowl and screw inward until it seats lightly. Back out three full turns as a preliminary idle adjustment.

6. Install main-well insert tubes (if used) into main well. Make sure they are seated into their recesses.

7. Drop steel pump-discharge check ball into discharge hole. Install discharge spring and T-shaped retainer. Top of retainer must be flush with top of casting when installed correctly.

8. Install venturi cluster and gasket. Install three cluster screws and lockwashers. Tighten evenly and securely.

NOTE: Center screw has smooth shank and uses a fiber gasket in place of lockwasher to seal pump-discharge passage.

9. Install two main metering jets. Tighten securely. Some type of screw holding tool is very helpful in starting the jets in their threaded holes.

10. Install power valve and gasket in bottom of fuel bowl using tool BT-3007 or a wide-bladed screwdriver. Tighten securely.

11. Install aluminum inlet-check ball in bottom of pump well if bowl has pump-inlet passage, insert pump-return spring into pump well. Make sure spring seats in well bottom.

12. Install pump-inlet screen in bottom of float bowl where used.

NOTE: Refer to parts lists for use of pump inlet-check ball and screen on a particular application.

8/Top view of RPD two-barrel showing cluster in place. Retaining screws pointed out by arrows.

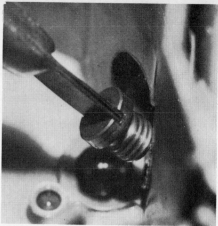

9/This shop-fabricated tool is excellent for starting metering jets and small screws. It is a piece of brass rod in which two pieces of 0.020" shim stock have been soldered. Saw a slot in the brass rod and shape shim-stock pieces by grinding to suit your needs. Similar screw drivers are commercially available.

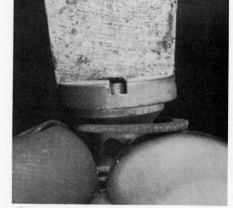

Small notch ground in tip of large screwdriver allows removing power valve without damage to center pin of valve.

Power valve at 1 inserts into hole in bottom of bowl. Make sure gasket is used as shown in right photo. Main jets are at 2. This 2GV model is used on the Chevrolet Vega.

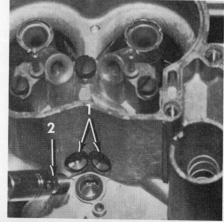

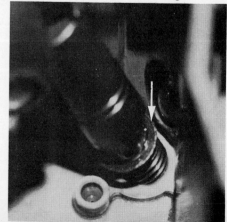

## AIR HORN ASSEMBLY

NOTE: Omit steps 1-8 if you have a 2G, 2GV or a 2GC with choke housing on the body.

1. Install choke housing and gasket to air horn with two attaching screws. Tighten evenly and securely.

2. On late units where choke trip lever is staked onto choke shaft, install upper choke lever and choke rod on choke shaft. Choke-lever tang should face towards trip lever and be positioned on top of trip lever.

3. Slide choke-shaft assembly through choke-shaft bores in air horn.

4. Install choke valve into slot in choke shaft. RP trademark should be on top side of valve with choke closed. Install two choke-valve retaining screws. Do not tighten.

5. Install choke piston on choke-piston lever-and-link assembly with piston pin. If choke piston has a flat on one side, this should face inward toward choke valve.

6. Install choke-piston lever and link assembly into choke housing. Align piston-lever flats with those on choke shaft. Install retaining screw in choke-shaft end.

7. To center choke valve and shaft, close choke valve, then place 0.020" feeler gage between choke trip lever and upper choke lever. Center shaft to maintain above clearance, then tighten choke-valve screws and stake securely.

8. On early units where the choke piston and lever assembly is riveted to choke shaft, install choke piston on choke lever and link. Flat side on choke piston should face inward toward air horn. Install choke shaft into air horn and rotate shaft to install choke piston in choke housing. Install choke valve, upper choke lever, trip lever and align as described in steps 4 and 7.

9. Lubricate pump shaft with light grease before installation. Install outer pump lever and shaft assembly and inner pump lever into air horn and tighten set screw. Check this shaft-to-bore clearance as it sometimes becomes excessive. Erratic pump action will accompany a wobbly shaft.

10. Attach pump-plunger assembly to inner pump lever with end of pump shaft pointing inward. Install pump retainer.

11. Install float needle seat and gasket into air horn. Use tool BT-3007 or wide-bladed screwdriver and tighten securely.

12. Install power-piston assembly into power-piston bore. Stake power-piston re-tainer lightly in place. A small plastic-handled screwdriver tapped firmly with a small hammer does a good job. Power piston should travel up and down freely after installation. Vega 2GV carburetors use a power-enrichment valve without the power piston.

13. Install new air-horn gasket on air-horn casting.

14. Install float needle into float-needle seat and float assembly to air horn. Secure in place with float hinge pin. On models using a float-needle pull clip, make sure it is aligned correctly in float arm.

15. Adjust float assembly as outlined in the rebuild instructions supplied in the carburetor rebuilding kit for your specific unit, or in the shop manual for your particular car. Drop the accelerator-pump return spring into the pump well.

16. Install air horn onto float bowl, making certain accelerator-pump plunger is aligned correctly in pump well. Lower cover gently straight downward to prevent damage to floats or power piston.

17. Install eight air-horn screws and lockwashers (if used). Tighten evenly and securely. Vega carburetors must have Loctite or similar screw-retaining compound to prevent loosening.

18. On 2GV only, install vacuum-break lever on choke-shaft end. Install attaching screw and tighten securely. Install vacuum-break unit on air horn using two attaching screws. Install vacuum-break line; retain with clip.

2/Choke-shaft levers on later-model air horns are staked permanently in place. Early ones had a screw. There must be a gap of approximately 0.020-inch between these two levers and the air-horn casting to ensure that the assemblies will move freely.

12/Top view of partially reassembled carburetor. Cluster, power valve and accelerator-pump-return spring are in place. Vega installation. Arrow indicates gasket-sealing bead.

NOTE: 4 & 5 apply to 2GC units only.
4. Install choke baffle plate into choke housing on models with the choke housing mounted to air horn.

NOTE: Some early carburetors use a stat coil torque-relief spring assembled on the choke shaft in the choke housing. This spring contacts a tang on the choke baffle plate when the choke valve is completely closed to 15° open. This prevents over-choking and loading when starting with a partly warm engine by offsetting the torque of the thermostatic coil. Hopefully, you illustrated this well with a sketch prior to disassembly.

The choke valve must be in the wide-open position when installing the baffle plate in the choke housing to prevent damaging the stat coil torque-relief spring. If the relief spring is not installed correctly the choke valve may be locked closed. Check to be sure choke valve is free.
5. Install choke stat cover, coil assembly and gasket using three attaching screws and retainers.

To adjust thermostatic coil, rotate cover until choke valve is just closed and index marks are aligned. Tighten stat cover screws securely after adjustment.
6. Install fuel strainer, inlet fitting and gaskets. Tighten securely. If an integral fuel filter is used, install pressure-relief spring, filter element with large open end toward inlet nut, filter gasket inside inlet nut. Then install inlet nut and gasket and tighten securely.

## CHOKE ASSEMBLY ON 2GC
NOTE: The following applies to the 2GC models with the choke assembly mounted on the throttle body.
1. Install intermediate choke shaft and lever into choke housing. Lever on shaft should hang downward between mounting bosses.
2. Install new gasket, then install choke housing on throttle body using two attaching screws. Tighten securely.
3. Assemble choke piston to choke piston link and lever assembly with piston pin. Then install choke piston and lever assembly into choke housing. Attach choke piston lever to intermediate choke shaft end

with retaining screw. Make sure flats in lever hole line up with flats on intermediate choke shaft. Tighten retaining screw securely.

## FINAL CARBURETOR ASSEMBLY
1. Install accelerator-pump rod into upper pump lever and throttle lever, retaining with clips provided.
2. Install idle-vent valve and shield, retain with attaching screw; tighten securely.
3. Install choke rod into fast-idle cam, then install fast-idle cam onto float bowl, retaining with fast-idle-cam attaching screw. Tighten securely. RP on cam should face outward.
4. Install intermediate choke rod to upper choke lever and intermediate choke lever, retain with clips.
5. Install baffle plate inside choke housing, then install stat cover and coil assembly using new gasket. Set index mark on the cover adjacent to the most prominent mark on the housing. You can tailor this setting as you see the need after a few days of operation. Install three cover screws and retainers. Tighten securely.

## CONCLUSION
This concludes the reassembly operation. Be sure to read the reinstallation section of chapter nine; along with installation hints it gives some safety procedures which are well worth learning and practicing.

After you have completed all operations including setting idle mixture and speed, drive the vehicle for a few days. If it drives to suit you, enjoy it—should some problem exist, obtain carburetor-adjustment details from your United-Delco dealer or from the shop manual for your car. Adjust your unit to proper specification. Many suggestions in the adjustment section of chapter nine relate carburetor malfunction to vehicle behavior. It suggests ways to tailor the carburetor to suit you if standard settings are not satisfactory. Most of these rules of thumb will apply to the 2G carburetor.

Should you encounter specific problems or have trouble obtaining parts, write Doug Roe Engineering.

1/Accelerator pump rod A is installed and secured with clip B.

3/Hold throttle partially open for best access to fast-idle-cam screw.

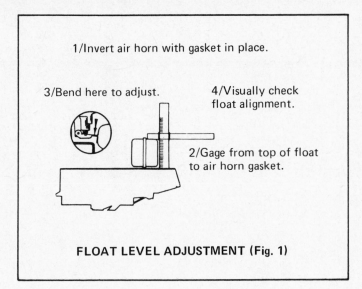

1/Invert air horn with gasket in place.

3/Bend here to adjust.

4/Visually check float alignment.

2/Gage from top of float to air horn gasket.

**FLOAT LEVEL ADJUSTMENT (Fig. 1)**

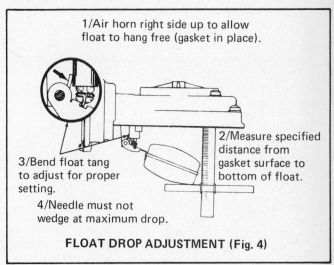

1/Air horn right side up to allow float to hang free (gasket in place).

2/Measure specified distance from gasket surface to bottom of float.

3/Bend float tang to adjust for proper setting.

4/Needle must not wedge at maximum drop.

**FLOAT DROP ADJUSTMENT (Fig. 4)**

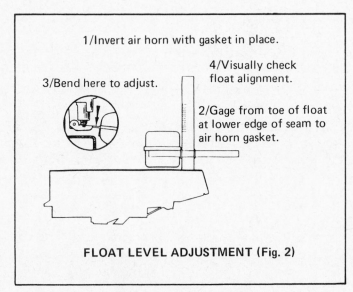

1/Invert air horn with gasket in place.

4/Visually check float alignment.

3/Bend here to adjust.

2/Gage from toe of float at lower edge of seam to air horn gasket.

**FLOAT LEVEL ADJUSTMENT (Fig. 2)**

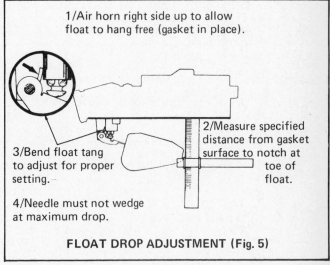

1/Air horn right side up to allow float to hang free (gasket in place).

2/Measure specified distance from gasket surface to notch at toe of float.

3/Bend float tang to adjust for proper setting.

4/Needle must not wedge at maximum drop.

**FLOAT DROP ADJUSTMENT (Fig. 5)**

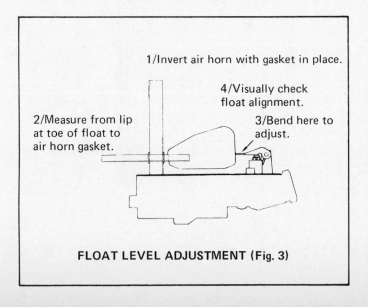

1/Invert air horn with gasket in place.

4/Visually check float alignment.

2/Measure from lip at toe of float to air horn gasket.

3/Bend here to adjust.

**FLOAT LEVEL ADJUSTMENT (Fig. 3)**

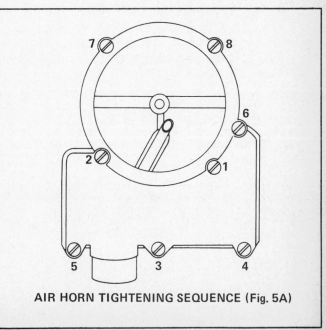

**AIR HORN TIGHTENING SEQUENCE (Fig. 5A)**

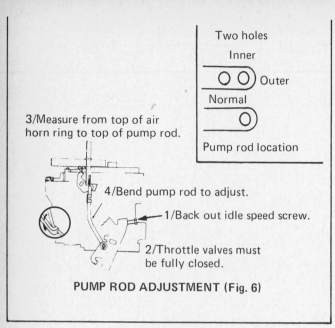

Two holes

Inner

Outer

Normal

Pump rod location

3/Measure from top of air horn ring to top of pump rod.

4/Bend pump rod to adjust.

1/Back out idle speed screw.

2/Throttle valves must be fully closed.

**PUMP ROD ADJUSTMENT (Fig. 6)**

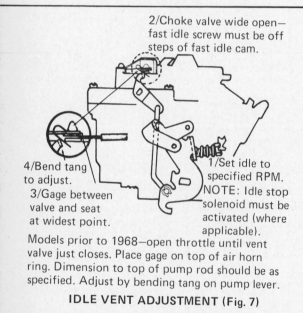

2/Choke valve wide open— fast idle screw must be off steps of fast idle cam.

4/Bend tang to adjust.

3/Gage between valve and seat at widest point.

1/Set idle to specified RPM. NOTE: Idle stop solenoid must be activated (where applicable).

Models prior to 1968—open throttle until vent valve just closes. Place gage on top of air horn ring. Dimension to top of pump rod should be as specified. Adjust by bending tang on pump lever.

**IDLE VENT ADJUSTMENT (Fig. 7)**

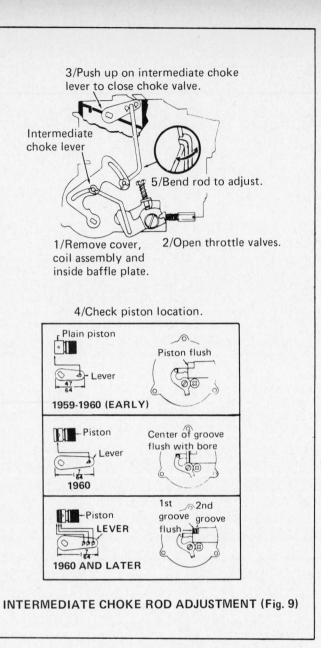

3/Push up on intermediate choke lever to close choke valve.

Intermediate choke lever

5/Bend rod to adjust.

1/Remove cover, coil assembly and inside baffle plate.

2/Open throttle valves.

4/Check piston location.

Plain piston

Piston flush

Lever

$\frac{47}{64}$

**1959-1960 (EARLY)**

Piston

Center of groove flush with bore

Lever

$\frac{7}{64}$

**1960**

Piston

1st groove flush

2nd groove

**LEVER**

$\frac{7}{64}$

**1960 AND LATER**

**INTERMEDIATE CHOKE ROD ADJUSTMENT (Fig. 9)**

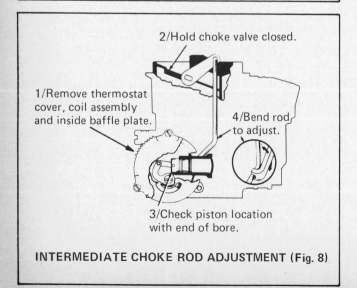

2/Hold choke valve closed.

1/Remove thermostat cover, coil assembly and inside baffle plate.

4/Bend rod to adjust.

3/Check piston location with end of bore.

**INTERMEDIATE CHOKE ROD ADJUSTMENT (Fig. 8)**

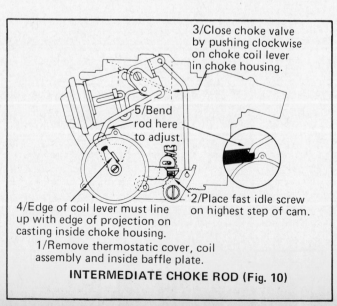

3/Close choke valve by pushing clockwise on choke coil lever in choke housing.

5/Bend rod here to adjust.

4/Edge of coil lever must line up with edge of projection on casting inside choke housing.

1/Remove thermostatic cover, coil assembly and inside baffle plate.

2/Place fast idle screw on highest step of cam.

**INTERMEDIATE CHOKE ROD (Fig. 10)**

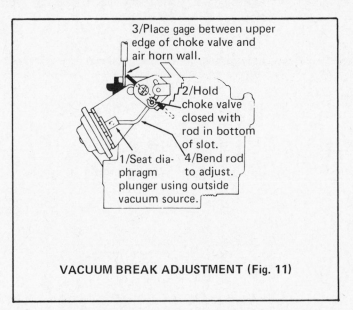

3/Place gage between upper edge of choke valve and air horn wall.

2/Hold choke valve closed with rod in bottom of slot.

1/Seat diaphragm plunger using outside vacuum source.

4/Bend rod to adjust.

**VACUUM BREAK ADJUSTMENT (Fig. 11)**

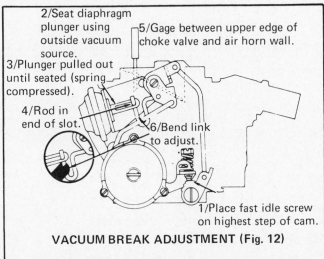

2/Seat diaphragm plunger using outside vacuum source.

3/Plunger pulled out until seated (spring compressed).

4/Rod in end of slot.

5/Gage between upper edge of choke valve and air horn wall.

6/Bend link to adjust.

1/Place fast idle screw on highest step of cam.

**VACUUM BREAK ADJUSTMENT (Fig. 12)**

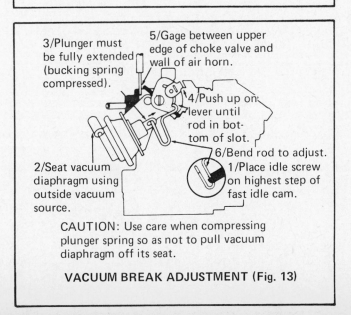

3/Plunger must be fully extended (bucking spring compressed).

5/Gage between upper edge of choke valve and wall of air horn.

4/Push up on lever until rod in bottom of slot.

2/Seat vacuum diaphragm using outside vacuum source.

6/Bend rod to adjust.

1/Place idle screw on highest step of fast idle cam.

CAUTION: Use care when compressing plunger spring so as not to pull vacuum diaphragm off its seat.

**VACUUM BREAK ADJUSTMENT (Fig. 13)**

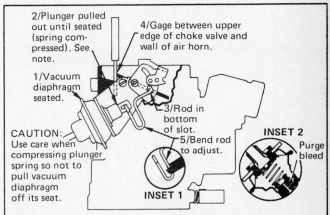

2/Plunger pulled out until seated (spring compressed). See note.

1/Vacuum diaphragm seated.

4/Gage between upper edge of choke valve and wall of air horn.

3/Rod in bottom of slot.

5/Bend rod to adjust.

CAUTION: Use care when compressing plunger spring so not to pull vacuum diaphragm off its seat.

INSET 1

INSET 2 Purge bleed

NOTE: If purge filter is used (see inset 2), remove vacuum break diaphragm hose and rubber-covered filter element from vacuum break tube and, using a small piece of tape, plug the small bleed hole. After adjustment, remove the tape, making sure the small bleed hole is open, and install the rubber-covered filter element over the vacuum break tube.

**AUXILIARY VACUUM BREAK ADJUSTMENT (Fig. 14)**

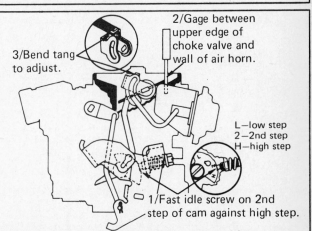

3/Bend tang to adjust.

2/Gage between upper edge of choke valve and wall of air horn.

L—low step
2—2nd step
H—high step

1/Fast idle screw on 2nd step of cam against high step.

It is important to position both slow idle and fast idle screws, as follows, before making choke rod adjustment.

Step 1—Models using single idle stop screw only—turn stop screw in until it just contacts bottom step of fast idle cam. Turn screw in one full turn. Models using both a slow idle and a fast idle screw—turn slow idle stop screw in until it just contacts stop. Then turn this screw in one full turn from this point. Next turn the fast idle screw in until it touches bottom step of fast idle cam.

Step 2—All models—place idle screw on second step of fast idle cam against shoulder of high step. While holding screw in this position, check clearance between upper edge of choke valve and air horn wall, as shown. Adjust to specified dimension by bending tang on choke lever and collar assembly, as shown above.

**CHOKE ROD (FAST IDLE CAM) ADJUSTMENT (Fig. 15)**

**NOTE: PERFORM ADJUSTMENTS IN PROPER SEQUENCE.**

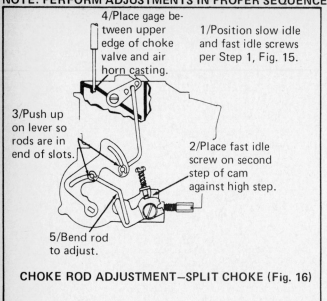

4/Place gage between upper edge of choke valve and air horn casting.

1/Position slow idle and fast idle screws per Step 1, Fig. 15.

3/Push up on lever so rods are in end of slots.

2/Place fast idle screw on second step of cam against high step.

5/Bend rod to adjust.

**CHOKE ROD ADJUSTMENT—SPLIT CHOKE (Fig. 16)**

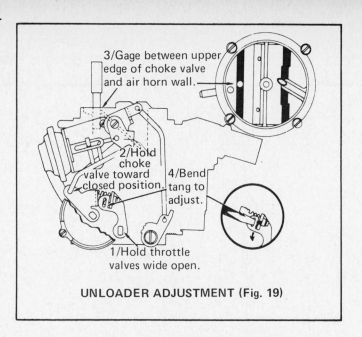

3/Gage between upper edge of choke valve and air horn wall.

2/Hold choke valve toward closed position.

4/Bend tang to adjust.

1/Hold throttle valves wide open.

**UNLOADER ADJUSTMENT (Fig. 19)**

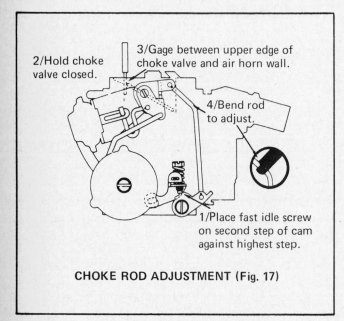

2/Hold choke valve closed.

3/Gage between upper edge of choke valve and air horn wall.

4/Bend rod to adjust.

1/Place fast idle screw on second step of cam against highest step.

**CHOKE ROD ADJUSTMENT (Fig. 17)**

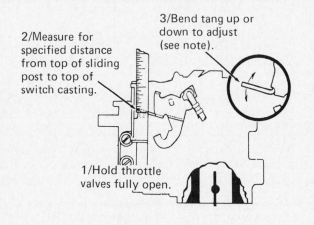

2/Measure for specified distance from top of sliding post to top of switch casting.

3/Bend tang up or down to adjust (see note).

1/Hold throttle valves fully open.

NOTE: On some models, loosen switch attaching screws and move switch up or down to adjust.

CAUTION: Be careful not to bump or bend lever after adjustment is made. Open and close throttle to make sure arm on pump lever does not bind post on vacuum switch.

**VACUUM SWITCH ADJUSTMENT (CENTER CARBURETOR) (Fig. 20)**

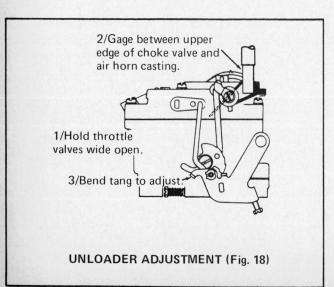

2/Gage between upper edge of choke valve and air horn casting.

1/Hold throttle valves wide open.

3/Bend tang to adjust.

**UNLOADER ADJUSTMENT (Fig. 18)**

**NOTE: PERFORM ADJUSTMENTS IN PROPER SEQUENCE.**

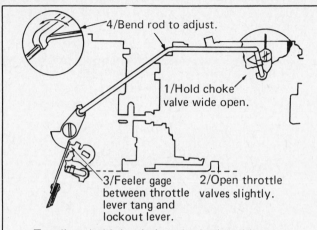

4/Bend rod to adjust.

1/Hold choke valve wide open.

3/Feeler gage between throttle lever tang and lockout lever.

2/Open throttle valves slightly.

To adjust, hold the choke valve in the wide-open position. With the throttle valves slightly open on the carburetor to which the diaphragm is attached, there should be a clearance, as specified, between the lockout lever and the throttle lever as shown. Measure clearance with a feeler gage and bend the lockout rod to adjust.

### CHOKE LOCKOUT ADJUSTMENT (Fig. 21)

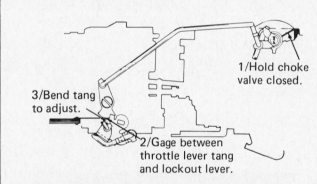

1/Hold choke valve closed.

3/Bend tang to adjust.

2/Gage between throttle lever tang and lockout lever.

To adjust, hold the throttle valves completely closed. With the choke valve on the center carburetor in the closed position and the choke lockout lever rod connected, bend the lockout tang on the throttle lever to obtain specified clearance between the lockout lever and tang on the throttle lever of the carburetor to which the diaphragm assembly is attached.

### CONTOUR ADJUSTMENT (Fig. 22)

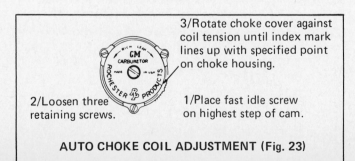

3/Rotate choke cover against coil tension until index mark lines up with specified point on choke housing.

2/Loosen three retaining screws.

1/Place fast idle screw on highest step of cam.

### AUTO CHOKE COIL ADJUSTMENT (Fig. 23)

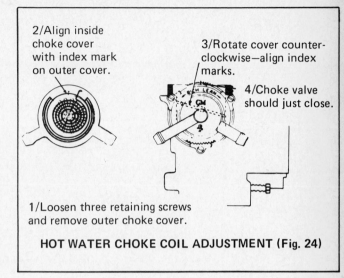

2/Align inside choke cover with index mark on outer cover.

3/Rotate cover counter-clockwise—align index marks.

4/Choke valve should just close.

1/Loosen three retaining screws and remove outer choke cover.

### HOT WATER CHOKE COIL ADJUSTMENT (Fig. 24)

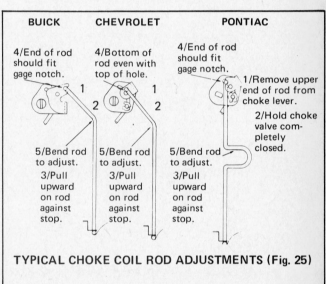

| BUICK | CHEVROLET | PONTIAC |
|---|---|---|
| 4/End of rod should fit gage notch. | 4/Bottom of rod even with top of hole. | 4/End of rod should fit gage notch. |
| | | 1/Remove upper end of rod from choke lever. |
| | | 2/Hold choke valve completely closed. |
| 5/Bend rod to adjust. 3/Pull upward on rod against stop. | 5/Bend rod to adjust. 3/Pull upward on rod against stop. | 5/Bend rod to adjust. 3/Pull upward on rod against stop. |

### TYPICAL CHOKE COIL ROD ADJUSTMENTS (Fig. 25)

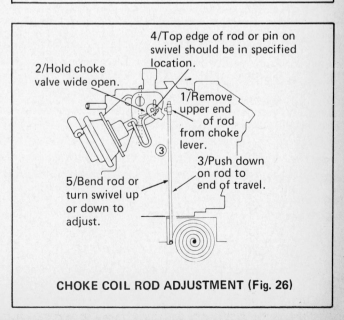

4/Top edge of rod or pin on swivel should be in specified location.

2/Hold choke valve wide open.

1/Remove upper end of rod from choke lever.

3/Push down on rod to end of travel.

5/Bend rod or turn swivel up or down to adjust.

### CHOKE COIL ROD ADJUSTMENT (Fig. 26)

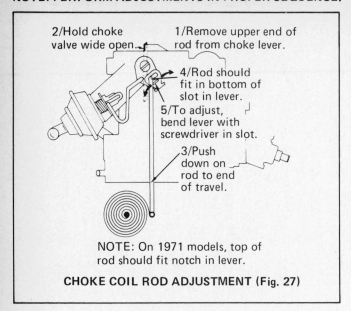

2/Hold choke valve wide open.

1/Remove upper end of rod from choke lever.

4/Rod should fit in bottom of slot in lever.

5/To adjust, bend lever with screwdriver in slot.

3/Push down on rod to end of travel.

NOTE: On 1971 models, top of rod should fit notch in lever.

**CHOKE COIL ROD ADJUSTMENT (Fig. 27)**

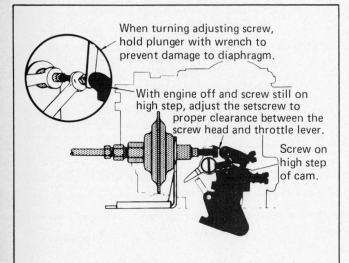

When turning adjusting screw, hold plunger with wrench to prevent damage to diaphragm.

With engine off and screw still on high step, adjust the setscrew to proper clearance between the screw head and throttle lever.

Screw on high step of cam.

1/Place fast idle screw on high step of cam (with engine off).

2/Using two wrenches, loosen and hold locknut while adjusting contact screw for specified clearance (see note).

NOTE: Slow and fast idle adjustments must be made first. Any time slow or fast idle adjustments are changed, readjust the throttle return check as above. On models having a specified engine speed, with engine at curb idle speed and transmission in neutral, disconnect and plug end of hose to throttle return check, using two wrenches, loosen and hold locknut while adjusting contact screw for specified engine RPM.

**THROTTLE RETURN CHECK ADJUSTMENT (Fig. 28)**

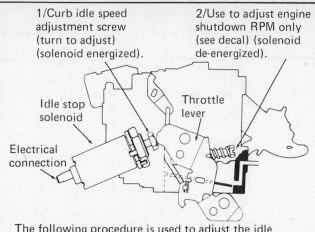

1/Curb idle speed adjustment screw (turn to adjust) (solenoid energized).

2/Use to adjust engine shutdown RPM only (see decal) (solenoid de-energized).

Idle stop solenoid

Throttle lever

Electrical connection

The following procedure is used to adjust the idle stop solenoid to control engine speed on a running engine. Follow instructions on vehicle tune-up sticker before proceeding:

1/With engine at normal operating temperature and idle stop solenoid energized (plunger stem extended), adjust plunger screw to obtain specified engine speeds.

Low idle adjustment
2/To set low engine idle speed, with the idle stop solenoid disconnected electrically, adjust idle speed screw on throttle lever to obtain specified engine speed.

**IDLE STOP SOLENOID ADJUSTMENT (Fig. 29)**

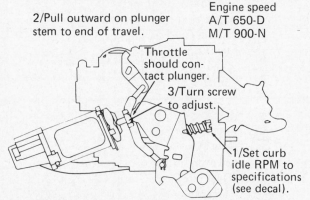

2/Pull outward on plunger stem to end of travel.

Engine speed
A/T 650-D
M/T 900-N

Throttle should contact plunger.

3/Turn screw to adjust.

1/Set curb idle RPM to specifications (see decal).

This adjustment is to be made only after: (1) Replacement of the solenoid, (2) Major overhaul of the carburetor is performed or (3) The throttle body is removed and replaced.

The following procedure is used to adjust the C.E.C. valve controlled engine speed on a running engine (in "neutral" for manual or in "drive" for automatic transmissions) with air conditioning off, distributor vacuum hose removed and plugged, and fuel tank hose from vapor canister disconnected. Follow instructions on vehicle tune-up sticker before proceeding.

1/Adjust curb idle speed to specifications (see decal).

2/Manually extend C.E.C. valve plunger to contact throttle lever and limit of its travel.

3/Adjust plunger length to obtain specified engine speeds.

**C.E.C. VALVE ADJUSTMENT (Fig. 30)**

# 4G Carburetors

**W**e can now look into the function of the first RPD four-barrel-carburetor design. This was the 4G—4GC carburetor used almost exclusively on General Motors cars from 1952 through 1967. Recently there has been a revival of interest in these carburetors for use on stock-model cars (1967 and earlier) being drag raced in stock classes.

With the trend toward lower and lower hood lines plus increased power output in the mid-50's, it became necessary to lower the carburetor and increase its capacity. This called for what was then a new approach in carburetors. The new approach came in the form of the four-barrel carburetor with two fuel clusters. The four-jet carburetor was essentially two 2-jet carburetors combined into a single casting. The primary or fuel-inlet side contains all six systems of carburetion as described in the carburetor systems chapter: float, idle, part-throttle, power, accelerator pump and choke. The secondary side supplements the primary side with separate float and power systems to add extra capacity for high-speed and power ranges.

Design-wise it was desirable to use as small a venturi as possible to keep high air velocity and allow efficient metering. This idea was carried even further in the current Rochester Q-jet design. It can easily be seen that the larger the carburetor bores, the lower the air velocity through each bore. By combining two carburetors into a single casting and providing a delayed opening of the second carburetor, the carburetor's capacity could be used as required. During normal operation the primary side does the majority of the work and venturi sizes are kept sufficiently small to provide efficient metering. At between 42 and 60 degrees primary-throttle opening, secondary throttles begin opening and the linkage is designed so that when the primaries reach full-open position the secondaries are also fully open. The secondary bores also contain a spring-loaded auxiliary throttle valve which provides further control for metering in the secondary side. This feature will be discussed later.

4GC is readily identifiable by a forest of vents atop the air horn. Note nipple for water hose on choke thermostatic coil housing. This was Rochester's first four-barrel and it was used from 1952 through 1967. It is an air-valve carburetor, but the air valve is hidden below the secondary venturi. In 1957 the bowl height was reduced to allow fitting the unit under lowered hood lines. Rubber boot over accelerator pump keeps dirt out of float bowl. Keep a good one installed. The accelerator pump is very adjustable with five holes in the lever. Pump lever also opens bowl vent at idle. Manual-choke models are designated 4G.

Ricky and Leo Klarr's "Lil ole Nova" runs a 4GC on their SS/LA with a 327 small block. The car has turned an 11.48-second ET and held the NHRA MPH record at 115.38 as of August 1973.

Lou Unser on his way up Pikes Peak at the 1961 Annual Hill Climb. He won it with his 4GC-equipped Chevrolet. Author Doug Roe blueprinted the carburetor. Hot Rod photo.

Chevrolet drivers won more points per entry and more points per event than any other entrant for five consecutive years in the Pure Oil Trials. Here NASCAR driver Marvin Panch is performing a passing maneuver in an early 1960 event. Rochester 4GC carburetors were commonly used on these entries.

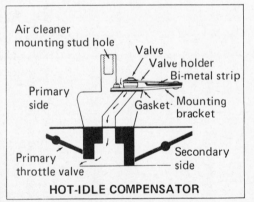

HOT-IDLE COMPENSATOR

## SYSTEMS DESCRIPTIONS

With minor exceptions, the 4G and 4GC carburetor systems function as described in chapter two.

### FLOAT SYSTEM

There are two float systems, one for each side of the carburetor. Floats are aided in their action by a torsion spring around the float hinge pin or a balance spring between the float hanger posts. Either of these arrangements helps to control float bouncing, regardless of float position. Some models had a vacuum-assisted-float arrangement which helped the float to close the needle valve during periods of minimum fuel demand. A coiled assist spring on the power-piston stem exerted an upward (closing) force on a tang on the primary float arm during periods of high manifold vacuum (above 9 inches Hg). When manifold vacuum dropped, the power-piston stem was pushed down to open the power valve, moving the assist cup (spring) to allow full float drop as required by fuel use. This system only aided closing force against the needle.

Primary and secondary needle/seat assemblies may differ: The secondary seat may be larger than that in the primary. This should be watched for during disassembly to aid in assembly. Some models use a ball-tipped needle to allow greater fuel flow with less float drop.

### IDLE SYSTEM

During the later years of existence the Pontiac-initiated RPD-design idle by-pass system showed up on many 4GC units. This was to combat contaminant buildup on the throttle blades which caused inconsistent idle. With this by-pass system the throttle blades are stopped in a near-closed position and the required idle air by-passes around them via drilled passages. The amount of air and thus engine-idle RPM is adjustable.

### PART-THROTTLE (TRANSITION)

Many Rochester carburetors utilize a lower idle air-bleed to supplement main and power system fuel and to cover up any "hole" after the throttle passes the off-idle holes/slots. In the case of the 4GC, off-idle to main-metering transition fuel is quite dependent on this.

The eyebrow nozzles protruding from the wall of the venturi-exit area in the primary bores (and in some sec-

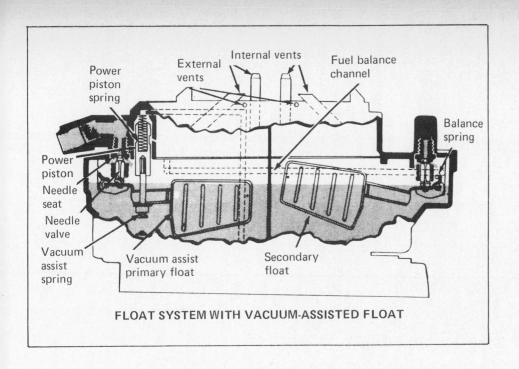

FLOAT SYSTEM WITH VACUUM-ASSISTED FLOAT

Power piston spring

Power piston

Needle seat

Needle valve

Vacuum assist spring

External vents

Internal vents

Fuel balance channel

Balance spring

Vacuum assist primary float

Secondary float

Air-horn assembly with: 1—fuel-inlet needles and seats, 2—fuel inlet, 3—power piston, 4—fuel-slosh baffles and secondary fuel passage.

Ten internal float-bowl vents are in the air horn. White arrow indicates location of "hidden" screw attaching air horn. Other screws are completely visible.

Primary float on a 4GC air horn. Be sure to reinstall float-assist spring (arrow). It is necessary to maintain fuel-level control. Note upper end of spring rests against seat (arrow). Float-drop adjustment tang also contacts seat in full dropped position.

**Cluster design**—Fuel-cluster design places the main-well tubes, idle tubes, mixture passages, air bleeds and the pump jets in a removable assembly called the *cluster*. Thus, when the cluster is removed for servicing, virtually all of the vital metering parts can be seen readily, cleaned and examined. The cluster fits on a platform in the float-bowl casting so the main-well and idle tubes are suspended in the fuel. A gasket between the cluster casting and the body platform insulates the main-well and idle tubes from engine heat. This reduces the effects of percolation on metering.

Inverted 4GC bowl assembly shows "eyebrow" lower idle-air bleed nozzles (arrows) that also supplement main-system fuel feed whenever off-idle air flows are exceeded. Note secondary air-valve assembly in place.

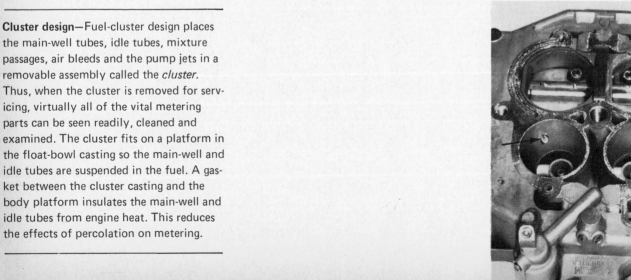

ondaries) were put in by RPD at no small expense to create higher depressions in the idle-feed down channels. Generally, this bleed-feed consists of a simple drilled hole.

As throttle blades are opened past the off-idle discharge holes, air velocity over these eyebrowed lower idle-air bleeds creates a vacuum and they begin to feed fuel, continuing to feed even after the main system starts. Their fuel source is the idle-system metering orifices from the fuel bowl. When this system has a strong signal and is feeding heavily, the idle discharge orifice and slots are dry.

## MAIN & ACCELERATOR-PUMP SYSTEMS

These operate as described in chapter two.

## POWER SYSTEM

The 4GC power system is similar to that described in the carburetor systems chapter. There is no secondary power system.

## SECONDARY SYSTEM

Fuel is fed through a metering system similar to that in the primary side. The damper or auxiliary throttle valve above the secondary throttle valves is spring-loaded and offset on the shaft so air velocity against the wider side will force it open. If all four bores were wide open at low engine speed, more air would be available than the engine could use. Instead of supplying power mixtures the carburetor would supply a lean mixture because of low air velocity. This spring-loaded valve is calibrated to open the secondary side only after air flow has reached sufficient velocity to provide good metering in all four bores. In this way power is maintained at low speed by providing good metering and at high speed by allowing complete opening of all four bores. This air valve is the forerunner of the one used on the Quadrajet and its mechanism similar to that used on the later design.

When the secondary throttle valves are opened and air velocity is not high enough to open the auxiliary valves, additional fuel is sometimes needed. This is supplied by either of two methods to offset the momentary leanness caused by air passing around the auxiliary valves just prior to opening of the auxiliary throttle valves:

1. Secondary feed tubes extend from the secondary idle passages down below the auxiliary throttle valves and feed a mixture

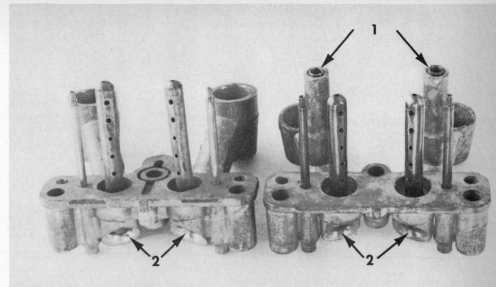

Primary (left) and secondary clusters for 4GC. Accelerating fuel nozzles 1 supply extra fuel as secondary air valves start to open. Note the idle tubes in both clusters. The fixed idle on the secondary system also used fuel from the secondary side to prevent gum formation in the event the secondaries were not being used. It also provided transition fuel through secondary feed tubes on some models. Opening into main discharge nozzles 2 is called an aspirator, hence the nozzles are called aspirated fuel nozzles.

as soon as the secondary throttles open but before the auxiliary throttles open. The tubes continue to operate after the main nozzle starts to flow.

2. Or, mixture down tubes extend from the mixture passage to below the auxiliary throttle valves. These tubes operate similar to the secondary feed tubes except fuel is obtained from the main nozzle tube. When the air flow is high enough to open the auxiliary valves, the down tubes no longer feed fuel because the low pressure point is now in the small venturi. With this feature, the correct air/fuel mixture can be supplied at any point during secondary throttle valve operation.

## CHOKE SYSTEMS

Several types of choke systems were used on the 4GC: conventional, choke modifier, split-linkage and hot water. These all operate similarly to the 2GC carburetors described on pages 73–75.

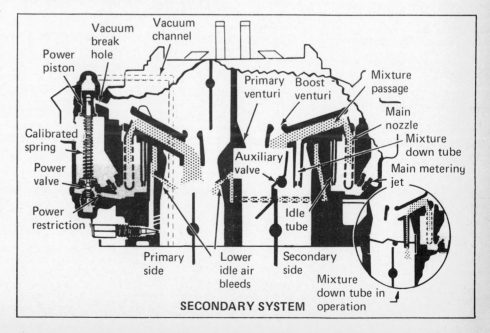

SECONDARY SYSTEM

Primary side of the carburetor showing: 1—main metering jets, 2—primary cluster 3—power valve, 4—fuel-level equalizing passage to prevent fuel overflow into throttle bores.

Secondary venturis showing: 1—secondary-metering jets, 2—cluster, 3—hot-idle compensator, 4—idle-air bleed control which allowed primary throttle blades to be almost completely closed at idle.

## SUMMARY

This 4GC carburetor had a good many pluses. Adequate fuel orifices and air bleeds allowed altering calibration, which made development work fairly rewarding. A very drivable carburetor usually resulted from these efforts.

The demand for larger capacity and even greater part-throttle metering control dictated the design of the Q-jet, which is RPD's sole four-barrel unit as of 1974. See chapter eight for complete service hints on the Q-jet.

Many 4GC units are still in service today. They can be found on 1952–67 vintage GM cars and trucks. The 4GC provides good service and is relatively easy to overhaul and maintain. Delco kits are available to restore most units to perfect working condition.

Economy tuning—If you decide to tune your 4GC for utmost economy and would like to take advantage of some of the hints provided in the economy and drivability section of chapter fourteen, you have a problem. No power valve springs or assemblies are supplied for the 4GC because it is now an obsolete carburetor. You'll have to make any modifications on the parts you have available or whatever you can find in rebuild kits.

Spring tension of auxiliary throttle-valves in secondary side can be adjusted with a slim screwdriver. Doug Roe uses a special screwdriver (arrow) with 10° markings on a circular handle. Turn the tool or screwdriver counterclockwise to add tension after neutral position is established. Add tension in 10° increments until desired performance is obtained. Don't forget to tighten the setscrew.

Underside of 4GC shows idle mixture screws in primary side 1, heat track in cast-iron throttle base 2, idle-air by-pass track into primaries 3, and idle-air by-pass adjusting screw 4. White arrow indicates one of the throttle-body-attachment screws.

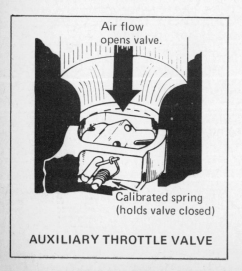

Air flow opens valve.

Calibrated spring (holds valve closed)

AUXILIARY THROTTLE VALVE

# Models 4G, 4GC—Exploded View

1. Intermediate choke rod clip (lower)
2. Intermediate choke rod
3. Intermediate choke rod clip (upper)
4. Choke shaft & lever assembly
5. Choke valve screw
6. Choke valve
7. Air horn assembly
8. Air horn screw
9. Choke lever and collar assembly
10. Choke trip lever
11. Choke trip lever screw
12. Pump shaft & lever assembly
13. Pump rod clip
14. Pump rod
15. Pump shaft & lever clip
16. Air horn gasket
17. Fuel inlet fitting
18. Needle & seat assembly (primary)
19. Power piston assembly
20. Pump plunger boot
21. Pump plunger clip
22. Pump plunger assembly
23. Needle & seat assembly (secondary)
24 & 24A Float assembly
25. Float balance spring & clip assembly
26. Float hinge pin
27. Venturi cluster screw & lockwasher
28. Venturi cluster (primary)
29. Venturi cluster (secondary)
30. Venturi cluster gaskets
31. Idle compensator assembly
32. Pump discharge guide
33. Pump discharge spring
34. Pump discharge check ball
35. Power valve assembly
36. Main metering jet (primary)
37. Main metering jet (secondary)
38. Pump inlet screen retainer
39. Pump inlet screen
40. Pump return spring
41. Pump inlet check ball
42. Float bowl assembly
43. Auxiliary throttle valve assembly
44. Throttle-body-to-bowl gasket
45. Idle speed screw
46. Idle speed screw spring
47. Throttle body assembly
48. Choke rod
49. Choke rod attaching clips
50. Fast idle cam
51. Fast idle cam screw
52. Fast idle screw spring
53. Fast idle screw
54. Throttle-body-to-bowl screw (large)
55. Throttle-body-to-bowl screw (small)
56. Idle mixture screw
57. Idle mixture screw spring
58. Choke cover & coil assembly
59. Choke cover & coil gasket
60. Choke baffle plate
61. Choke cover attaching screws
62. Choke cover retainers (toothed)
63. Choke piston lever screw
64. Choke piston lever and link assembly
65. Choke housing attaching screw
66. Choke cover retainer
67. Choke housing plug
68. Choke Piston
69. Choke piston pin
70. Choke housing assembly
71. Intermediate choke shaft & lever
72. Choke housing gasket
73. Carburetor-to-manifold gasket
74. Idle vent valve assembly
75. Filter element relief spring
76. Filter element
77. Filter element gasket
78. Fuel inlet fitting gasket
79. Fuel inlet fitting

**NOTE: PERFORM ADJUSTMENTS IN PROPER SEQUENCE**

1/Invert air horn with gasket in place.

2/On models with round or "D" type floats—gage from gasket surface to top of each float.

3/Bend float arm to adjust.

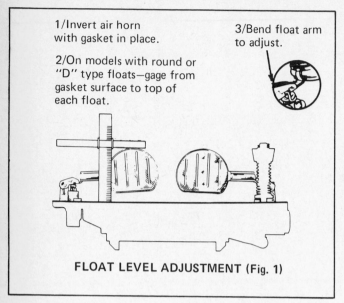

**FLOAT LEVEL ADJUSTMENT (Fig. 1)**

1/Invert air horn with gasket in place.

3/Bend pontoon lightly to adjust.

2/Measure from gasket surface to center of dimple.

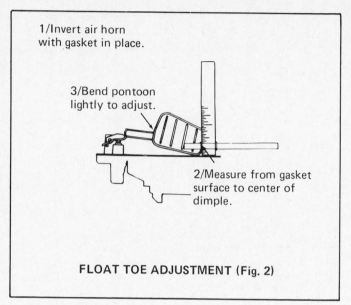

**FLOAT TOE ADJUSTMENT (Fig. 2)**

1/Invert air horn with gasket in place.

2/Gage from gasket surface to top heel of each float.

3/Bend float arm to adjust.

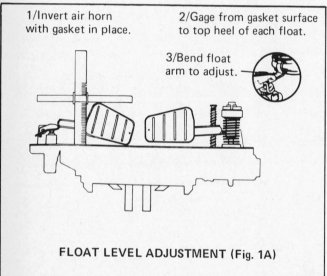

**FLOAT LEVEL ADJUSTMENT (Fig. 1A)**

1/Invert air horn with gasket in place.

3/Bend pontoon lightly to adjust.

2/Float toe should be flush when sighting across air horn casting when adjusting wedge floats without dimples.

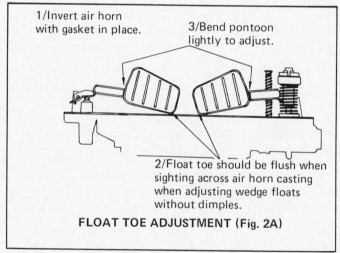

**FLOAT TOE ADJUSTMENT (Fig. 2A)**

1/Air horn inverted with gasket in place.

2/Gage from gasket surface to top of float.

3/Bend float arm to adjust.

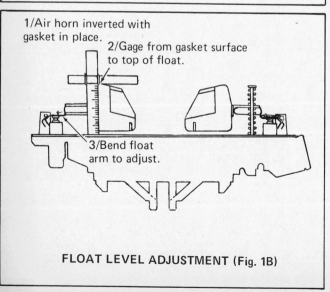

**FLOAT LEVEL ADJUSTMENT (Fig. 1B)**

2/Center float pontoons in gasket cutout.

3/Floats should be parallel—bend float arms to adjust.

1/Align holes in gasket with holes in air horn casting.

2/Center float pontoons in gasket cutout.

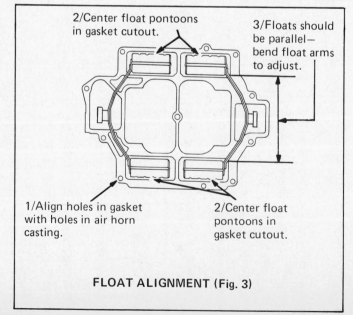

**FLOAT ALIGNMENT (Fig. 3)**

**NOTE: PERFORM ADJUSTMENTS IN PROPER SEQUENCE**

1/Air horn right side up to allow float to hang free (gasket in place).

4/Needle must not wedge at maximum drop.

3/Bend float tang to adjust.

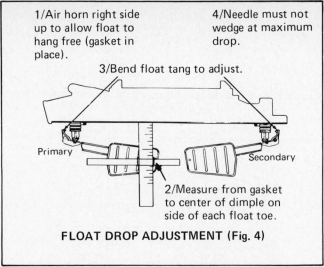

2/Measure from gasket to center of dimple on side of each float toe.

**FLOAT DROP ADJUSTMENT (Fig. 4)**

1/Air horn right side up to allow float to hang free (gasket in place).

4/Needle must not wedge at maximum drop.

3/Bend float tang to adjust.

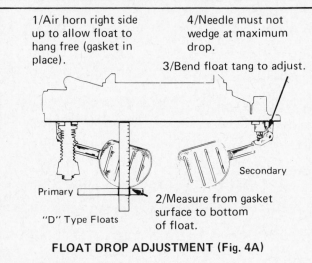

2/Measure from gasket surface to bottom of float.

"D" Type Floats

**FLOAT DROP ADJUSTMENT (Fig. 4A)**

1/Air horn right side up to allow float to hang free (gasket in place).

4/Needle must not wedge at maximum drop.

3/Bend float tang to adjust.

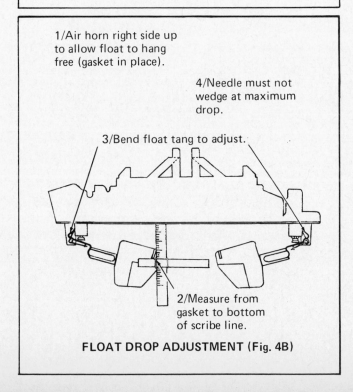

2/Measure from gasket to bottom of scribe line.

**FLOAT DROP ADJUSTMENT (Fig. 4B)**

1/Hold power piston in full up position (lightly move float assembly to be sure there is no bind).

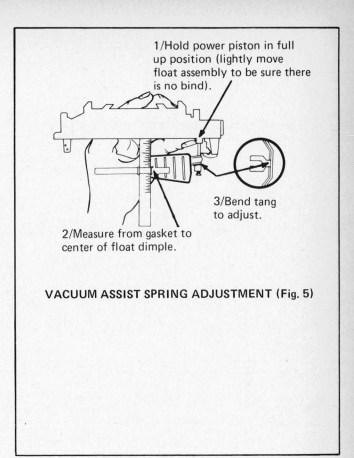

3/Bend tang to adjust.

2/Measure from gasket to center of float dimple.

**VACUUM ASSIST SPRING ADJUSTMENT (Fig. 5)**

3/Measure from top of air horn to bottom of plunger shaft.

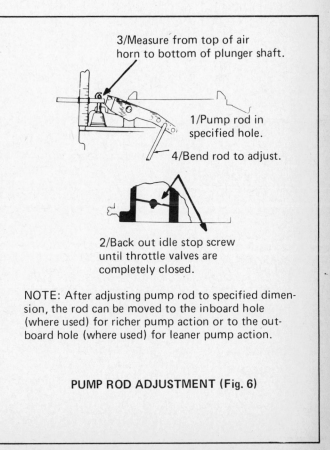

1/Pump rod in specified hole.

4/Bend rod to adjust.

2/Back out idle stop screw until throttle valves are completely closed.

NOTE: After adjusting pump rod to specified dimension, the rod can be moved to the inboard hole (where used) for richer pump action or to the outboard hole (where used) for leaner pump action.

**PUMP ROD ADJUSTMENT (Fig. 6)**

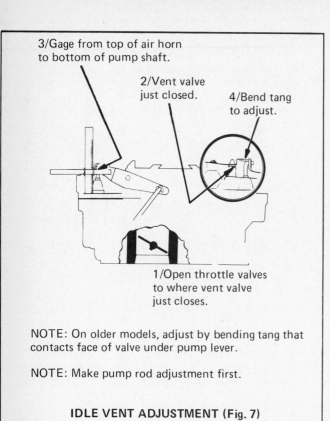

3/Gage from top of air horn to bottom of pump shaft.

2/Vent valve just closed.

4/Bend tang to adjust.

1/Open throttle valves to where vent valve just closes.

NOTE: On older models, adjust by bending tang that contacts face of valve under pump lever.

NOTE: Make pump rod adjustment first.

**IDLE VENT ADJUSTMENT (Fig. 7)**

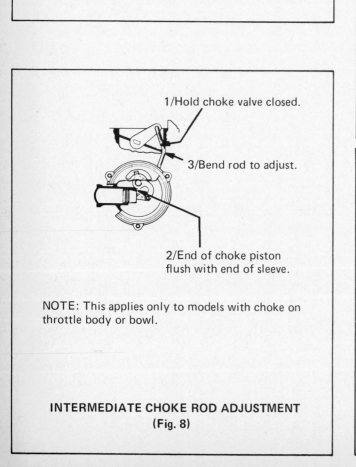

1/Hold choke valve closed.

3/Bend rod to adjust.

2/End of choke piston flush with end of sleeve.

NOTE: This applies only to models with choke on throttle body or bowl.

**INTERMEDIATE CHOKE ROD ADJUSTMENT (Fig. 8)**

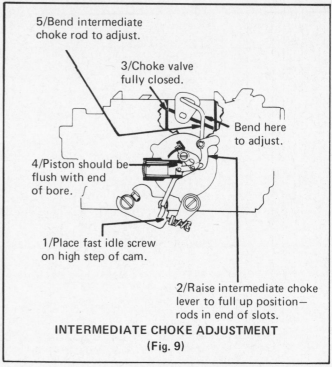

5/Bend intermediate choke rod to adjust.

3/Choke valve fully closed.

Bend here to adjust.

4/Piston should be flush with end of bore.

1/Place fast idle screw on high step of cam.

2/Raise intermediate choke lever to full up position— rods in end of slots.

**INTERMEDIATE CHOKE ADJUSTMENT (Fig. 9)**

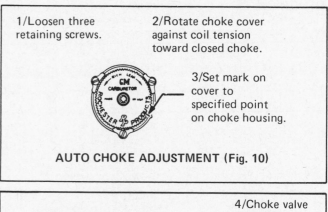

1/Loosen three retaining screws.

2/Rotate choke cover against coil tension toward closed choke.

3/Set mark on cover to specified point on choke housing.

**AUTO CHOKE ADJUSTMENT (Fig. 10)**

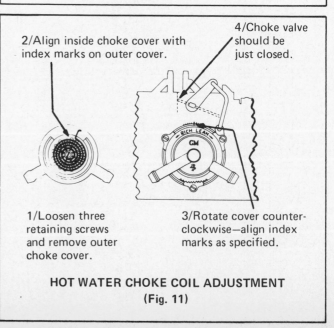

2/Align inside choke cover with index marks on outer cover.

4/Choke valve should be just closed.

1/Loosen three retaining screws and remove outer choke cover.

3/Rotate cover counter-clockwise—align index marks as specified.

**HOT WATER CHOKE COIL ADJUSTMENT (Fig. 11)**

**NOTE: PERFORM ADJUSTMENTS IN PROPER SEQUENCE**

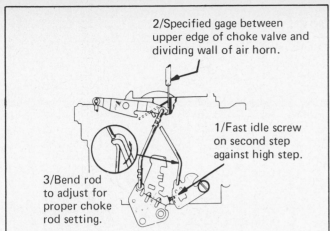

2/Specified gage between upper edge of choke valve and dividing wall of air horn.

1/Fast idle screw on second step against high step.

3/Bend rod to adjust for proper choke rod setting.

It is important to position both slow idle and fast idle as follows before making choke rod adjustment:

Step 1
Models using single idle stop screw: Turn stop screw in until it contacts bottom step of fast idle cam. Then turn screw in one full turn.

Models using separate fast idle screw: Turn slow idle stop screw in until it touches stop, then turn one full turn. Then turn the fast idle screw in until it touches bottom step of fast idle cam.

Step 2
After positioning slow idle and fast idle screws as described above, position idle screw on second step of fast idle cam against the shoulder of the high step, then check clearance between upper edge of choke valve and air horn wall. Bend choke rod, to adjust.

**CHOKE ROD ADJUSTMENT (Fig. 12)**

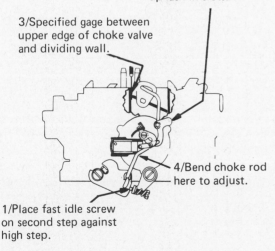

2/Move intermediate lever to full up position to take up lash in slots.

3/Specified gage between upper edge of choke valve and dividing wall.

4/Bend choke rod here to adjust.

1/Place fast idle screw on second step against high step.

**CHOKE ROD ADJUSTMENT (Fig. 13)**

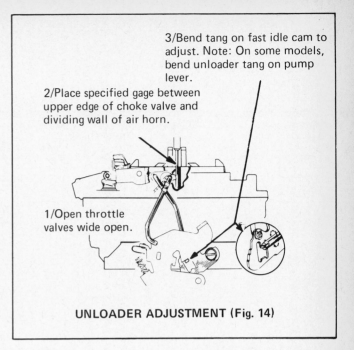

3/Bend tang on fast idle cam to adjust. Note: On some models, bend unloader tang on pump lever.

2/Place specified gage between upper edge of choke valve and dividing wall of air horn.

1/Open throttle valves wide open.

**UNLOADER ADJUSTMENT (Fig. 14)**

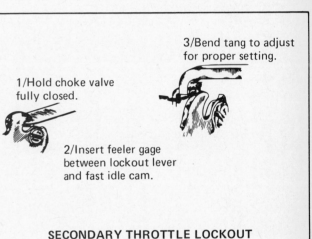

3/Bend tang to adjust for proper setting.

1/Hold choke valve fully closed.

2/Insert feeler gage between lockout lever and fast idle cam.

**SECONDARY THROTTLE LOCKOUT ADJUSTMENT**
**(Fig. 15)**

1/Hold choke valve wide open.

2/Insert feeler gage between lockout lever and fast idle cam.

3/Bend tang to adjust.

**SECONDARY THROTTLE LOCKOUT CONTOUR ADJUSTMENT**
**(Fig. 16)**

# Quadrajet Carburetor Design

The first-design Quadrajet carburetor was a revolution in General Motors induction systems. This section deals with this unit, first released to the public in 1965, and with design changes through 1974 units.

## GENERAL DESCRIPTION

The Quadrajet carburetor as we know it today is a very versatile unit serving millions of people. This speaks well of RPD people who put the original thinking into its design during the early '60s.

This section is intended to describe the systems of the original units and comment on changes that transpired, bringing us to current Quadrajet configurations.

The original concept of the Quadrajet carburetor was a new approach to the four-barrel incorporating unique ideas. Its design is easily understood for servicing and trouble-shooting. Its versatility and principles of operation make it adaptable to engine displacements ranging from small to very large without major design changes.

Three separate Quadrajet models are designated 4M (basic carburetor equipped for manual choke operation), 4MV (automatic-choke model for use with a manifold-mounted thermostatic choke coil), and 4MC (automatic choke model with choke housing and thermostatic choke coil mounted on the side of the float bowl). Except for the choke systems, all models utilize basically the same operation principles.

The Quadrajet carburetor has two stages of operation. The primary (fuel-inlet) side has small bores with a triple-venturi arrangement. Its metering principles are similar to most carburetors using the venturi principle. The triple-venturi stack, plus the small primary bores, gives a stable and fine fuel control in the idle and economy ranges of operation. This lends itself to good part-throttle metering control for best drivability with modern-day low-emission air/fuel mixtures which are typically lean at low air flows. Fuel metering in the primary side is accomplished with tapered metering rods positioned in metering

Buick's 1974 Q-jet features Teflon coating on both the pump lever and the primary throttle shaft for extremely smooth operation of these parts. Pull-over enrichment and dual power piston springs are included as described in the text of this chapter. All 1974 Rochester carburetors have limited-adjustment limit caps on the idle-mixture adjusting screws.

1974 Chevrolet Q-jet with idle-stop solenoid. This model has detail design improvements over previous models including Teflon-coated pump lever for smooth operation of the pump rod in its hole in the lever. The lower throttle lever has a spun-in plastic bushing for the 1974-required extra throttle-return spring. A heavier pump-return spring also helps to ensure positive throttle closing.

Top view of a first-design Quadrajet. Note the small, very efficient primary venturis and the big, big secondaries. The small centrally located bowl reduces fuel loss by evaporation during hot engine shutdowns.

Close-up of the primary venturi/nozzle arrangement.

jets by a power piston which is responsive to manifold vacuum.

The secondary side has two very large bores with large air capacity to meet present and future large engine demands. This supplements air/fuel flow through the primary bores when engine demand requires the added air-flow capacity. The air-valve principle is used in the secondary side for metering control. The air-valve mechanically positions tapered metering rods in orifice plates to control fuel flow from the secondary nozzles in direct proportion to air flow through the secondary bores.

The centrally located fuel reservoir avoids fuel-slosh problems which cause engine turn cut-out and delayed fuel flow to the carburetor bores. The float system uses a single float pontoon for ease in servicing.

First-design units had a pressure-balanced float needle valve to overcome problems encountered with high fuel-pump pressures and to allow a small float to control fuel "shut-off" through the large fuel-inlet-needle seat. It had a synthetic tip designed to reduce dirt-caused flooding problems. Service problems in the field and perhaps pressures from skeptics caused RPD to revert back to a conventional type needle and seat in 1967 models.

A sintered-bronze fuel filter was mounted in the fuel-inlet casting which is an integral part of the float bowl. The filter was easily taken out for cleaning or replacement. It removed dirt particles

The vacuum piston supporting the primary metering rod hanger is housed in the round cylinder protruding from the fuel-bowl floor. The tapered portion of the metering rods is in the orifice area of the metering jets.

as small as 80 microns (0.003") in diameter. These filters were later replaced by more efficient paper filters.

The primary side has six fuel-control systems: float, idle, main-metering, power, accelerator pump and choke. Some models incorporate a pullover enrichment system for added enrichment at high air flows. The secondary side has one metering system which supplements the primary main metering system and receives fuel from a common float chamber.

The following text describes all fuel-control systems in detail to simplify servicing and troubleshooting.

## FLOAT SYSTEM

Quadrajet's centrally located float reservoir with a single pontoon float and float needle valve is unique. The fuel bowl is centered between the primary bores and adjacent to the secondary bores. This design assures adequate fuel supply to all carburetor bores, giving excellent performance with respect to car inclination or severity of turns. This float system has proven to be very good for off-road-vehicle use. Bumps or severe jolts upset the fuel level in most carburetors, causing erratic metering and performance. See off-road preparation in chapter fourteen for further comments.

The solid float pontoon is made of closed-cell plastic material. Because it is lighter than a brass pontoon, its added buoyancy allows use of a smaller float to maintain constant fuel levels.

The first-design float needle valve was a completely new design. It was diaphragm-assisted so that additional float buoyancy was not needed to hold the float needle valve closed to shut off fuel flow—even with rising fuel pressures.

The first-design float-valve seat is a brass insert pressed into the bottom of the float bowl. The large-diameter seat is not removable for servicing because the first-design valve tip creates negligible seat wear. Care should be used during servicing so that the seat is not nicked, scored, or moved. The float-valve seat is factory staked and tested and should not be restaked in the field.

A sintered-bronze fuel-inlet filter is used with a pressure-relief spring. The relief spring allows fuel-pump pressure to force the filter off its seat if it should

Primary metering jets 1 and secondary metering jets 2 are well placed in this centrally located compact fuel bowl. Idle/off-idle fuel is metered from the primary jet wells via idle tubes 3. Accelerating wells for the secondary side are at 4.

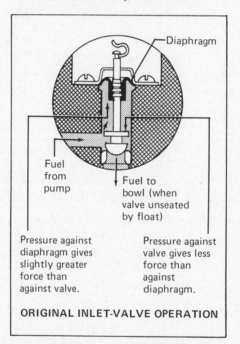

ORIGINAL INLET-VALVE OPERATION

Pressure against diaphragm gives slightly greater force than against valve.

Pressure against valve gives less force than against diaphragm.

Diaphragm

Fuel from pump

Fuel to bowl (when valve unseated by float)

Bottom and side view of first-design Quadrajet float.

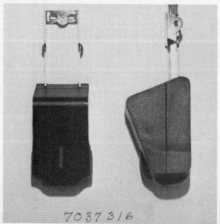

7037316

This unique inlet-valve design had a number of distinct advantages, but it died in infancy—perhaps prematurely. With today's emission tests, maximum fuel-level control and vapor containment are needed. The initial design offered both of those features.

clog, thereby ensuring fuel flow to the carburetor. A guide cast into the filter housing prevents installing the filter the wrong way.

The original Q-jet float system consisted of a float chamber, plastic pontoon float assembly, float hinge pin and retainer combination, a float valve and diaphragm assembly, float valve diaphragm retainer and screws, and a float-valve pull clip. The original type float system operates as follows.

Fuel from the fuel pump enters the fuel-inlet passage passing through the filter element and into the float-inlet-valve chamber. Fuel under pressure enters the chamber above the float valve tip and its pressure on the tip tends to force it downward or closed. The same pressure in the float-valve chamber also acts upon the small diaphragm fastened to the top of the float-valve stem. Pressure on this diaphragm tends to force the float valve upward to pull the float valve off its seat. Because the valve is pressure-balanced, even very high fuel pressures do not tend to unseat the valve. Very light float forces (small movements) operate the valve. Because there is such a large seat area, bowl filling or level restoration occurs very quickly.

Fuel pressure acting on the diaphragm and the float link work together to raise the valve tip off its seat, allowing fuel to pass by the open valve and enter the fuel passage leading to the float chamber. This fuel passage leads upward to a point just above normal

# Model 4MV Quadrajet—Exploded View

1 Choke shaft & lever assembly
2 Air valve lockout lever roll pin
3 Air valve lockout lever
4 Choke rod clip (upper)
5 Choke valve
6 Choke valve screw (2)
7 Air horn—long screw
8 Secondary metering rod holder & screw
9 Air horn—short screw
10 Air horn—countersunk screw (2)
11 Dash pot lever roll pin
12 Dash pot actuating
13 Air valve rod clip
14 Air valve rod
15 Idle vent valve lever
16 Pump actuating lever
17 Pump lever roll pin
18 Idle vent valve screw
19 Idle vent valve
20 Idle vent valve (thermostatic type)
21 Air horn assembly
22 Secondary metering rod (2)
23 Early dashpot assembly
24 Pump assembly
25 Pump return spring
26 Air horn gasket
27 Float assembly
28 Float assembly hinge pin
29 Primary metering rod retainer spring
30 Float bowl insert
31 Idle compensator cover screw (2)
32 Idle compensator cover
33 Idle compensator assembly
34 Idle compensator seal
35 Choke rod
36 Secondary bores baffle
37 Choke rod lever (lower end)
38 Primary metering rod (2)
39 Primary main metering jet (2)
40 Power piston assembly
41 Power piston spring
42 Pump discharge ball retainer
43 Pump discharge ball
44 Pull clip float needle (early)
45 Float needle diaphragm retainer
     screw (early)
46 Float needle assembly retainer (early)
47 Float needle and diaphragm assembly
     (early)
48 Needle and seat assembly (standard)
49 Float bowl assembly
50 Vacuum break control screw
51 Vacuum control hose
52 Vacuum break control assembly
53 Vacuum break control rod
54 Vacuum break rod clip
55 Vacuum break diaphragm
56 Fast idle cam
57 Secondary lockout lever
58 Fuel inlet filter nut
59 Filter nut gasket
60 Fuel filter gasket
61 Fuel inlet filter
62 Filter relief spring
63 Idle adjusting screw spring
64 Idle adjusting screw
65 Throttle-body-to-bowl gasket
66 Throttle body assembly

67 Fast idle lever
68 Cam and fast idle lever attaching screw
69 Cam and fast idle lever spring
70 Fast idle screw spring
71 Fast idle adjusting screw
72 Fast idle cam follower lever
73 Idle mixture needle spring (2)
74 Idle mixture needle (2)
75 Throttle-body-to-bowl attaching screw
76 Primary throttle lever
77 Throttle lever attaching screw
78 Pump rod clip
79 Pump rod

liquid level and the fuel spills over into the float bowl.

As incoming fuel fills the float bowl to the prescribed fuel level, the float pontoon rises and forces the fuel-inlet valve closed, shutting off all fuel flow. As fuel is used from the float chamber, the float drops and allows more incoming fuel to enter the float bowl until the correct fuel level is reached. This cycle continues, constantly maintaining the required fuel level in the float bowl.

A float-needle pull clip fastened to the float-valve stem hooks over the center of the float arm above the inlet valve. It assists in lifting the float valve off its seat whenever there is low fuel-pump pressure and the fuel level is low in the bowl.

1967 and later Q-jets used a conventional Viton-tipped needle and brass seat arrangement for the inlet-valve assembly. The seat screws into the fuel bowl and the needle is a separate item. Operation of the inlet needle and seat in conjunction with the float is as described in the second chapter.

A plastic filler block in the top of the float chamber just above the float valve prevents fuel slosh on severe brake applications. The filler maintains a more constant fuel level to prevent stalling during this type of maneuver. This filler also reduces total fuel capacity and thereby reduces the total vapors which can emit into the atmosphere.

The original float chamber was vented both internally and externally. Internal vent tubes are in the forward and aft sides of the primary bore section just above the float chamber. The elongated slots for the secondary metering rods also provide a sizable vent area. The internal vent balances incoming air pressure beneath the air cleaner with air pressure acting on fuel in the bowl. Therefore, a balanced air/fuel mixture ratio can be maintained during part-throttle and power operation because the same pressure acting upon the fuel in the float bowl is balanced with the air flow through the throttle bores. The internal vent tubes allow fuel vapors to escape from the float chamber during hot-engine operation. This prevents fuel vaporization from causing excessive pressure buildup in the float bowl which can cause excessive fuel spillage into the throttle bores. During hot engine soak, an external idle-vent valve vents fuel vapors from the float bowl to

This plastic filler displaces a large void in the top-forward portion of the bowl, preventing fuel from rushing to the area during brake stops. In certain performance operations it directs inlet fuel to the metering jet area more directly.

The most prominent internal vent is the forward vertical stack (white arrow) shown on this first-design Q-jet. Inspection to the rear of the vertical choke-blade housing will reveal other openings to the fuel bowl. These openings are also internal vents. Most Quadrajets from '65 thru '70 had an external vent—opened during closed throttle by a tang off the pump lever (black arrow).

the atmosphere. This problem exists primarily during warm summer weather. This vent was connected to the throttle and only opened at curb idle. It closed as the throttle was opened farther.

When the throttle valves are in the idle position, an actuating link on the pump lever strikes the spring arm on the idle-vent valve on the top of the primary side of the air horn, opening the valve. Fuel vapors are vented externally, so they cannot enter the throttle bores to be drawn into the engine. This prevents rough engine idle and excessively long hot-engine starting.

A temperature-controlled vent valve was used on some pre-emission Q-jets. A heat-sensitive bi-metal strip held the vent valve in position beneath the idle-vent-valve arm. The valve is held on its seat at temperatures below 75°F. At higher underhood temperatures the bi-metal unseats the valve so fuel vapors caused by hot engine operation escape to improve hot starting and idling at these temperatures. Closing the valve below 75°F. retains fuel vapors in the carburetor to supply extra fuel for good cold-engine starting. During hot-engine

operation, except during idle, the valve is closed by a spring steel vent arm operating off the wire lever on the pump lever. Opening the throttle from idle closes the vent valve, regardless of underhood temperature.

Some Q-jets have a vacuum vent switch valve in the air horn to vent vapors to the air cleaner during engine operation and to the charcoal canister when the engine is off. A spring beneath a vacuum diaphragm piston forces the piston upward to close off the air-cleaner vent and to open the canister vent. The air-cleaner vent is held open during heavy acceleration or low manifold vacuum by an actuating arm on the pump lever. This maintains internal carburetor pressure balance at all times when the engine is running.

Newly designed systems have since eliminated this means of venting during hot soak because of vapor loss to our atmosphere. Current systems plumb these vapors to a charcoal-filled canister where they are retained until the engine is run. This purges the vapors and takes them to the combustion chamber. See chapter fifteen for more details.

## IDLE SYSTEM

The Quadrajet has an idle system on the primary side of the carburetor to supply the correct air/fuel mixture ratios during idle and off-idle operation. The idle system is used during this period because not enough air flows through the venturis to obtain efficient metering from the main discharge nozzles.

The idle system is used only in the two primary bores. Each bore has a separate and independent idle system consisting of idle tube, idle passage, idle-air bleed, idle-channel restriction, idle-mixture-adjustment needle, curb-idle-discharge hole and off-idle slot.

During curb idle the throttle valve is held open slightly by the idle-speed adjusting screw. The small amount of air passing between the primary throttle valve and bore is regulated by this screw to give the desired idle speed. Fuel is added to the air by the direct application of intake-manifold vacuum (low pressure) to the idle-discharge hole below the throttle valve. With the idle-discharge hole in a very low-pressure area and fuel in the bowl vented to atmosphere (high pressure), the idle system operates as follows.

Fuel is forced from the float bowl down through the primary main metering jets into the main fuel well. It passes from the main fuel well into the idle passage where it is picked up by the idle tube. Fuel metered at the idle-tube tip passes up through the idle tube. At the top of each idle tube the fuel is mixed with air supplied through an idle air bleed. The air/fuel mixture then crosses over to the idle down channel where it passes through a calibrated idle-channel restriction.

It then passes down the idle channel past the lower idle-air bleed hole and off-idle-discharge port, just above the primary throttle valve where it is mixed with more air. The air/fuel mixture then moves down to the curb-idle-discharge hole where it enters the carburetor bore and mixes finally with air passing around the slightly open throttle valve. It then enters the intake manifold and is conducted to the engine cylinders as a combustible mixture.

The adjustable idle-mixture needle controls the amount of fuel mixed with the air going to the engine. Turning the mixture screw clockwise (inward) decreases the fuel discharge to give a leaner idle mixture. Turning the mix-

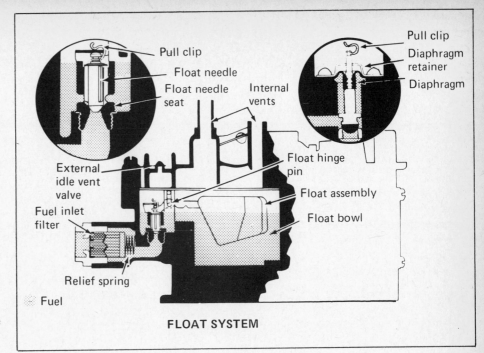

**FLOAT SYSTEM**

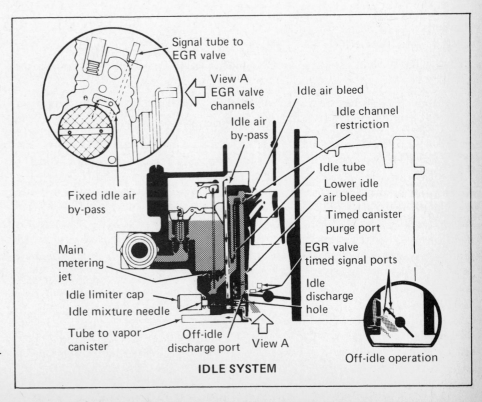

**IDLE SYSTEM**

ture screw counterclockwise (outward) increases fuel discharge to richen the idle mixture.

Note the idle air bleed above each idle tube. This bleed also serves as a siphon break and a calibration aid. The air bleed leads from the top of this channel into the air horn above the venturi. The bleed aids the escape of any fuel vapors in the idle tube channel during hot idle and prevents them from mixing with fuel being picked up by the idle tube. This gives a more consistent idle-fuel mixture during periods of hot-engine idling.

Some Q-jets have a fixed idle-air-by-pass system. An air channel leads from the top of each primary bore in the air horn to a calibrated hole below each primary throttle, or to the throttle-body gasket surface. At normal idle, extra air passing through these channels supplements the air passing by the slightly opened primary throttles. This reduces the amount of air going past the throttles so they can be nearly closed at idle. This reduces the amount of air flowing through the carburetor venturis so the main nozzles will not feed (drip) during idle operation. Because the venturi system in the primary bores is so efficient, this idle-air-by-pass system is used on some applications where large amounts of idle air are required to maintain the idle speed.

On some emission-controlled applications the idle-mixture discharge holes were reduced in size to prevent a too-rich idle adjustment (rich roll) if the mixture needles were turned far out beyond normal mixture requirements. On some models the idle-discharge holes were increased in size when idle air bleed sizes were enlarged as part of recalibration to combat fuel-percolation-caused problems during high-temperature operation.

Another feature sometimes added to emission carburetors is an adjustable idle air bleed system. A separate air channel leads from the top of the air horn to the idle-mixture cross channel. A tapered-head adjustment screw at the top of the channel is used to control the amount of air bleeding into the idle system. This bleed is factory-adjusted when the carburetor is flow checked and a triangular spring clamp is forced over the vent tube to discourage field adjustments.

## OFF-IDLE OPERATION

As the primary throt-

The curb-idle adjustment feed orifice 1 and part of the off-idle feed slot 2 are shown with the throttle blade slightly open. Fixed idle air by-pass port is at 3.

There is one idle-mixture adjusting screw for each primary throttle bore.

tle valves are opened from curb idle to increase engine speed, additional fuel is needed to combine with the extra air entering the engine. This is supplied by the slotted off-idle discharge ports. As the primary throttle valves open, they pass by the off-idle ports, gradually exposing them to high engine vacuum below the throttle valves. This delicately metered system serves a critical drivability area.

Without correct fuel-flow response to engine demand, you have sags and backfires when tipping in the throttle from idle. The accelerator pump

helps here unless throttle movement is very slow.

Further opening of the throttle valves sufficiently increases air velocity through the venturi to cause low pressure at the lower idle air bleeds. As a result, fuel begins to discharge from the lower idle-air-bleed holes and continues to do so throughout part-throttle operation to WOT ranges, supplementing the main-discharge-nozzle delivery.

The idle-discharge holes and off-idle discharge ports continue to supply sufficient fuel for en-

White arrows indicate bushed air bleeds, known as the "small-bleed" type. Black arrows show the "large-bleed" type which is used on some Chevrolet models. Don't think the air-bleed jets have fallen out if you happen to be working on a "large-bleed" unit. It was built that way purposely and the rest of the metering components are designed to match.

gine requirements until air velocity is high enough in the venturi area to obtain sufficient fuel flow from the main metering system.

## HOT-IDLE COMPENSATOR

The hot-idle compensator is in a chamber at the rear of the carburetor float bowl adjacent to the secondary bores or in the primary side of the float bowl. It offsets richening effects caused by excessive fuel vapors during hot-engine operation.

The compensator is a thermostatically controlled valve consisting of a heat-sensitive bi-metal strip, a valve holder and bracket. The valve closes off an air channel leading from a vent in the air horn to a point below the secondary throttle valves.

The compensator valve is normally held closed by the bi-metal strip's tension. During extreme hot-engine operation, excessive fuel vapors entering the engine manifold cause too-rich mixtures, resulting in rough engine idle and stalling. At a predetermined temperature, when extra air is needed to offset the richening effects of fuel vapors, the bi-metal strip bends to unseat the compensator valve. This opens the air channel to allow enough air to be drawn into the engine manifold to offset the richer mixtures and maintain a smooth engine idle. When the engine cools and the extra air is not needed, the bi-metal strip closes the valve and operation returns to normal mixtures.

The compensator valve assembly is held in place by a dust cover over the valve chamber. A seal is used between the HIC valve and the float-bowl casting on HIC types B and C shown in the accompanying illustration.

## OTHER "IDLE-CONNECTED" SYSTEMS

As described in the other functions section of chapter two, there are vacuum advance, exhaust-gas recirculation and timed canister purge ports in the throttle body. A calibrated hole in the throttle body provides constant bleed purge of the canister whenever the engine is running.

## MAIN METERING SYSTEM

The main metering system supplies fuel to the engine from off-idle to wide-open-throttle operation. The Quadrajet has two small primary bores which feed fuel and air during this range, metering fuel according to the venturi prin-

These pieces making up the hot-idle-compensation assembly are mounted in a recessed area at the rear of the carburetor bowl. Some HIC valves are in the primary side of the float bowl.

Q-jet throttle body and main body casting passages: idle by-pass delivers air from above venturi through fixed orifices in throttle body (not all models) 1, idle down channel 2, timed spark (horizontal slot above primary throttle) 3, manifold-vacuum port 4, and power system manifold-vacuum port 5.

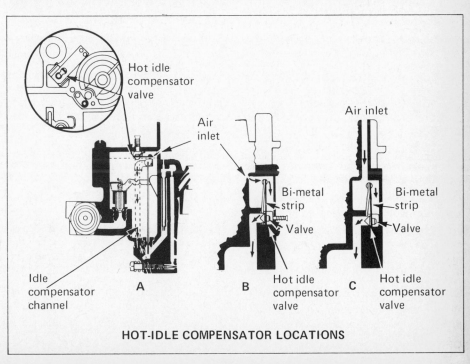

**HOT-IDLE COMPENSATOR LOCATIONS**

109

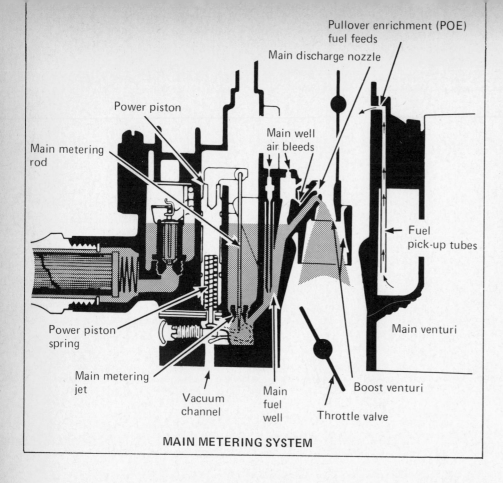

**MAIN METERING SYSTEM**

Labels (clockwise): Power piston · Main metering rod · Main metering jet · Power piston spring · Vacuum channel · Main fuel well · Throttle valve · Boost venturi · Main venturi · Fuel pick-up tubes · Main well air bleeds · Main discharge nozzle · Pullover enrichment (POE) fuel feeds

Adjustable-part-throttle system raises power piston so primary metering rods are precisely positioned in the main jets to provide the required part-throttle fuel mixture. This is primarily done for emission performance. Adjustable lever at 1 (barely visible) provides stop for the power-piston tip, 2. Adjusting screw for the APT lever is hidden by plug 3 after adjustment is made with computer-controlled equipment at RPD.

ciple. This design allows the use of multiple venturis for fine, stable metering control during light engine loads.

The system is in operation at all times when air flow through the venturi is high enough to maintain efficient fuel flow from the main fuel-discharge nozzles. It begins to feed fuel when the idle system can no longer meet the engine requirements.

The main metering system consists of main metering jets, vacuum-operated metering rods, main fuel well, main well air bleeds, fuel-discharge nozzles, and triple venturis. The system operates as follows.

During cruising speeds and light engine loads, engine manifold vacuum is high. Manifold vacuum holds the main-metering rods down in the main-metering jets against spring tension. Manifold vacuum is supplied to the vacuum-operated power piston through a channel connected to the primary main-metering rods. Fuel flow from the bowl is metered between the metering rods and the main-jet orifice.

As the primary throttle valves are opened beyond the off-idle range to allow more air to enter the engine mani-

fold, air velocity increases in the carburetor venturi. This causes a drop in pressure in the large venturi which is increased many times in the pair of boost venturis. Because the lowest pressure (vacuum) is in the smallest boost venturi, fuel flows from the main discharge nozzles.

Fuel flows from the float bowl through the main-metering jets into the main fuel well and is mixed with air from the bleed vent at the top of the main well and side bleeds. One side bleed leads from inside the area above the venturi and another from the cavity around the main fuel nozzle into the main well. Fuel in the main well is mixed with air from the main well air bleeds and then passes through the main discharge nozzle into the boost venturi. At the boost venturi the fuel mixture then combines with the air entering the engine through the throttle bores and passes through the intake manifold and on into the engine cylinder as a combustible mixture.

The main metering system is calibrated by air bleeds and by tapered and stepped metering rods in the main jets. A vacuum-responsive power piston positions the metering rods in the jets to provide fuel as

required by various engine loads. This is described in the following section on the power system.

Two types of primary main metering rods are used in Q-jets. 1967 and earlier models use rods with a single taper at the metering tip; 1968 and later models use rods with a double taper at the metering tip, identified with the suffix B after the part number. Both types of rods use a part number identifying the diameter of a specific point on the tapered portion. Specifications for primary metering rods are in chapter fourteen.

**Adjustable part-throttle feature—** This is used on emission-controlled Q-jets to maintain very close fuel-mixture tolerance for part-throttle operation. A pin pressed into the base of the adjustable-part-throttle (APT) power piston protrudes through the float bowl and gasket to contact an adjustable link in the throttle body. The APT primary main-metering rods are tapered at the upper metering end so fuel flow is controlled by the depth of the taper in the main metering jet orifice. During production, the APT screw is turned in or out to place the taper at the exact point in the jet orifice which provides the desired air/fuel mixture. Once set the APT screw is capped to discourage field adjustments. Tapered metering rods used with the APT feature have the suffix B after the number stamped on the side of the rod.

APT as used on the Q-jet gives good part-throttle performance because all of the fuel is being supplied through the main metering jets.

**Pull-over enrichment**—Some Q-jets are equipped with a fuel pull-over enrichment circuit to supplement fuel feed from the primary main-discharge nozzles. These supplementary fuel feeds allow lean air/fuel mixtures during part-throttle operation while providing the extra fuel needed at higher speeds for good performance.

There are two calibrated holes, one in each primary bore, either just *above* or just *below* the choke valve. During high air flows low pressure created in the air horn pulls fuel from the calibrated holes. The system starts feeding fuel at approximately eight pounds of air per minute and continues to feed at higher flows.

On those models with the two calibrated holes just *below* the choke valve, the pull-over enrichment system feeds additional fuel when the choke is closed for good cold-engine starting. Calibrated air bleeds in the air horn are used with this system, called a *choke enrichment* system (described in the choke section later in this chapter).

## POWER SYSTEM

The power system provides extra mixture enrichment for power requirements under heavy acceleration or during high-speed operation. The richer mixture is supplied through the main metering system in the primary and secondary sides of the carburetor.

The power system in the primary side consists of a vacuum piston and spring in a cylinder connected by a passage to intake-manifold vacuum. The spring beneath the vacuum-operated power piston tends to push the piston upward against manifold vacuum.

During part-throttle and cruising ranges, manifold vacuums are sufficient to hold the power piston down against spring tension to hold the larger diameter of the main-metering-rod tips in the main-metering-jet orifices. Mixture enrichment is not necessary at this point. As engine load is increased to a point where

Some Q-jets have a pull-over enrichment system consisting of tubes which pick up fuel directly from the bowl (white arrows). These connect with feed holes (black arrows) in the air horn. When velocity increases in the primary side (high speed or medium-to-heavy accelerations), a depression created at these holes pulls fuel into the primaries. This allows extremely lean metering at low speeds to pass emission-certification requirements, yet provides adequate fuel for good high-speed performance. Wide boss around center bowl vent indicates carburetor equipped with pull-over circuit. This is also called "high-speed fuel feed."

extra mixture enrichment is required, manifold vacuum weakens and spring tension overcomes the vacuum pull on the power piston. The tapered primary-metering-rod tip moves upward in the main-metering-jet orifice so a smaller diameter portion of the metering rod tip is in the jet. This allows more fuel to pass through the main metering jet to richen the mixture flowing into the primary main wells and out the main discharge nozzles.

When the load is reduced, manifold vacuum rises. Mixture enrichment is no longer needed so manifold vacuum overcomes the power-piston-spring tension, returning the larger portion of the metering rod into the metering-jet orifice and reducing fuel flow back to normal economy ranges.

The main metering and power systems are integrated. The system acts as a main metering system under light loads (manifold vacuum above 7 inches Hg). This places the straight part of the larger diameter or the tapered part of the main metering rod in the jet.

The taper acts as a transition to the full-power mixture. Under power situations, the rod is pulled up so only the smaller straight portion of the rod extends into the jet orifice.

On some models two power-piston springs are used beneath the power piston. The smaller (primary) power-piston spring seats in the center of the piston and bottoms on the float-bowl casting.

A larger diameter (secondary) spring surrounds the smaller spring. It exerts added pressure on the bottom of the power piston to provide the required mixture ratios at heavy engine loads. The dual-power-piston-spring feature assists in providing the air/fuel ratios required to meet emission and power requirements. The two springs convert the power-piston/metering-rod combination into a two-stage power system. This was done to make some engine applications pass certain critical portions of emission certification.

## SECONDARY SYSTEM

As engine speed increases, the primary side of the carburetor can no longer meet the engine's air-flow demands and the secondary side of the carburetor begins to operate. The secondary section contains throttle valves, spring-loaded air valves, metering orifice plates, secondary metering rods, main fuel wells with air-bleed tubes, fuel-discharge nozzles, and accelerating wells and tubes. The secondary side operates as follows.

When engine demand requires more air/fuel than the primary bores can supply, the primary throttle lever begins to open the secondary throttle valves with its connecting linkage to the secondary-throttle-shaft lever. As air flow through the secondary bores creates a low pressure (vacuum) beneath the air valve, atmospheric pressure on top of the air valve forces it open against spring tension. This allows the required air for increased engine speed to flow past the air valve.

As the air valve begins to open, its upper edge passes the accelerating-well ports, exposing them to manifold vacuum. The ports immediately start to feed fuel from the accelerating wells and continue to feed fuel until the accelerating well fuel is depleted.

The accelerating ports prevent momen-

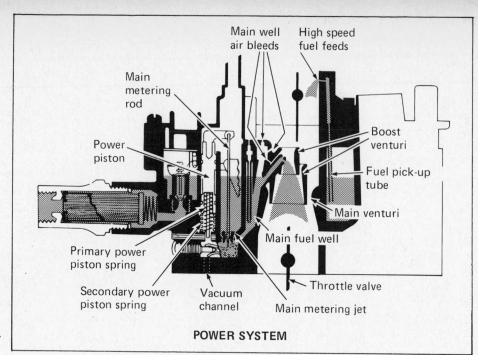

**POWER SYSTEM**

The baffle in the secondary side tends to channel a large portion of the air past the nozzles until the air valve is fully open. This creates a low-pressure area, causing fuel to flow from the nozzles. Some consider this arrangement to be similar to a crude venturi.

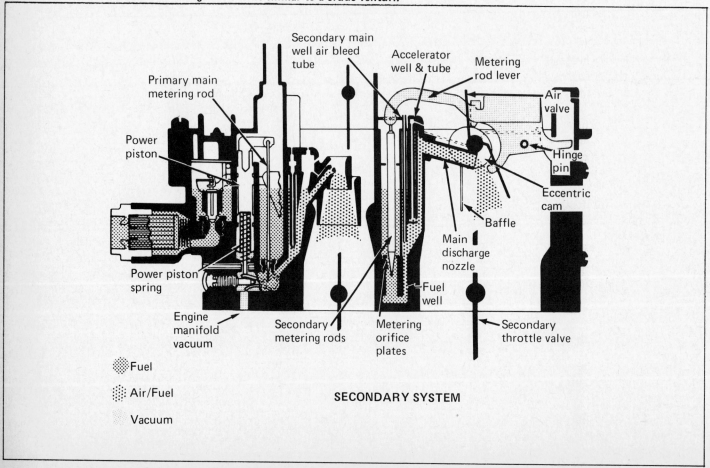

Fuel

Air/Fuel

Vacuum

**SECONDARY SYSTEM**

tary leanness as the valve opens and the secondary nozzles begin to feed fuel. Some Q-jets do not include the accelerating well feature.

The secondary main-discharge nozzles (one for each secondary bore) are just below the air valve and above the secondary throttle valves. As they are in the area of lowest pressure, they begin to feed fuel as follows.

When the air valve begins to open it rotates a plastic cam attached to the main air-valve shaft. The cam pushes on a lever attached to the secondary main-metering rods, pulling the rods out of the secondary-orifice plates. Fuel from the float chamber flows through the secondary-orifice plates (jets) into secondary main wells, where it mixes with air from the main-well tubes. The air-emulsified fuel mixture travels from the main wells to the secondary-discharge nozzles and into the secondary bores. Here the fuel mixture is mixed with air traveling through the secondaries to supplement the air/fuel mixture delivered from the primaries and goes on into the engine manifold and to the engine cylinders as a combustible mixture.

As the throttle valves are opened farther and engine speeds increase—increased air flow through the secondary side opens the air valve to a greater degree—which in turn lifts the secondary metering rods farther out of the

As the secondary air valve starts to open, the forward edge of the blades passes the secondary fuel discharge ports, exposing them to a reduced pressure area. This pulls a shot of fuel which eliminates lean sag as the air valve starts to open. A few models of the Quadrajet did not have this feature.

A spare plastic cam is here to show its shape. The metal pick-up lever between the air valves rides on the cam (arrow). The secondary metering rod hanger mounts to the center lever. When the air valve opens, the high part of the cam lifts the lever—and the secondary metering rods for automatic enrichment at high air flows.

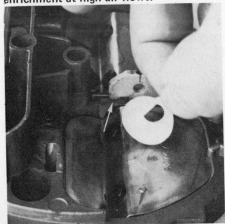

orifice plates. The metering rods are tapered so that fuel flow through the secondary-metering-orifice plates is directly proportional to air flow through the secondary bores. In this manner correct air/fuel mixtures supplied from the secondary bores are maintained by the depth of the metering rods in the orifice plates in relation to the air-valve position. This relationship is factory-adjusted to meet the air/fuel requirements for the specific engine model. No further adjustment should be required for normal road-use applications.

When the air valve has lifted the secondary metering rods to their maximum height (at or near full air-valve opening) the small diameter tip of the rods is in the orifice plates. This provides a richer full-power mixture for high-RPM operation. Secondary metering rod specifications are in chapter fourteen.

Four other features are incorporated in the secondary metering system as follows:

1. Main-well bleed tubes extend below the fuel level in the main well. These bleed air into fuel in the well to emulsify the fuel with air for good atomization as it leaves the secondary-discharge nozzles.

2. Secondary metering rods (through 1972) have a slot milled in the side to ensure adequate fuel supply in the secondary fuel wells. These slots were considered necessary because the secondary metering rods are nearly seated against the secondary metering orifice plates when the air valve is closed. During hot-engine idle or hot soak the fuel could boil away to empty the fuel well—at least partially. The slots allowed enough fuel to by-pass the orifice plates so the main wells are filled during this period. This ensured immediate fuel delivery from the secondary fuel wells at all times. In 1973 it was found that the slots could be eliminated without detrimental effects.

3. A directional baffle plate in each secondary bore extends up and around the secondary fuel-discharge nozzles. The baffle aids in directing the air flow for improved distribution under certain conditions.

4. An air horn baffle is used on some models to prevent incoming air from the air cleaner from forcing fuel down in the secondary wells through the bleed tubes. This prevents secondary nozzle lag on heavy acceleration.

When an early-model air horn is lifted off and turned bottoms-up, many items are exposed. Refer to this photo as you read about secondary operation.
1  Air valve damping piston
2  Fuel feed tubes for secondary accelerating well discharge holes
3  Secondary main well bleed tubes
4  Secondary metering rods

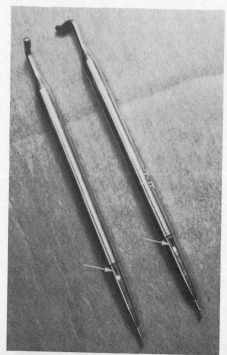

The secondary rods are precision pieces. Since the first design many various tapered ends have been developed to meet the fuel requirements of specific engines. A total listing of these can be found in high-performance chapter fourteen. Arrows point out the slots referred to in the text.

## AIR-VALVE DASHPOT

In first-design Q-jets the secondary air valve has an attached piston (shown in an accompanying drawing) which acts as a damper to prevent valve oscillation due to engine pulsations. The damper piston operates in a well filled with fuel from the bowl. Piston motion is retarded by fuel which must by-pass the piston when it moves up in the fuel well. The piston is attached loosely to a plunger rod. The rod has a rubber seal which holds the damper piston onto the plunger rod and also acts as a valve. The valve seats on the piston when the air valve opens and the piston rod moves upward, closing off the piston's center area and slowing down the air valve opening to prevent secondary-discharge-nozzle lag.

1967 and later Q-jets have an air-valve dash pot operated off of the choke vacuum break diaphragm unit. The air valve is connected to the vacuum break by a rod. Whenever manifold vacuum is above approximately five to six inches Hg, the vacuum break diaphragm is seated. The plunger is held fully inward against spring tension. At this point the vacuum break rod is in the forward end of the slot in the air valve lever and the air valve is closed.

During acceleration or heavy engine loads when the secondary throttle valves are opened, the manifold vacuum drops. A spring in the vacuum break diaphragm overcomes the vacuum pull and forces the plunger and link outward, permitting the air valve to open when the combination of depression under the air valve and velocity through it combine to force the air valve open against the tension of its torsion spring. Air-valve opening rate is controlled by the calibrated restriction in the vacuum inlet to the diaphragm. This provides the damping action to delay air-valve opening enough so fuel flow can begin at the secondary-discharge nozzles.

## ACCELERATOR-PUMP SYSTEM

When the throttle is opened rapidly for quick acceleration, the air flow and manifold vacuum change almost instantaneously. Fuel is heavier and tends to lag behind the air flow, causing a momentary leanness. The accelerator pump provides the extra fuel necessary for smooth operation during this time.

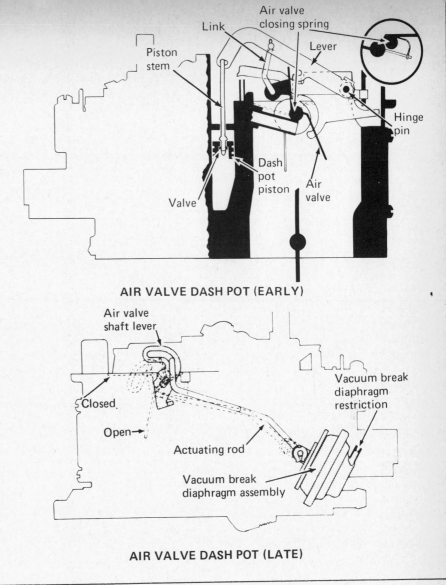

AIR VALVE DASH POT (EARLY)

AIR VALVE DASH POT (LATE)

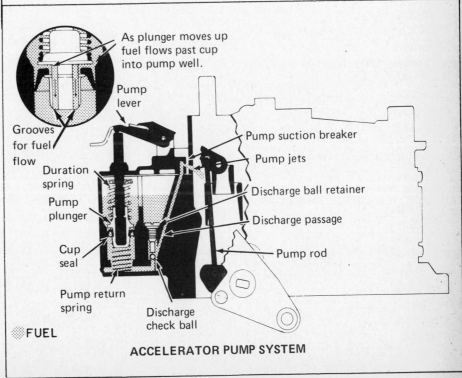

ACCELERATOR PUMP SYSTEM

The accelerator-pump system is in the primary side. It consists of a spring-loaded pump plunger and pump-return spring operating in a fuel well. The pump plunger is operated by a pump lever connected directly to the throttle lever by a pump rod.

When the pump plunger moves upward in the pump well during throttle closing, fuel from the float bowl enters the pump well through a slot in the top of the pump well. The floating synthetic-type pump cup moves up and down on the pump-plunger head. When the pump plunger is moved upward, the flat on the top of the cup unseats from the flat on the plunger head to allow free movement of fuel through the inside of the cup into the bottom of the pump well. This floating cup also vents any vapors which may be in the pump well so a charge of liquid fuel is maintained in the fuel well beneath the pump plunger head.

When the primary throttle valves are opened the connecting linkage forces the pump plunger downward. The pump cup seats instantly and fuel is forced through the pump discharge passage, where it unseats the pump discharge check ball and passes on through the passage to the pump jets in the air horn. Here it sprays into the venturi area of each primary bore.

It should be noted that the pump plunger is spring-loaded. The upper duration spring is balanced with the bottom pump-return spring so that a smooth sustained charge of fuel is delivered during acceleration.

The pump-discharge check ball seats in the pump-discharge passage during upward motion of the pump plunger so air will not be drawn into the passage. Otherwise, a momentary acceleration lag could result.

During high-speed operation, a vacuum exists at the pump jets. A cavity just beyond the pump jets is vented to the top of the air horn, outside the carburetor bores. This acts as a suction breaker so that when the pump is not in operation fuel will not be pulled out of the pump jets into the venturi area. Thus, a full pump stream is ensured when needed and there is no fuel "pull-over" from the pump-discharge passage.

## CHOKE SYSTEM

The variations of design used to get the choke on, to modulate its action

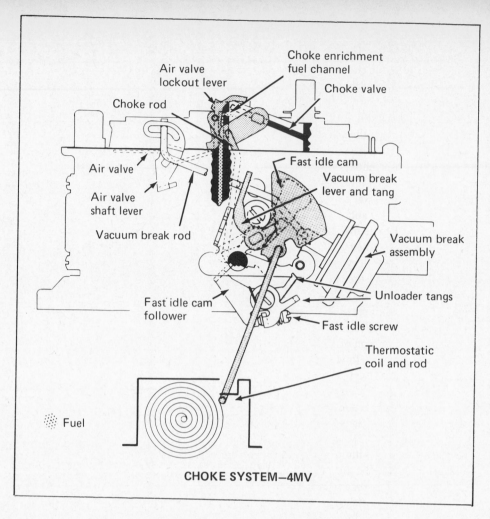

**CHOKE SYSTEM—4MV**

according to engine load and temperature and to get the choke off when the engine warms up are many and complex. Study the accompanying drawings and you should get some general idea of what the designers have been trying to do. I'll agree with your wonderment at the many designs which have been used thus far. There will undoubtedly be others because getting the choke off quickly is one of the important requirements in passing total emissions tests where the engine must be cold-started.

Adjustments to the various choke systems are described at the conclusion of the following service chapter.

The choke valve is on the primary side of the carburetor. It provides the correct air/fuel mixture enrichment for quick cold-engine starting and during the

warm-up period. The air valve is locked closed until the engine is thoroughly warm and the choke valve is wide open.

The choke system consists of a choke valve in the primary side of the air horn, a vacuum-diaphragm unit, fast-idle cam, connecting linkage, air-valve-lockout lever and a thermostatic coil. Some applications may use a split choke pick-up spring or a vacuum-break modulating spring. The thermostatic coil is usually in the engine manifold and it is connected to the intermediate choke shaft and lever assembly. Choke operation is controlled by the combination of intake-manifold vacuum, the offset choke valve, temperature and throttle position.

The thermostatic coil in the intake manifold (or on the carb body in the case of Oldsmobile units) is calibrated to hold the choke valve closed when the engine is cold.

NOTE: To close the choke valve manually, the primary throttle valves have to be partially opened to allow the fast-idle-cam follower to by-pass the fast-idle-cam steps and come to rest on the fast-idle-cam's highest step.

When the choke valve is closed, the air-valve lockout lever is weighted so that a tang on the lever catches the upper edge of the air valve and keeps it closed. Or, depending on the model, the secondary throttles may be locked out when the choke is closed. Both types of lockout are used widely through the Q-jet line.

During engine cranking, air pressure against the top side of the choke valve causes it to open slightly against tension of the thermostatic coil. This restricts air flow through the carburetor to provide the required rich starting mixture. When the engine starts and is running, manifold vacuum applied to the vacuum-break diaphragm unit on the float bowl opens the choke valve to a point where the engine will run without loading or stalling. At the same time, the fast-idle-cam-follower lever on the primary-throttle-shaft end drops from the highest step on the fast-idle cam to the second step as the throttle is opened. This gives the engine sufficient fast-idle throttle and correct fuel mixture for running until the engine begins to warm up and heat the thermostatic coil. As the thermostatic coil becomes heated, it relaxes its tension and allows the choke valve to open further because of intake air pushing on the off-set choke valve. Choke-valve opening continues until the thermostatic coil is completely relaxed at which point the choke valve is wide open.

The vacuum-break modulating spring allows the vacuum break to vary choke-valve position according to ambient temperature. The vacuum-break modulating spring connected to the vacuum-break link allows varying choke openings depending on the closing force of the thermostatic coil. As the closing force of the coil increases (cool weather) the link moves in the slotted lever until the modulating spring overcomes the coil force, or the link is in the end of the slot. This results in less vacuum break (choke valve more closed) during cooler weather and more vacuum break (choke valve more open) during

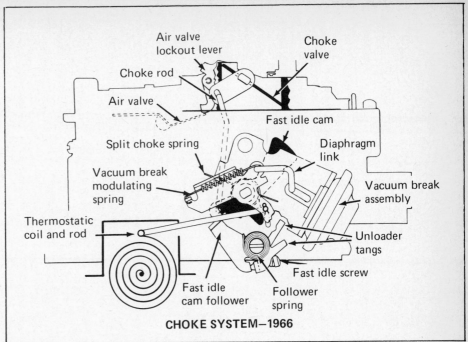

CHOKE SYSTEM—1966

The vacuum-break modulating spring feature was used on the 1966-69 Oldsmobile Q-jets.

warmer weather.

**Split-choke (early MV models)**—This feature operates during the last few degrees of choke-thermostat rotation. It maintains fast-idle speed long enough to keep the engine from stalling but allows the use of a choke coil which lets the choke valve open quickly. The operation of the split-choke feature is controlled by a torsion spring on the intermediate-choke-lever shaft. As explained earlier, air pressure action on the offset choke valve tends to force the choke valve open against tension of the choke thermostatic coil. In the last few degrees of thermostatic-coil opening motion a tang on the intermediate-choke lever contacts the end of the torsion spring. This keeps the fast-idle-cam-follower lever on the last step of the fast-idle cam longer to maintain fast idle until the engine is thoroughly warm. The spring works against the thermostatic coil until the coil is hot enough to pull on

the intermediate choke lever and overcome torsion-spring tension.

When the engine is thoroughly warm, the choke coil pulls the intermediate choke lever completely down and allows the fast-idle cam to rotate so that the cam follower drops off the last step of the fast-idle cam, allowing the engine to run at normal idle speeds. When the choke shaft lever moves toward the up position, the end of the rod strikes a tang on the air-valve lockout lever. As the rod moves to the end of its travel, it pushes the lockout tang upward and unlocks the air valve.

Models used on applications where little or no air flow can be tolerated from the secondary throttle bores during engine warmup, a secondary throttle valve lockout in place of the air-valve lockout feature.

A lockout lever on the float bowl is weighted so a tang on

the lower end of the lever catches a lock pin on the secondary throttle shaft to hold the secondary throttle valves closed. As the engine warms up, the choke valve opens and the fast idle cam drops. As the cam drops the last few degrees it strikes the secondary lockout lever, pushing it away from the secondary valve lockout pin to allow the secondary valves to open.

**Spring-assist choke closing system**—Some Chevrolet 4MV carburetors use a torsion-type assist spring on the intermediate choke shaft. It exerts pressure on the vacuum-break lever to force the choke valve toward a closed position. Torsion-spring tension is overcome by the choke thermostatic coil located on the engine manifold which, during the engine warmup period, will pull the choke valve open. The addition of a torsion spring assists in closing the choke valve to ensure good engine starting when the engine is cold.

Along with the choke closing assist spring, 1970 and 1971 Chevrolet 4MV models use the fast idle cam "pull-off" feature.

When the engine starts and is running, manifold vacuum is applied to the vacuum break diaphragm and the diaphragm plunger moves slowly inward to open the choke valve. As the diaphragm plunger moves inward, a tang on the plunger contacts the end or "tail" of the fast idle cam to "pull off" the cam from the high step to the lower second-step setting.

Some Chevrolet 4MV models have the choke closing assist spring on the vacuum break plunger stem (instead of on the intermediate choke shaft). The closing assist spring tension is added to the closing tension of the remote choke thermostatic coil to assist in closing the choke valve during engine starting. The choke closing assist spring only exerts pressure on the vacuum-break link to assist in closing the choke valve during engine starting. When the engine starts and the choke vacuum break diaphragm seats, the closing spring retainer hits a stop on the plunger stem and no longer exerts pressure on the choke valve.

With this arrangement, the tang on the vacuum-break-diaphragm plunger is removed to eliminate the fast-idle-cam "pull-off" feature. Also, the slot for free travel of the air-valve-dashpot link is on the vacuum-break plunger instead of on the air-valve-shaft lever.

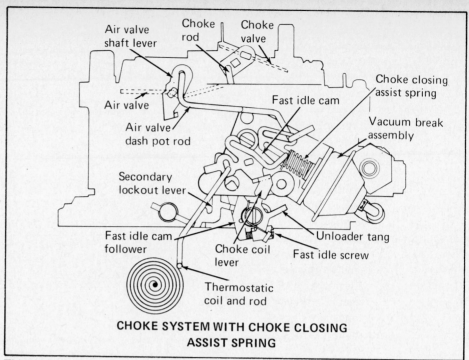

**CHOKE SYSTEM WITH CHOKE CLOSING ASSIST SPRING**

**This choke arrangement is found on Q-jets as used on 1972-74 Chevrolets. On 1968-71 Chevrolets the choke closing spring is wrapped around the choke lever pivot. This scheme was used in conjunction with a feature called "fast-idle-cam pull-off."**

**Choke features are often used in combinations**—The schematic drawings of choke mechanisms are presented on these pages to illustrate some of the many ways Rochester engineers have used to modulate choke operation. You will find features used in various ways, some of which are not shown in these drawings.

**Delayed vacuum-break system**—Models using this system operate as follows:

When the engine is started cold, vacuum applied to the choke vacuum break diaphragm unit opens the choke valve against tension of the thermostatic choke coil to a point where the engine will run without loading or stalling. An internal air bleed check valve inside the diaphragm unit delays the choke valve from opening too fast. When the engine starts, vacuum acting on the internal check valve bleeds air through a small hole in the valve which allows the vacuum diaphragm plunger to move slowly inward. This gives sufficient time to overcome engine friction and wet the engine manifold to prevent a lean stall. When the vacuum break diaphragm is fully seated, which takes a few seconds, the choke valve will remain in the vacuum-break position until the engine begins to warm and relax the thermostatic coil. Some models include a rubber-covered filter over the purge bleed hole in the vacuum tube to the rear vacuum break. The rubber-covered filter must be removed and the bleed hole closed with a piece of tape when making system adjustments. In addition to the internal-bleed check valve, some car applications have a separate vacuum-delay tank. It is connected "in series" to a second vacuum tube on the vacuum diaphragm unit to further delay the choke vacuum-break-diaphragm operation.

The check valve in the choke vacuum diaphragm unit is designed to "pop" off its seat and allow the diaphragm plunger to extend outward when the spring force against the diaphragm is greater than the vacuum pull. This will give added enrichment as needed on heavy acceleration during cold driveaway by allowing the choke coil to close the choke valve slightly. Some 4MV models use a calibrated restriction in the vacuum inlet to the vacuum break diaphragm unit in place of the internal air bleed check valve. Similar to the internal air bleed check valve, the calibrated restriction delays the supply of vacuum to the diaphragm unit to retard opening of the choke valve for good engine starting.

**Bucking spring**—A spring-loaded plunger used in the vacuum break unit on some 4MV models offsets choke thermostatic coil tension and balances the greater opening of the choke valve with choke-coil tension. This enables further refinement of

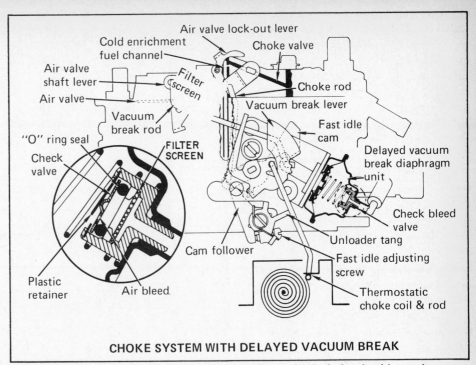

**CHOKE SYSTEM WITH DELAYED VACUUM BREAK**

1969-72 Pontiacs have this choke arrangement. Some also include a bucking spring.

Although not all models used the choke enrichment feature, this schematic is typical for 1969-74 Cadillacs and for some 1969-72 Pontiacs.

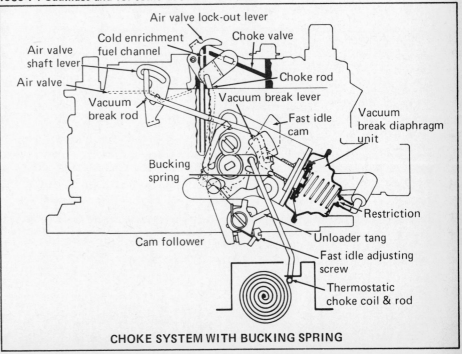

**CHOKE SYSTEM WITH BUCKING SPRING**

air/fuel mixtures because the coil, which senses engine and ambient temperatures, will allow the choke valve to open gradually against "bucking spring" tension in the diaphragm plunger head. In very cold temperatures, the extra tension of the thermostatic coil overcomes the diaphragm-plunger (bucking) spring to provide less choke valve opening and slightly richer mixtures. In warmer temperatures, the thermostatic coil has less tension and cannot compress the spring as much. A greater choke valve opening then gives slightly leaner mixtures.

**Dual delayed-vacuum-break system**—Other Quadrajet carburetors include a front or main vacuum-break diaphragm unit with a rear or auxiliary vacuum-break diaphragm unit. The dual delayed-vacuum-break system operates as follows: During engine cranking, the choke valve is held closed by thermostatic-coil tension, restricting air flow through the carburetor to provide a richer starting mixture.

When the engine starts and is running, manifold vacuum is applied to both vacuum break diaphragm units mounted on the side of the float bowl. The front or main vacuum break diaphragm opens the choke valve to a point where the engine will run without loading or stalling. As the engine is wetted and friction decreases after start, an internal bleed in the auxiliary vacuum break unit causes a delayed action to gradually open the choke valve a little further until the engine will run at a slightly leaner mixture to prevent loading.

The auxiliary vacuum break unit includes a spring-loaded plunger. The purpose of the spring is to offset choke thermostatic coil tension and balance the greater opening of the choke valve with tension and balance the greater opening of the choke valve with tension of the choke coil. This enables further refinement because the coil, which senses engine temperature, will allow the choke valve to open gradually against spring tension in the diaphragm plunger head.

**Early 4MC models (choke stat housing on carburetor)**—have a choke mechanism which includes an adjustable vacuum-break diaphragm. These operate similarly to the 4MV models, with one major exception.

When the engine starts and is running, manifold vacuum is applied to the vacuum diaphragm unit. There is an adjustable plastic plunger

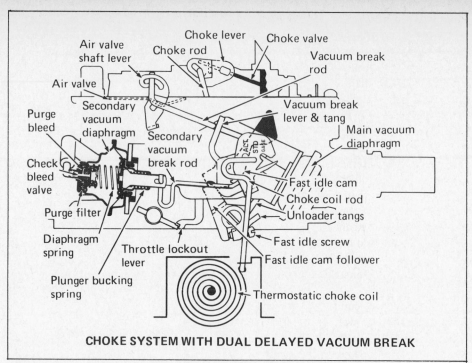

**CHOKE SYSTEM WITH DUAL DELAYED VACUUM BREAK**

1970-74 Buick Q-jets typically use the dual delayed vacuum break set up pictured here. Use of this more complex choke mechanism will probably increase in future years as the total emission requirements become even more stringent.

1966 Buicks used this type choke on their Q-jets.

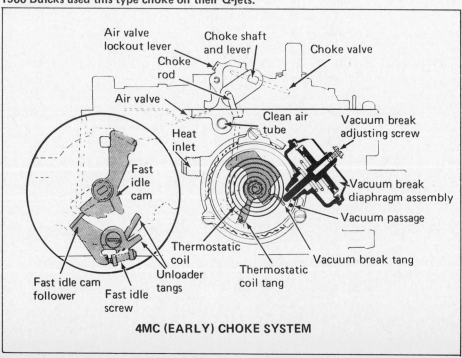

**4MC (EARLY) CHOKE SYSTEM**

on the vacuum-break diaphragm. Engine vacuum pulls inward on the diaphragm and this plunger strikes the vacuum-break tang inside the choke housing which, in turn, rotates the intermediate choke shaft and through connecting linkage opens the choke valve.

**Later 4MC models (choke stat housing on carburetor)**—These use a delayed-vacuum-break system. An internal bleed check valve in the vacuum-break diaphragm delays diaphragm action a few seconds before it becomes seated. This allows the engine manifold to be wetted and engine friction to decrease enough so that when the vacuum-break point is reached, the engine will run without loading or stalling.

When the choke valve moves to the vacuum-break position, the fast-idle-cam follower lever on the end of the primary throttle shaft drops from the highest step to the next lower step when the throttle is opened. This gives the engine sufficient fast-idle speed and correct fuel mixture for running until the engine begins to warm up and heat the thermostatic coil in the choke housing. Engine vacuum pulls heat from the manifold heat stove into the choke housing and gradually relaxes choke-coil tension, allowing the choke valve to continue opening through inlet air pressure pushing on the offset choke valve. Choke valve opening continues until the thermostatic coil is completely relaxed, at which point the choke valve is wide open and the engine is thoroughly warm.

Both early and late 4MC models use an air-valve lockout lever which operates similarly to 4MV models.

**Vacuum re-indexing unit**—Starting in 1974, Cadillac models had a vacuum-operated diaphragm plunger on the remote choke coil assembly to improve cold-engine starting. The assembly is called a vacuum reindexing unit. It applies additional pressure on the choke coil which, in turn, increases the pressure holding the choke valve closed when starting the engine. This provides the correct air/fuel mixtures for good engine starting, yet allows the choke to open quickly for good economy and reduced emissions.

The vacuum re-indexing unit is a spring-loaded diaphragm operated by direct manifold vacuum. When the engine is not running, no vacuum is applied to the diaphragm. The spring on top of the diaphragm

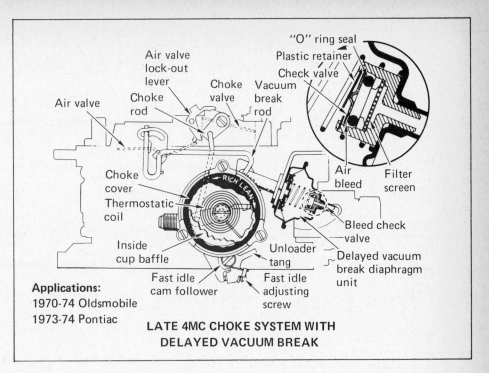

**Applications:**
1970-74 Oldsmobile
1973-74 Pontiac

**LATE 4MC CHOKE SYSTEM WITH DELAYED VACUUM BREAK**

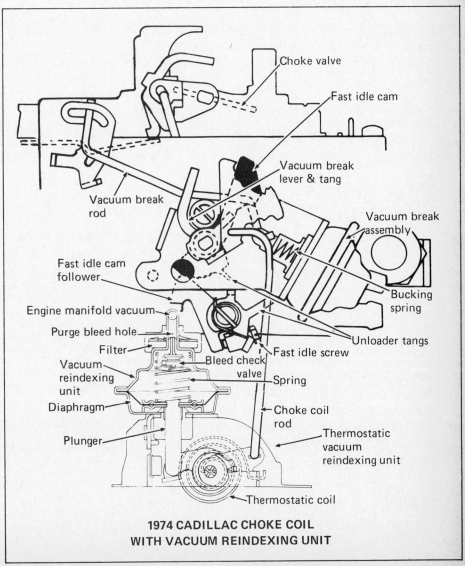

**1974 CADILLAC CHOKE COIL WITH VACUUM REINDEXING UNIT**

pushes the diaphragm and plunger downward. The end of the plunger strikes a lever attached to the center choke coil shaft, rotating the choke coil counterclockwise and increasing the pressure on the coil which, in turn, increases the pressure holding the choke valve closed for good engine starts.

After the engine starts, manifold vacuum acts on top of the vacuum diaphragm, pulling the diaphragm plunger upward against spring tension until it clears the lever on the remote choke coil shaft. The choke coil then operates in the normal manner allowing the choke valve to completely open as the thermostatic coil warms up.

The bleed check valve in the top of the diaphragm unit retards upward movement of the diaphragm plunger to allow gradual reduction of choke coil pressure so the engine will not stall lean. A small purge hole in the vacuum tube allows a small amount of clean filtered air to be drawn into the vacuum channel to remove any engine vapors or dirt which might tend to plug the internal check valve filter or small bleed hole in the check valve.

The unit can be checked for operation by making sure the plunger is extended (full downward) with the engine off and slowly moves to the full-up position when the engine is started and running. The vacuum unit should not be immersed in any type of carburetor cleaner. If the unit becomes inoperative, replace the assembly.

**Unloader**—4MV and 4MC choke systems have an unloader mechanism to open the choke valve partially should the engine become loaded or flooded during the starting period. To unload the engine, depress the accelerator pedal so the throttle valves are held wide open. A tang on a lever on the choke side of the primary throttle shaft contacts the fast-idle cam and forces the choke valve slightly open through the intermediate choke shaft. This allows extra air to enter the carburetor bores and pass on into the engine manifold and engine cylinders to lean out the fuel mixture so the engine will start.

**Choke enrichment system**—Some Quadrajet models use a system which provides fuel supplementing that from the primary main discharge nozzles for good cold-engine starting. Two calibrated holes, one in each

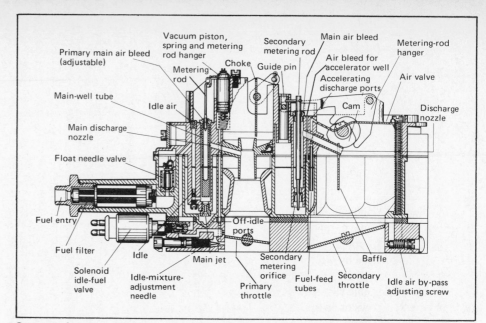

Cutaway drawing of German Solex for 1974 Mercedes six- and eight-cylinder engines. The primary main air bleed is adjusted by the vacuum piston to provide an air/fuel mixture compatible with throttle opening and engine load. Quadrajet ancestry of many design features is easy to see. Some features which do not show in this drawing are a secondary lockout arrangement and an air-valve dashpot just like those on the Q-jet.

primary bore, are located in the air horn just *below* the choke valve to supply added fuel for cold enrichment during the cranking period. The extra fuel is supplied through channels to the secondary accelerating well pick-up tubes to allow fuel at closed choke to be drawn from the secondary accelerating wells in the bowl. Also, during warm-engine operation, the two calibrated holes in the air horn are a pull-over enrichment system feeding a metered amount of fuel at higher air flows to supplement fuel flow in the primary bores to provide the extra fuel needed at higher engine speeds. This is mentioned earlier in the main metering system discussion.

Other Quadrajets use a fuel pull-over enrichment system similar to the choke enrichment system except that the two calibrated holes, one in each primary bore, are just *above* the choke valve to supply added fuel *only* during higher air flows. The calibrated holes above the choke valve do not feed fuel at closed choke during engine cranking.

**DESIGN CHANGES**

This concludes our explanation and illustrations of the Quadrajet. Nearly every portion of the carburetor has been subtly changed since the 1965 intro-

duction. A few of the more significant ones are detailed in the preceding text. They include a choke thermostatic coil housing introduced by Oldsmobile (4MC). The air-valve damping design was changed to a vacuum-diaphragm unit (incorporating the choke vacuum break on some models), and the fuel inlet reverted to a conventional needle/seat assembly. The *basic* design has not been changed.

All in all, it must be said the design concept was unusually good. It was far ahead of its time. This has been substantiated by increased demand year after year. And, it should be noted that many Quadrajet features have been incorporated into carburetors of other manufacturers. Most recent of these as this book went to press was the special Solex design for the 1974 Mercedes sixes and V-8's.

The combination of venturi-primaries with air-valve secondaries is a good solution to the air-flow and emission-control problems of modern engines. Rochester saw the need in the early sixties when they started designing the Q-jet. Other makers waited until nearly the end of the decade to begin offering such sophisticated fuel-control mechanisms.

# Acknowledgments

Should you delve into carburetion beyond service operations, you will become aware it is a study all its own. No other component on the vehicle is so influenced by weather, fuel, vapors, heat, forces, hydraulics and many seemingly unpredictable variables. There are no fixed rules, laws or formulas that solve all of what we term "carburetion problems."

In more than 20 years of concerted effort I have been confronted by and solved thousands of problems—some little, some big. Regretfully, a few have gone unsolved. Only after these years of effort in the field in becoming acquainted with general trends and actual happenings could I relate the best information I knew in this book. Much of this information can be attributed to many individuals who contributed directly or indirectly to the writing of this book.

A special thanks goes to a few who probably were not aware of my writing this book. They are former supervisors and fellow workers who in various ways contributed so much to my understanding of carburetion and its function on the automobile. Regretfully the majority cannot be listed; but, if in the past years you and I sweated out some problems together, I am thankful for the association.

Lou Phillips, now retired from Stromberg Carburetor (Bendix Aviation Corp.), some 20 years ago set me on a course of methodical test procedures that became habit and a means of instilling confidence in tests that remains with me today.

The carburetor engineering management of Rochester Products hired me as a shop technician in 1956 and had confidence enough to assign me challenging jobs that led to an engineering position in carburetor development. Art Winters, Al Kehoe and Russ Sanders allowed these rewarding opportunities to happen.

Bob Sternaman, to whom I was assigned on my first liaison engineering job as a carburetor consultant to Chevrolet Motor Division, provided an atmosphere of trust and confidence that dictated meticulous and positive approaches to problems.

The majority of carburetor design and development personnel during the Rosenberger-Roensch era (early '60s) were great and sharp. To those outside Chevrolet the referenced era was a period of time when economy and performance were key words and emission design and development were infants. Rosenberger and Roensch at that time were two of the key Chevrolet managers—since retired. Of this group a special thanks goes to Bob Aldrich who provided me with confidence and responsibility which netted me experience attainable no other way.

Walter MacKenzie and Vince Piggins of the GM product performance group for whom I worked in the '60s assigned me to competitive events around the nation. (GM participated in racing back then.) Through these time-limit (sometimes crash!) programs I learned new methods and procedures in testing that brought about expediency in obtaining accurate results. This is a most valuable ingredient in any test program when time or money is a factor. As you know in the work world, time is money.

Several people have assisted in one way or another since the conception of the book.

Al Kehoe, Rochester's Chief Engineer, helped us get underway with essential information and technical details.

Dick Klotzbach of RPD's Service Department was most helpful in providing materials, photos and finally valid comments on written material. Race enthusiast Tom Toal, also of RPD, shared his personal knowledge of building Quadrajets for performance. He has built many for drag and oval-track winners. Another RPD Service Engineer, Bill Hobbs, assisted us by clarifying important details now included in the emission chapter.

Bob Prior, RPD's Desert Test Facility operations manager, provided photos, answers and encouragement.

Jim McFarland of Edelbrock Equipment Co. pleasantly surprised us with photographs of Q-jets used in performance applications.

Jim Williams, Technical Products Manager at Chevrolet Public Relations, provided numerous photos, some of which took a lot of digging by his secretary Gloria Jezewski.

Gerry Holmes of Oldsmobile's Public Relations staff dug 'way back in their files for pictures of the three two-barrel manifold and of the Pike's Peak run. R. W. Emerick, Pontiac's Director of Public Relations, gave us important photos showing manifold construction and their 455 Super Duty engine.

Bill O'Neill, Public Relations Supervisor at Cadillac, helped with shop manual illustrations for the emissions chapter.

John Dinan of AC supplied fuel pump and air-filter photos and drawings to help in those areas.

Richard Watts II, tuner of Allan Barker's potent Corvette, contributed some of his experiences to the road-racing section of chapter fourteen.

There are names like Louis Cuttitta, Jack Layton, Jack Jaquette and Dave Brown—each one a racer, car builder or performance enthusiast and a friend from my GM days. All gave some of their personal time to read and comment on various portions of the book.

Thanks goes to Mike Urich for materials used from the *Holley Carburetors* book which he co-authored with Bill Fisher.

And, to my immediate family—a giant thank you for tolerating the constant paper-picture mess scattered throughout our house for almost two years!

Doug Roe Engineering personnel Jim LeRoy (slalom/road racer), Max Cookman (drag racer) and Jim McKinney (off-road racer) contributed assistance in too many ways to tell.

Bill Fisher himself deserves credit for building a successful organization by demanding accurate material for his books. His expertise in automotive technology has contributed heavily to getting our thoughts said right.

—*Doug Roe*

NOTE: Although Bill Fisher co-authored this book, any references to personal experiences, to "this writer" or specific recommendations are by Doug Roe.

123

# Quadrajet Service

**Chevrolet Blazer truck used to illustrate
this Quadrajet service section.**

## INTRODUCTION

Many well-designed pieces of
equipment get chastised unfairly by errone-
ous comments such as "too complex" or
"too complicated." In the case of carburetors, extremely harsh and unfair judgements
are cast on certain units which rightfully
stand among the greatest. The Quadrajet is
just such a carburetor. Great in many respects, but all too often "put down" or
"bad-mouthed" by those who don't understand it.

In many circles the Q-jet is considered difficult to service. It does not have
the "blessing" of most hotrodders. It is seldom considered a "super" carburetor. Like
all other carburetors it catches the derogatory accusations of every owner and mechanic who has a problem and does not correctly diagnose ignition and other potential
trouble areas with equal enthusiasm.

This writer
feels that racers should take a long look at
the Q-jet or be prepared to lose trophies
unnecessarily. It's great in many race applications.

It is also my strong conviction that
the general user should not spend money
to replace original-equipment carburetors.
The carburetor is seldom the culprit in
tune-up problems and you'll rarely come
out ahead with a replacement unit as opposed to servicing the one you already
own . . . if you do it correctly. To support
this, remember the factory-installed carburetor on your vehicle has thousands and
thousands of man-hours in its development
and calibration. No replacement manufacturer can afford to spend a fraction of this
time per vehicle or engine option. Assuming
the proposed replacement unit is basically
good, the best you can hope to do is get a
good carburetor calibrated to do a fair job.
Check the list of engines it is recommended
for. Do you really think it could be near-
perfect for all of them?

Pursue this thinking in
another light. Literally millions of carburetors are bolted onto new vehicles at the
factory and without any more than an occasional idle adjustment, many of these original-equipment units have been intact and
working good when the vehicle took its
place in the wrecking yard. Carburetors
are designed to give good long-term
service.

How many sets of spark plugs go
60,000 to 100,000 miles? How many
points, condensers or spark-plug wires last
three to ten years? The motoring public
rightfully accepts periodic expenditures on
these general tune-up items. Unfortunately,
motorists are all too often talked into unnecessary $40 to $100 carburetor conversions.

There are times when carburetor service and overhaul are necessary. Consider this
costly task only after a thorough check of
simpler things. Volumes have been written
on the shaky practices of some service sta-
tions and repair shops but the car owner
remains vulnerable to large unnecessary
expense.

We want this chapter to save you
frustration and unnecessary expense. It is
written and illustrated so you can master
Q-jet carburetor service. If you have a fair
knowledge of automotive mechanics the
explanations will seem unnecessarily detailed. To those who want to accomplish a
feat completely out of their realm, the
trivials expounded on will be confidence
builders.

With the following text and pictures
as a guide, a housewife, a 13-year-old girl,
a young secretary, a college student and
numerous other inexperienced people did
flawless Q-jet carburetor service jobs. Their
average time to perform a complete job—
from carburetor removal to completion—
was eight hours. Not an easy task for a
complete novice; but if you save $80 on
repair bills and get satisfaction in new ac-
complishments, that's $10 per hour pay
for a meaningful day. The next time, you
can do a similar job much faster and with
confidence.

Read this section and carefully
scan the illustrations. Better yet, plan on a
day to become a Q-jet expert. Now may be
the time to go through one you have. Perhaps you can help someone else.

If there are
variances in hose-wire connectors etc. on
your vehicle, tag them or mark them plain-
ly for reinstallation. Strips of masking tape
marked with a marking crayon, or Dymo
label tape wrapped back on itself works
fine.

Several Q-jet carburetors are shown here
to illustrate variances in fuel-line entry,
choke components, linkages and outside
"dress" pieces.

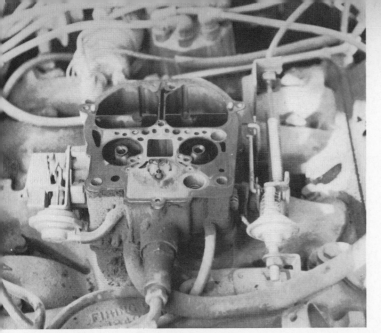

1969 Olds Toronado Q-jet being serviced on the vehicle. This approach is recommended to pinpoint problems and repair known ills within the fuel-bowl or air-horn area. Read the cleaning section before starting carburetor repairs. This engine would have been easier to work on had the engine been cleaned at a 25¢ car wash.

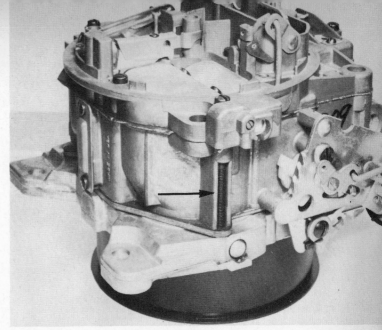

Exposed screw threads on both sides identify 1965-66 Q-jet without removing air cleaner.

All references to the left and right side of the car or engine will be as shown in this drawing.

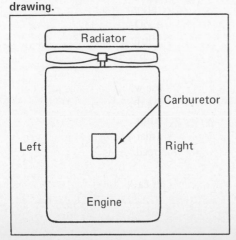

**Just a minute**—Before you grab your tools and charge into a Q-jet service job, scan these short cut suggestions. Solving your problem may require only a few minutes of time. We promise you a savings and it is further intended that you end up with a satisfactory job for least effort.

1. It is very seldom necessary to completely disassemble a carburetor and soak it in a special carburetor cleaner as the majority of manuals and printed instructions direct you to do. If inspecting every piece and having the entire unit spotless inside and out is the satisfaction *you* want, then prepare to proceed through the total exercise. Estimated time—novice 8 hours/experienced 2 to 4 hours.

Locate the following items and commence when time permits.

    Carburetor kit
    Base gasket (carburetor-to-manifold)
    Carburetor cleaner
    Tools

Most carburetor ailments are minor. If appearance isn't your primary goal then proceed with one of the following exercises.

2. In most cases you can cut the repair time in half and reduce the overall cost $10 to $15. To do this you remove the carburetor, but clean it with less costly materials than the strong cleaners sold specifically for that purpose. Doing it this way *eliminates the need to remove the very complex* and awkward-to-handle choke mechanism. Although the text is complete in getting you through this awesome task, it is a bear. Avoid it if at all possible.

Estimated time—novice 5 hours/experienced 1 to 3 hours.

3. In the event you have pretty well pinpointed your problem and it is accessible by removing the air horn only, that's all you should do.

Estimated time—novice 3 to 5 hours/ experienced 1 to 2 hours/pro 30 minutes. Inclement weather, working environment and physical aptitude may be factors causing you to use the 2nd method; even for minor repairs. If not, this way saves you a costly $4 carburetor-to-manifold gasket as well as a need for an expensive carburetor cleaner. One main drawback must be considered. It is tough to lean over a fender or radiator and grill if you're not used to it. Perhaps you can start the task by just pulling the top and if the working position gets too difficult or bad weather closes in or whatever, go ahead and pull the carburetor off as suggested in item 2.

1967 Chevrolet has an air-deflector plate (arrow) beyond the choke-blade housing. This was popular on several models. The plate directs air flow to stabilize bowl pressures during certain heavy-throttle maneuvers.

1965-66 Chevrolet secondary air valves on first-design units were dampened by a lever-arm fuel-piston assembly. This arm (arrow) identifies first-design units. A plastic fast-idle cam (arrow at lower left) was used in 1965-67. 1968 and later cams were made of steel because the plastic had a tendency to deteriorate.

1969 Pontiac Q-jet. Pontiac nearly always has a large fuel-inlet fitting. Note dust cover over idle-vent valve. These covers helped to keep dust out of the power-piston area but the real fix came in 1970 when idle vents were eliminated to meet emission regulations.

1972 Oldsmobile has the entire choke assembly integrated with the carburetor. All other Quadrajet users mount the choke thermostat (arrow) on the manifold and connect it to the choke blade by a small rod.

1967 Cadillac Quadrajets were furnished with no throttle lever. Threaded holes and alignment pins (arrows) allowed the Cadillac-supplied lever to be secured properly.

Here are the tools normally required for complete Q-jet service. A ratchet and 1/2" socket can be used in place of the 1/2" box-end wrench shown. A medium-sized phillips screwdriver or impact driver will also be needed if the throttle body and bowl are to be separated.

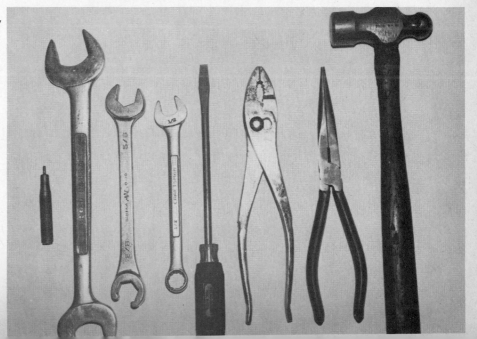

## REMOVING THE Q-JET

1. For ease in removing the carburetor and maintaining tidy working conditions, wash the engine at a 25¢ car wash or with a suitable engine cleaner prior to starting the job. Do this with the air cleaner and all components in place to prevent water entry into the carburetor or engine.

2. Before removing the air cleaner, reach down and remove the vacuum line from tube extending out from carb as shown. Leave it attached to the air cleaner.

3. Don't worry if the steel line pulls out of the carburetor. It is a press fit and it's common for the hose to stick tight on the tube. Separate the hose and tube and drive the tube back in with modest tapping as illustrated. Note the screw placed in the tube. It prevents damage as you tap it back in place.

4. Now remove the wing nut and lift upward on the cleaner. The heat-stove pipe which connects the snorkle to the exhaust manifold is a slip fit. Ease it apart as the cleaner is lifted upward. Some vehicles have a flex tube in place of the steel pipe shown. In some cases it's easiest to take the pipe loose at the heat stove (bottom end). Gentle turning and pulling will separate the two allowing air-cleaner removal. Some heat pipes are clamped and must be loosened first.

5. Next remove the vacuum hose located at upper front of the carburetor. Break the hose loose by prying with a screwdriver if it does not pull off easily.

6. To remove the PCV valve line (large hose) reach in with needle-nose pliers and open the squeeze clip as you pull gently on the hose. Pry with a screwdriver if it does not come loose easily. Hoses get hard and brittle with heat and age. Replace defective or damaged hoses before completing the job. Note where the hose was connected (both ends!), then take the hoses with you so your dealer or auto-parts-house counterman can give you the right size and length.

7. To remove the throttle cable from the carburetor, reach in with a screwdriver and pry off the "spring" socket which connects the cable to the throttle lever.

NOTE: On some cars the throttle linkage cannot be pried off—you may have to unscrew a nut holding the fitting to the throttle arm.

8. To remove the transmission kick-down cable from the carburetor lever remove E-clip and slip elongated cable end from throttle linkage. There will be considerable differences in transmission shift levers and cables from one vehicle to another. None will be difficult to handle. Make a mental note or preferably a sketch if differences in hookup are noted. This caution will be beneficial during reinstallation.

1

2

3

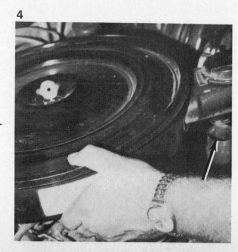

4

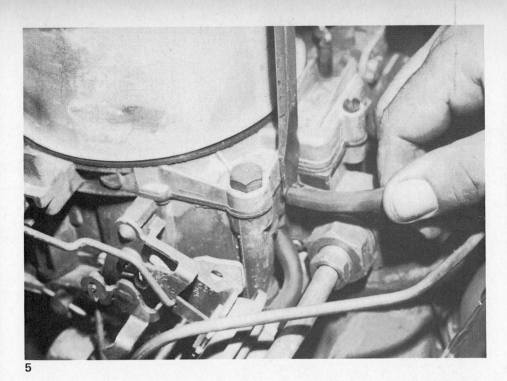

5

6

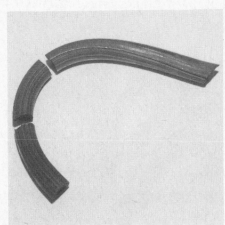

6A/Hoses can look good and be so brittle from heat they break like uncooked spaghetti.

8A/Enlarged photo of the E clip shows three retaining ears.

7

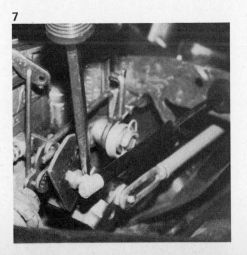

8

129

9/Arrow indicates tang to be sprung outward off rod end so clip can be pulled free.

9. Remove clip from upper end of choke-actuating rod. A screwdriver or needle-nose pliers will handle various retainers found on these rods. Push rod from hole in lever and let it stand free. There is no need to remove thermostat housing or rod from the engine. Cadillac and perhaps other applications do not have a clip at the top of the choke rod.

10. To remove fuel line, hold large nut with a 1-inch-end wrench and fuel-line nut with a fuel-line wrench. Because the fuel-line nut is large and often over-tightened it may be difficult to loosen. If this is the case hold the 1-inch carburetor fitting securely so it does not turn. Put a good wrench on the fuel-line nut (line wrench preferred) and tap it with firm blows in a counterclockwise direction to loosen. Sharp blows will loosen a nut or fitting easier than a strong-arm pull.

11. With the fuel-line nut backed out loosen the 1-inch fuel-inlet nut while the carburetor is still secured to the engine. Also, while the carburetor is secured to the engine there is a screw that should be loosened; refer to photo 11 and 11A. It is usually secured with a thread-lock substance and is literally impossible to get out. Do it now if you intend to remove the outside choke mechanism. Remove two front bolts (1/2-inch wrench fits capscrew hex head).

12. Two nuts hold the rear of the carburetor to the manifold. Remove each of these. The nut on the left rear secures that corner of the carburetor and also the throttle-cable bracket (arrow). Various pieces are secured by these retaining nuts. Hardware such as cable brackets, etc. vary from one make and model vehicle to another. In most cases logic will dictate how to handle the situation.

13. As we continue with the Chevy Blazer Q-jet carburetor removal you will note the throttle-return spring (arrow) also hooks into the throttle-cable bracket. This does not have to be removed until the bracket is free. At this time, release the spring from the carburetor throttle linkage and let it stay with the bracket which remains on the cable.

10

11/Be sure your screwdriver grips this screw firmly. If it does not loosen with a good firm turning effort, put a drift punch or a 5/16 bolt against the screw head and hit it firmly with a hammer. This should jar it loose. An impact screwdriver from Sears or a motorcycle shop is even better.

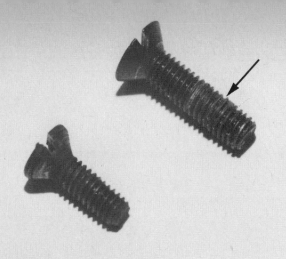

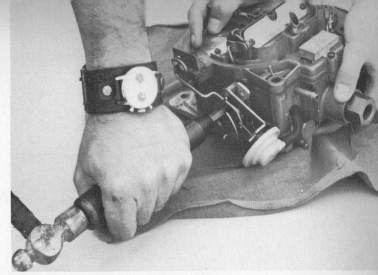

11A/The longer screw holds the choke mechanism. Note the locking substance clinging in the threads (arrow). It holds! The short screw holds a solenoid bracket used on some model carburetors. It has no thread-locking compound and should give you no trouble.

11B/An impact driver is a useful tool for removing stubborn screws such as this one which attaches the choke assembly. Similar tough-to-remove screws are found in the throttle bases.

11C

11D

12

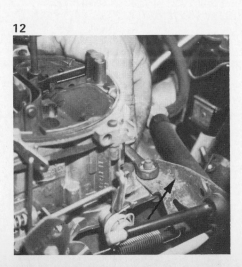

12A

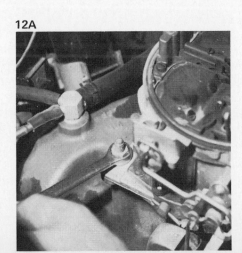

13

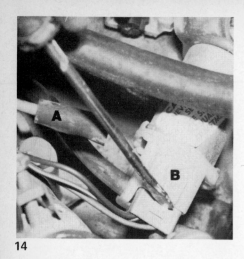

14

15

14. The vacuum line to distributor A and wire plug B should be removed from solenoid. Solenoid bracket may be removed later during carburetor disassembly.

15. Lift the carburetor off next. Heat, pressure and time often bond the two together quite firmly. If you have to tap the carburetor to loosen it from the gasket, tap upward with a plastic hammer against a boss or other substantial (strong) part of the carburetor body. Do not tap against any linkage, lever or bracket. It is not necessary to pry against the base of the carburetor and this is not recommended. After removal of the carburetor it is a good time to inspect for possible intake (air) leaks around the carburetor-base-to-manifold gasket. With months or years of service it is not uncommon to have gasket deterioration. A leak in this area can cause idle problems, exhaust popping noises, light throttle tip-in "sags," strange whistle noises and generally poor low-speed operation. This gasket deterioration is often accelerated by incorrectly torqued (loose) carburetor retaining nuts or capscrews. The accompanying photos illustrate a good example of one that leaked. This one caused popping noises in the exhaust during deceleration from cruise speed to a stop.

16. The manifold opening should be covered immediately after carburetor removal to prevent foreign objects from entering. A clean shop cloth folded to size works fine. If you are working outdoors where wind might shift the cover, place a heavy object to secure it.

17. Place carburetor on a holding fixture for ease of handling and to avoid damage to the throttle plates (blades).

16

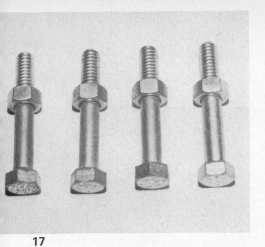

17

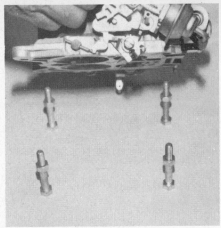

17A

17B/There are many ways to suspend the carburetor so throttle levers or throttle blades are free to move open or closed. Four 5/16 bolts and nuts work fine. Two more nuts installed and tightened on rear bolts would reduce wobble.

17C/This manufactured fixture with rigid legs could easily be duplicated with a piece of plywood, four studs, some nuts and a little imagination. If you get serious about carburetor tuning, buy or build a fixture.

1/This short screw attaches the solenoid.

## AIR HORN REMOVAL

1. Remove idle-stop-solenoid attaching screw. Some units will not have this solenoid, but you will have the following options for those that do.

Option 1: With the attaching screw removed, the air horn can be lifted off with the bracket secured to it. This procedure is advisable unless you intend to submerge the major components in a metal cleaner. My choice would be to leave it all intact and clean the air horn and attached pieces with a brush and mild cleaner.

Option 2: If you choose to dip these pieces in a carburetor or similar commercial cleaner, you must remove the solenoids, switches and other non-metal parts. Whether you choose to remove the necessary non-submersible items from the bracket or remove the bracket is dependent on your tools and ability. Solenoids, switches, etc. are generally held by a thin nut which may or may not be secured with a locking device. In most cases removing the specific item from the bracket will be easier than removing the bracket. Observe whether or not the switch or solenoid is adjustable to a depth within the bracket. Handle accordingly. Measure everything and make notes.

Option 3: Should you decide to get the bracket out of the way—proceed as follows. Use a screwdriver blade to spread the retaining tangs crimped beneath the square mounting boss. Wedge a suitable prying tool—such as the medium-sized screwdriver shown here—between the bracket and top of square boss. Pry upward until the lever slot matches the fixed retention key and pull it off.

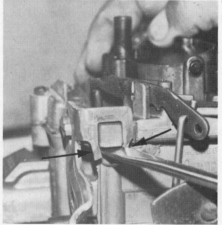

Option 3/Spreading tang with screwdriver blade.

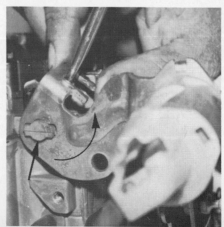

Option 3/Turn bracket counterclockwise to get it to position where you can slide it off the retention key (arrow).

2. Internal choke blade linkage disconnects by removing clip. Pull the rod free of the lever. Sometimes a little prying force is needed to separate these two pieces. Next, free rod from the choke lever hidden in the cavity. The rod is not secured in the lower lever. Just turn it 90° while applying pressure to separate them. Cutaway view shows how rod fits into lower link. This link would normally be secured on a flat-sided shaft which connects to the external choke components. There is minimal clearance for the rod to slip free of the link, so wiggle the rod as you pull up and turn it. This action encourages separation of the two parts. Observing the cutaway photo again, note closeness of the rod to the inner wall. It has worn a scuff pattern. When secured at the top, the rod cannot turn and thus will not come loose at the bottom.

3. When the rod is loose, lift it up out of the cavity.

## BASIC WARRANTY POLICY

Rochester Products Division manufactures carburetors but sells them only to Delco and to the car divisions of General Motors.

Delco warrants each Rochester product or parts thereof, to be free from defects in workmanship and material if product or parts are properly installed, adjusted to specifications as required and subjected to normal use and service.

Delco's total obligation under this warranty is limited to repair or replacement (at the option of Delco) of any Rochester product or part found defective within 90 days or 4000 miles from date of purchase by the original purchaser. Labor for removing and replacing is not covered by this warranty.

The car divisions of General Motors offer specific warranties covering Rochester carburetors and other parts utilized in their automobiles and trucks and sold as replacement parts through their dealers.

**Conditions not covered by warranty—** Failure caused by the following conditions voids warranty on Rochester carburetors: poor gasoline quality, dirt or other contaminants, gums or varnish, water or corrosion, modification, improper fuel-inlet pressure, installation damage, improper adjustment, improper installation, faulty repair.

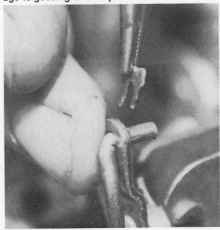

2/First step in removing choke-blade linkage is getting this clip off rod.

2A/Pull rod free of lever.

2B/Cutaway carb reveals lever you are trying to wiggle the rod out of. Normally the lever is on the D end of choke shaft. Note rod scuff pattern (arrow).

3/Taking rod out of cavity.

**Type A Pump Link**

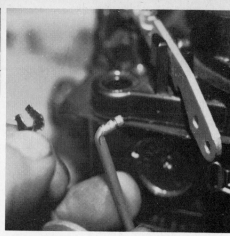

4. There are two types of pump-rod and pump-lever attachments. As you read the following, note which type you have and proceed accordingly. Remove the pump rod or lever—depending on which model you have. Study the accompanying photos to determine procedure. This is one of the most important points in Q-jet servicing, so read and reread this section and study the photos until you know it "by heart."
Type A uses a clip to attach the pump link to the pump arm. If your carburetor is this type, you are in luck because you can save 15 minutes in disassembling and reassembling your Q-jet. With this type you do not need a punch to drive out the hinge pin. Simply remove the clip and separate the rod from the pump arm.
Type B pump link attaches to the pump arm with a bent end protruding through the link. There is NO EASY WAY to disassemble this carburetor without following a special procedure which I will describe. The suggestions will save you a considerable amount of awkward effort in removing and replacing the air horn on models with this bent-rod-type assembly (arrow). When this exists you must remove the pump arm. *There is no alternative method.* Don't look for short cuts because there aren't any.
5. Using a 1/32" drift punch, drive pump-lever pin through the lever toward the choke. Secure a nail file or similar stopping object approximately 0.040-inch thick between the vertical choke housing and the pin. I suggest you use a small piece of tape to hold the stopper as illustrated. Continue to drive the pin against this object. If the lever does not come free, the stopping object is too thick. Use a slightly thinner one. If you drive the pin tightly against the air-horn casting, it is extremely difficult to start the pin back through the hole during reassembly. This is why I consider the stopping object to be absolutely essential.

**Type B Pump Link**

4/Type B pump link is more difficult to remove because pin (white arrow) has to be driven from lever. Black arrow indicates bent end on pump rod which makes Type B harder to disassemble.

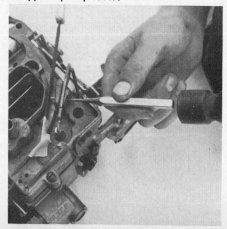

5/Drift applied to pivot pin allows removal of type B pump lever.

5A/Note how thin file was taped to air horn as a stopper to prevent pivot pin from ending up against air horn.

6. Next remove secondary metering rods and their holding fixtures by taking out the screw.

7. Rods will lift out with the bracket; refer to this on reassembly because the rods must hang in the bracket as illustrated. Rod ends face toward the center.

8. Next remove nine air-horn-to-bowl screws. Most screws are tightened quite firmly with a power tool during assembly. Many are secured with a locking compound on the threads, so be sure your screwdriver has a good square end to loosen the screws without damage to the slotted head. Super-bargain screwdrivers made of soft metal will cause you a lot of grief if you try to use them. Good tools are essential here. Two of the nine attaching screws are hidden. They are positioned as illustrated. Reach down along the choke blade to get at them. These are the screws which frustrate the majority of mechanics who are unfamiliar with the Q-jet.

On all models starting in 1967 there are two long screws, five medium-length, and two countersunk screws. 1965-66 Q-jets have four long screws, three medium-length screws and two countersunk screws. Exposed threads of these long screws on either side of carburetor is one of the many ways to identify 1965-66 units without removing air cleaner.

6/This screw attaches secondary-metering-rod hanger.

7/Secondary metering rods lift out with the bracket.

Nine screws attach the Q-jet air horn (arrows).

8C/The two screws which separate the knowledgeable Q-jet mechanic from the beginner. Arrow indicates second "hidden" screw.

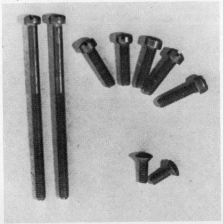

8A/Air-horn screws for 1967-73 models.

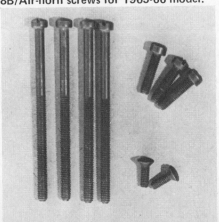

8B/Air-horn screws for 1965-66 model.

9/Gasket should stay on bowl as air horn is removed.

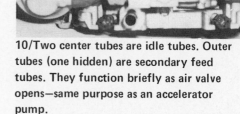

10/Two center tubes are idle tubes. Outer tubes (one hidden) are secondary feed tubes. They function briefly as air valve opens—same purpose as an accelerator pump.

11/Tip air horn 90° to allow removing secondary dampening link on 1967-73 models.

11A/ 1965-66 model with fuel dampened piston linked to secondary air valve.

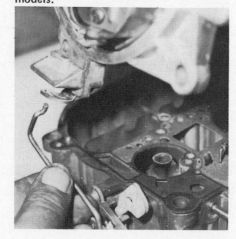

9. Remove air horn by lifting straight up. The air-horn gasket should remain with the carburetor bowl for removal. If the gasket clings to the air horn, use a thin object or your fingers to pry them apart gently. If you have a replacement gasket the same care will not be needed. Even with care, the gasket may be damaged so it is a good idea to have another gasket handy. Do not re-install a broken or obviously damaged gasket because the carburetor is very dependent on well-sealed passages, channels and fuel compartments.

10. As the air horn is lifted, keep it centered until the protruding tubes (arrows) clear their cavities. Mishandling can bend these non-serviceable items. Note the air horn is being tipped toward the choke-mechanism side. The accelerator-pump plunger may stay in its well instead of lifting out with the air horn as illustrated.

11. Continue to tip the air horn as it is lifted until it is at approximately a 90° angle to the carburetor. Remove the secondary damping link out of the air-valve lever. Early models with a fuel-damped air valve do not have this link. Set the air horn aside, turning it upside down to minimize the risk of bending any of the projecting tubes. Lift the pump plunger up out of its well if it did not come out with the air horn.

12. Remove air-horn gasket by freeing it from the alignment pins A and then from the rest of the carburetor body. Pick up gasket as shown and carefully remove it from around the metering rods B. Note how gasket removal starts at the primary (front) end of the carburetor. It is easiest if the gasket is first freed from the alignment pins at the rear. This allows it to slide from beneath the metering rods.

12/Arrow A indicates alignment pins. Lift gasket off these before trying to slide it from beneath metering rods at B.

## FLOAT BOWL DISASSEMBLY

NOTE: As these small items are removed, place them in a suitable container or tray. These small parts must be reinstalled and may not be included in a rebuilding kit.

1. Next remove plastic filler A which covers float assembly. Lift out pump-return spring B.

2. Now remove the power piston and primary metering-rod assembly. To do this without tools, push the piston assembly down until it bottoms—then release finger quickly to the side. Spring pressure from below will push the piston up against its retainer with some *striking force*. Repeat this very quickly five or ten times and the assembly will free itself. *This is the preferred method* unless a defective sticky piston dampens the striking action.

3. In the event the piston is stuck, use needle-nose pliers to pry the assembly out of its well. Do not exert side forces. Occasionally, attempts to pry the piston upward out of its well result in damage. If the metering-rod holding bracket separates from the piston as shown in photo 3A, carefully remove the piston by using needle-nose pliers and *patience*. Join the two pieces back together after a thorough cleaning. A few drops of epoxy will assure a lasting repair. The power piston must be 1.950—1.955 inches long overall, not including the APT pin at the bottom of the piston. Measure and reassemble to this length.

4. Clean the power piston as shown in photo 4.

1

3/Try to pry a stuck piston out of its well with needle-nose pliers.

2/Push piston assembly to bottom, then let your finger slide off the piston so the spring shoves the piston up against the retainer. Do this five or ten times and the assembly usually frees itself.

**Non-serviceable item**—This term means that Rochester does not sell the item as a separate part. They are also listed as NSS, meaning Not Serviced/Supplied Separately. If you damage a non-serviceable part you must either repair it yourself, steal an equivalent from a spare or junked carburetor, or buy a new assembly which includes the injured or worn-out part. NLA is also used (no longer available). This should explain why I've suggested caution in handling many of the pieces as you take your carburetor apart. Don't hesitate to protect the pieces so they will not be damaged as they await cleaning and/or reassembly. Care can save your hard-earned dollars.

3A/If the bracket separates from the piston, epoxy can be used to make an effective repair. Make sure hanger inserts into piston to its original depth as this affects metering-rod position.

Measure 1.950—1.955 inches from the hanger top to the piston bottom (not including APT pin) to reassemble power piston which has pulled apart.

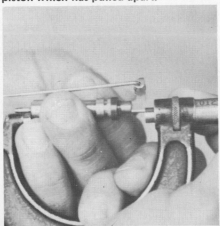

4/A good way to clean the power piston is with crocus cloth. It must be very clean. Do not use sand or emery paper as it will leave scratches.

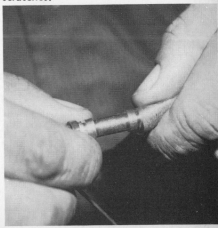

139

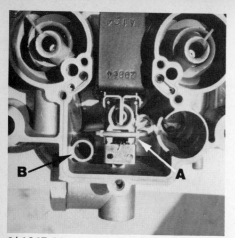

6/ 1965-66 type—As the float is lifted the D-shaped hinge pin A will rise enough to be slid to the side and out of the float arm. Note stack B where fuel enters the bowl from the diaphragm-inlet needle.

6A/Slide the float forward from under the retaining clip (arrow). Do not lift up against the clip on 1965-66 models.

7/Some 1967 and later models have a long D-shaped hinge pin. The air horn on all Quadrajets holds the float hinge pin down firmly.

5. The float will be removed next. Observe the illustrations to determine which type needle-seat assembly you have and proceed accordingly.

6. On 1965-66 models with a diaphragm fuel-inlet assembly, lift the float upward until the D-shaped hinge pin can be slid out to one side. Next slide the float toward the front until it slips clear of the small retaining clip; then lift it out. Attempts to lift it upward without care will damage the clip or diaphragm assembly.

7. On all 1967 and later models with a needle-seat type fuel inlet, just lift the D-shaped pin upward with fingers or needle-nose pliers and the needle and float will come up with it. Before the needle drops off, observe how the needle-to-float clip (arrow) is mounted, as you will have to get it back this way during reassembly. The clip (arrow) should remain on the needle when it is installed to prevent needle-valve hang-up and resulting problems. Note photo 7C.

Remove the seat from the carburetor with an appropriate width screwdriver. The gasket under the seat is often difficult to remove without damaging it. Be careful if you don't have a replacement.

In the event you have the 1965-66 carburetor with a diaphragm fuel-inlet assembly, a little more disassembly is necessary. With the float removed, loosen the two small screws (arrows) and lift out the diaphragm

7A/With your fingers or a pair of pliers lift the float hinge pin straight up on 1967 and later models. The long hinge pin is retained in slots. The slots are different depths, so the hinge pin must not be turned end for end. Observe it closely.

7B/Some late models retained the original short D-shaped hinge pin. The stamped "upsets" (arrows) in the pin arch hold it from side movements.

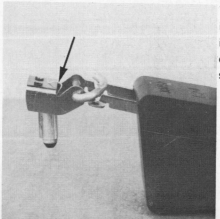

7C/Hang the needle pull clip on the float as illustrated. On some floats it can be hung in the pattern holes or from the front. On others this will create problems. Do it the safe way.

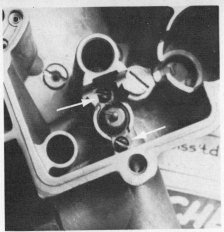

7D/Diaphragm-inlet-needle assembly used on 1965-66 models is retained with two small screws (arrows). Handle this assembly carefully.

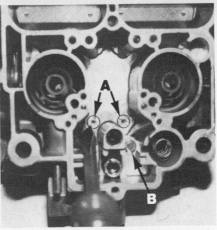

8/Primary metering jets A. Screw B holds pump-discharge assembly.

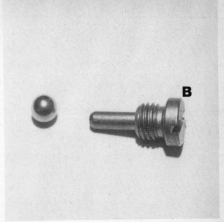

8A/Pump check ball and retainer. No spring is used.

assembly. If you intend to reuse the old diaphragm, handle it carefully and avoid using strong cleaners such as carburetor-cleaning solvents. Wash it off with kerosene or a similar safe cleaner when you are ready to reassemble.

8. Remove primary metering jets A by inserting a suitable screwdriver squarely in the jet and exerting a quick counterclockwise turn. The brass jets will be damaged if the screwdriver slips. Remove pump-discharge retainer screw B. A check ball should be visible in the well. This is easily dumped out by turning the carburetor upside down. Be careful not to lose it unless you have another. They come in some kits.

9. It is optional whether you remove air-directional baffle from secondary side of bowl. If it is tight, it's best to leave it in place. They may vary in shape from one side to another, so be sure they go back as they were.

Deflectors, such as the one on the bottom side of an air valve shown in 9A, vary in shape and size. They are developed to attain best distribution for a specific engine combination. This is another reason why off-brand or "universal" replacements are seldom the answer. See the distribution section in the chapter on the carburetor and engine variables.

10. Do not remove the secondary-metering orifices (arrows). These special stainless-steel washers are staked permanently in place.

9/Baffle in secondary side of bowl casting.

9A/Deflector on air valve.

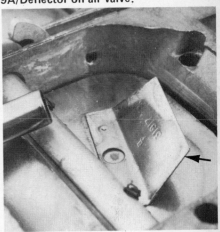

10/Secondary metering orifices are not removable.

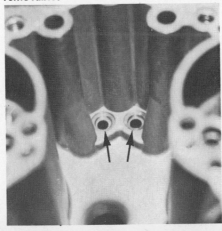

10A/Cutaway secondary-jet area shows wafer-thin stainless-steel secondary metering orifice (arrow). Donut-shaped piece is crimped tightly in place above it as a retainer.

11. Turn the carburetor upside down and remove the two (or three) phillips-head screws fastening the throttle body. Because these are steel screws in aluminum, they sometimes seize and require great force to break loose. An impact-driver as shown previously may be the *only* way to get the screws out.

12. Lift the throttle body off the carburetor and remove the idle-mixture screws. On 1971 and later carburetors the idle-limiting caps must be removed before the idle-mixture screws can be backed out.

13. Remove the hot-idle compensator. Some Quadrajets do not have this device. Also remove the circular cork gasket located in the circular well under the bi-metal valve. This compensator bleeds clean air into the intake manifold to compensate for vapors boiling out of the carburetor at high temperatures.

14. Remove the choke diaphragm vacuum hose.

15. If there is some problem in the choke assembly, remove the single screw fastening the choke assembly and slide the assembly off the carburetor. Watch for a secondary lockout lever which may fall off with the choke assembly. It may stay on its shaft, ready to fall off when you next tilt the carburetor.

16. Turn the carburetor body upside down to recover the hidden choke link which will fall free when the choke assembly is pulled off.

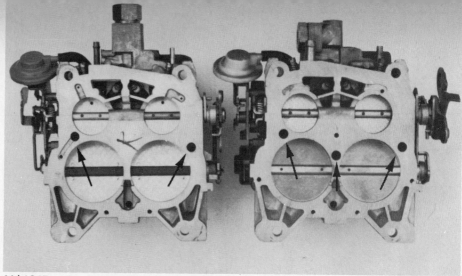

11/ 1967 and later Q-jets have two phillips screws holding throttle body as shown on left. 1965-66 first-design units had three retaining screws as shown on right.

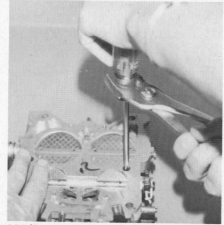

11A/Recommended method of removing seized throttle-body screws. Note second person holding the carburetor. Some of these require a firm effort. Or, use an impact driver.

13/Hot-idle compensator with the cover removed. This device is usually located at the rear of the carb. See page 109 for other locations.

12/Idle-mixture needle with spring and limiter cap.

13A/Bimetal valve and cork gasket.

12A/Removing idle-limiting caps from 1971 and later Q-jet. Be careful not to bend the idle needles.

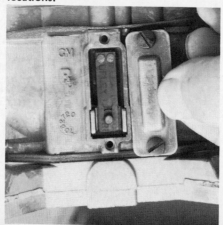

14/15/Typical Q-jet choke assembly with diaphragm vacuum hose and fastening screw.

Buick choke assembly with separate choke vacuum break and air-valve damper.

Oldsmobile choke assembly with integral thermostat.

15/When choke mechanism is removed this secondary lockout lever will be free to fall off (if carb has this type secondary lockout).

This is the most common choke arrangement with a single vacuum diaphragm used as the vacuum break and also serving to control the air-valve opening rate.

**Don't take the choke apart unless you have to**—Inspect the choke assembly very carefully before removal. This is the most complex mechanical linkage on the carburetor and will give you fits during reassembly. Don't be afraid to make drawings and compare your linkage to the ones pictured. There are only three major variations in standard Q-jet choke assemblies. Examples of each are pictured.

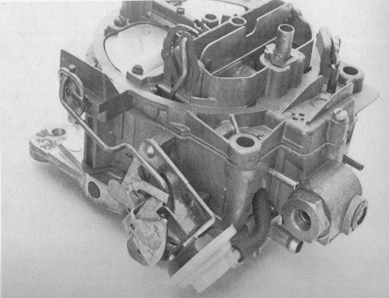

## CLEANING & INSPECTION

1. If you have elected to do a *complete* carburetor cleaning job for experience, thoroughness or the cosmetics of it; gather your submersible parts and put them in a cleaning basket.

CAUTION: Any rubber parts, plastic parts, diaphragms, pump plungers should not be immersed in carburetor cleaner. However, the Delrin cam on the air-valve shaft will withstand *normal* cleaning in carburetor solvents.

2. After rinsing blow out all passages in castings with compressed air. Do not pass drills through jets or passages. Use proper safety precautions whenever air or water pressure is used. Wearing safety glasses is *always* a good idea.

Work area cleanliness and organization will be beneficial to a good job. If you live in Southern California, visit the Carb Shop in Costa Mesa and you will observe the use of a clean towel for each tear down. The white towels help confine small parts and keep them visible to the carburetor technician. This meticulous approach allows them to guarantee carburetor rebuilds for six months. Care and know-how spells good work.

If you choose the economy carburetor-cleanup-and-overhaul method the same general rules of cleanliness apply. Cleaners such as kerosene, Stoddard solvent and mild (safe) paint thinners, etc. allow you to clean up most delicate parts along with the metal pieces. A small open can, a brush, a small scraping tool and some elbow grease let you do a fine job of washing. For dissolving deposits use lacquer thinner, toluol, MEK, Gum-Out, Chem-Tool, etc. These must be used in a well-ventilated area away from fire. Use with brush on the large surfaces to dissolve deposits, use a syringe (common ear syringe works fine) to shoot the solution through passages. Wear safety glasses and keep your hands out of any of the dissolving-type chemicals.

Should you anticipate stale fuel deposits and carbon, a special cleaning kit may be advisable before attempting the service job. One way or another you can do a good job without exotic equipment.

Only a few orifices and passages are subject to contaminant problems. Be sure the bleeds and vents are open in the air horn. You can see through most of these. If in doubt run a small tag wire through holes to check their opening. Orifices and bleeds in the air

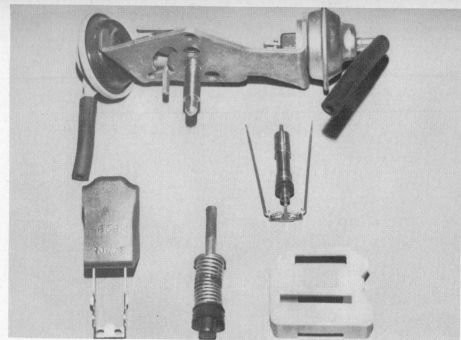

1/Do not put these parts or gaskets into carburetor cleaning solutions.

1A/The throttle body fuel bowl, air horn (shown) and small metal parts can go in this type cleaner basket.

2/The Carb Shop in Costa Mesa lays small non-submersible parts on a clean towel for easy accessibility. Two Rochester H Corvair carburetors are in the cleaning basket.

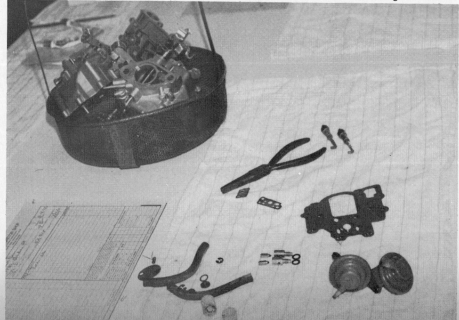

horn are not sensitive to minor changes that might be caused by inserting a piece of tag wire.

The fuel bowl contains most of the critical metering orifices. Be careful about probing with wire or drills. Squirt liquid through obvious fuel channels. This includes jets, fuel-inlet needle orifice seat, pump channel and idle channels A.

Be sure the power-piston well B is clean. A twisted rag or a roll of crocus cloth will do a good job.

Move the throttle blades full open and closed to be sure no dust or foreign material is binding the shafts.

Other than that, only the off-idle slots and idle-mixture needle orifice feed hole (arrows) must be open and clean. Check with needle screwed out.

Regardless of method or materials used to clean your carburetor, it is a good plan to wash the pieces thoroughly in a strong hot detergent for the final cleaning.

With everything tidy-clean we should now inspect a few items before starting reassembly.

3. Inspect idle-mixture needles for damage. Quite often a needle has been turned into the seat too tightly and damaged. The limiter caps placed on today's carburetors prevent well-meaning mechanics from doing this. Pre-emission carburetors are most likely to have damaged idle-mixture needles and seats.

2A/Some newspaper, an old dish or open can, a paint brush and some safe cleaner make a suitable do-it-yourself economy cleaning kit.

2B/Doug Roe Engineering sells a do-it-yourself carbon-gum removal kit for those without access to compressed air or easy access to suitable cleaners. Ear syringe forces concentrated cleaners into critical passages to assure full openings.

2C/"Improvise" or borrow your son's bicycle pump. Somehow cleaning can be done well without a ton of money or special equipment.

2D/Idle channels A and power-piston well B must be clean and free from deposits.

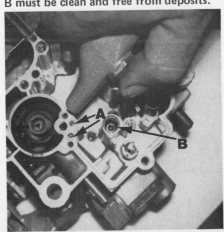

2E/Use compressed air to blow through the off-idle and curb-idle passages (arrows).

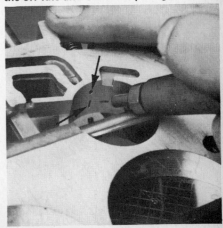

4. Examine float, needle or diaphragms for wear or damage. Replace if necessary. Be sure power piston is cleaned. Crocus cloth may be necessary if it is discolored or contaminated.

5. Inspect upper and lower surfaces of carburetor castings for visual damage.

6. Inspect holes in levers for excessive wear or out-of-round conditions. If considerable wear is noted readjust corresponding links and rods. If wear compensation can not be made, the unit will have to be replaced. This need is rare with most carburetor units.

7. Examine fast-idle cam for wear or damage. Inspect early plastic models carefully for visible cracks. Replace if cracked or badly crazed from heat and other under-hood elements. Metal cams were utilized starting in 1967.

8. Check air valve for binding conditions. If air valve is visibly damaged, air-horn assembly should be replaced. In most instances it will take four to six weeks for your dealer to get these parts. A plastic cam and air valve spring kit is available to fit all Q-jets (7035344).

9. Check all throttle levers and valves for binds or other damage. Lay the pieces out carefully after inspection. When all pieces meet your approval reassembly can be started.

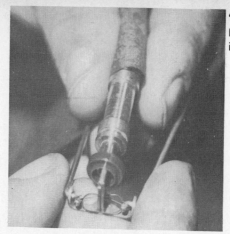

4/The power piston can be cleaned or polished with the primary metering rods intact if care is used.

7/Heat and under-hood environment including ozone can damage these cams. Sometimes the crack will form in the shaft-bushing area (arrow). Inspect and replace as necessary.

7A/All 1968 and later Q-jet choke cams have been made of metal. A plastic fast-idle cam used 1965-67 had a tendency to deteriorate. The plastic cam is shown in a photo on page 126.

## CHOKE MECHANISM

The Q-jet choke mechanism is relatively maintenance-free. Do not remove it unless a problem is evident. It is almost never advisable to go through the time-consuming exercise unless you are "dead set" on gunking the steel parts in carburetor cleaner or if there is a faulty part.

Three areas deserve occasional observance and light maintenance.

1. The external linkage to the choke shaft can bind up as a result of under-hood contaminants. Vapors and dust form a deposit that needs to be occasionally flushed and blown out of the pivot-bearing surfaces. The choke-shaft-bearing surface on either side of the air horn can bind up with gum from evaporated fuel. A spray can or ear syringe with varnish-dissolving solvent will clean it in a matter of minutes.

2. Plastic vacuum-break-diaphragm assembly units normally give long service. Now and then a plastic housing is broken by carelessness or becomes crazed badly by under-hood environment. Diaphragms can fatigue from age and use.

A visual check tells you the apparent condition of the plastic body.

3. Another simple check tells you the diaphragm condition. If the carburetor is several years old and the exterior plastic looks crazed and generally rough it would be advisable to replace the vacuum break unit at your first opportunity.

Remove vacuum hose from the housing. Now push the lever into the diaphragm housing, until it bottoms. With it held firmly, close off the vacuum opening with a finger. Now release the lever. It should move out a fraction of an inch—and hold that position while your finger seals the vacuum opening. If the lever is pushed out quickly by the spring, the diaphragm is ruptured or unsealed. If the lever creeps out slowly there is a small leak. In either event replace the unit. If the diaphragm is good and holds, move your finger away and allow the lever to return to a fully extended position.

The 1973 Buick uses a purge bleed hole in the auxiliary vacuum break tube. It must be covered or the diaphragm will not hold.

1/The most common problem with the Q-jet chokes is binding shafts. The choke blade shaft and choke link shaft must be cleaned occasionally with gum-dissolving solvent. Do not lube these areas with an oily substance as such lubricants collect dust and may bind the shaft instead of keeping it free.

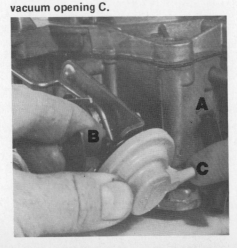

3/First step in checking diaphragm is to remove vacuum hose A, push lever B into housing and hold your finger over the vacuum opening C.

2/Plastic vacuum-break-diaphragm housing can be visually inspected on vehicle. Do this periodically.

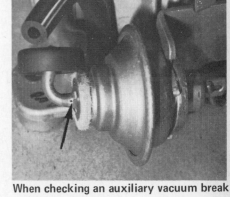

When checking an auxiliary vacuum break (mostly on Buicks through 1974), pull rubber purge-filter cover out of the way as shown. Cover bleed hole (arrow) with a piece of tape before checking holding ability of vacuum-break diaphragm.

3A/Release lever B. If diaphragm is good the lever should move out only a fraction of an inch and stay there so long as your finger seals the vacuum opening.

4/Check size of restriction if you replace vacuum-break unit. It should be similar to the one which came on your carburetor.

5/ 1965-66 secondary air-valve damping arm. Link A connects to air valve B. Lever extends down to piston in well C.

5A/Cutaway showing piston in well. Fuel in the well causes the damping effect required.

6/Typical choke thermostat cover. A screw (arrow) retains this one in place on the manifold.

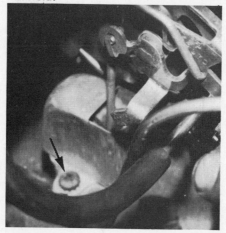

4. If you purchase a new vacuum-break unit observe the lever to be sure it is dimensionally the same as the old one. Be sure the vacuum restriction size (arrow) is similar. These are sized quite accurately so be sure they at least appear identical. A household item works well for a ball-park measurement.

Put the point of a small sewing needle in the hole and with your thumb nail hold a depth mark on the needle. Try it in the other unit to see how the depth compares. The orifice is generally 0.010" to 0.015". Occasionally one is 0.020-inch. Just be sure the old and new are *very close* in size. This tiny restriction limits the speed at which the choke blade is pulled to its first position (vacuum-break setting) after a cold start. This can be critical to stall after start, etc. If it does not pull the choke blade partially open after cold start extreme richness can result. This ailment is evidenced by black smoke, engine stalling from richness, poor gas mileage and eventual damage to rings and cylinder walls from excess fuel diluting cylinder lubricant. This orifice has a second, equally important duty. It regulates the air-valve opening speed.

If the unit is not functioning correctly, secondary-throttle operation will be bad. With no damping effect on the air valve, the engine will sag, hesitate and perhaps backfire every time you go to a heavy throttle such as during a passing maneuver. These annoyances are not designed into the carburetor.

7/Oldsmobile choke has thermostat in housing attached directly to the carburetor body. This is also used on 1966 Buicks and some Pontiacs.

Again I recommend, check the diaphragm fairly often—it's easy to do on the car.

5. Air-valve-opening speed on 1965-1966 models is controlled by a lever-piston damping device. More details on secondary-air-valve opening speed are given in the high-performance chapter fourteen.

6. Bi-metal choke thermostat. Most of these are mounted on the intake manifold and covered by a can. A great deal of design and test is required to determine the correct bi-metal material, its length and the number of coils. These three items determine the torque applied to close the choke blade and the rate at which this torque decreases with temperature at the thermostat (bi-metal). Generally all you have to do is be sure the coil housing is tight to the manifold and the can is in place. Some cans are held with a screw and some just clip on with indented tabs. Also check the rod to be sure it is in place and not binding or rubbing the can or a hose. Hold the throttle partly open with one hand and move the choke blade from full open to full closed several times. It must be free. The method of setting correct rod length is covered in the adjustment section at the end of this chapter.

7. A few Q-jets (Oldsmobile applications, typically) have the choke stat in a housing attached to the carburetor main body. As with the other type choke, be sure the parts are correctly attached and free from bind or contact with other objects which could keep them from working correctly.

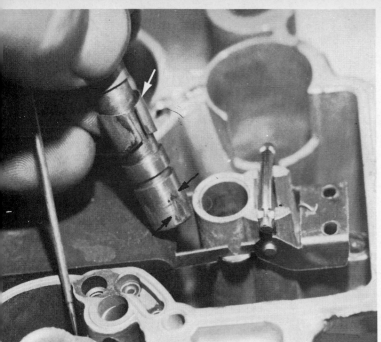

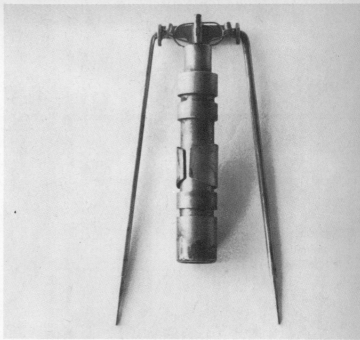

1/Early-model power piston was restrained in its bore by a metal clip (white arrow). Always inspect the power piston for deposits or marks indicating scuffing or binding against bore (arrows).

1A/Dust particles on early-model power pistons. This is common because there is no seal and dust enters through the idle vent.

## POWER PISTON

1. Carefully inspect the power piston for deposits or marks indicating scuffing or binding against its bore. This is a common problem with early-model idle-vent-equipped Q-jets. In dry areas of the country, the idle vent hole allows dust entry to the power pistons wall area. Early-model power-piston assemblies were restrained in the well by a metal spring clip, which afforded no sealing of the piston from dust. Most later models used a combination seal/retention collar to hold the piston. This helps keep out contaminants. When the idle-vent valves (external venting) were eliminated in 1970 the power-piston dust problem essentially ended.

2. Power-piston-sticking problems occasionally show up in today's carburetors. This is a result of fuel contaminants and foreign materials forced into the piston area by engine-backfire pressures. Today's extremely lean part-throttle calibrations combined with chokes that come off too early to meet emission standards, promote backfiring. The manifold and power-piston well are joined by a short vacuum passage. A portion of a backfire's flame front can enter the piston area through this vacuum passage. Most of the time only fuel and air are blown into this area. Occasionally a troublesome power piston is found with exhaust contaminants on its outer surface. This simply means that carbon and/or flame *did get there.* Eventually this will hang up the power piston. A cleaning procedure is outlined in the assembly portion of this section. If the piston sticks in the down position, you have no primary power system. The engine will falter and hesitate during medium-throttle accelerations. It will be extremely hesitant during engine warm up for the first 2–3 miles after each cold start. If the piston sticks in the up position, you have a rich power mixture for all driving conditions. Gas mileage will fall off as much as 50%. Low-speed throttle response will be very sluggish. Exhaust will be black and smoky.

Many a $35.00–$100.00 repair or carburetor replacement job has been sold on this one problem. A Q-jet trouble-shooting pro can remove the air horn and fix it in 30 minutes. A novice following instructions from this section can do it in less than two hours. I'd estimate 25% of all Q-jet problems are power-piston-caused. I designed a kit specifically engineered to fix this problem without taking off the air horn. This kit illustrates short cuts and provides all materials to do the repair, including the key-like tool I recommend using. In most cases a novice can free power pistons in less than 15 minutes. Doug Roe Engineering markets these kits.

You can make the special key-like tool from the accompanying photo. As an individual, you will save many dollars and much frustration. As a repairman you can do your customers a great service by making one of the keys for your tool kit. By inserting the specially designed notched key down the forward vent stack you can hook onto the power piston/metering rod assembly crossbar. With very modest up and down forces the power piston should travel approximately ¼". If any drag is noted, spray in some carbon-dissolving solvent (Gum-Out, Chem-Tool, etc.), then work the piston up and down. Repeat this until it is completely free. Starting the engine and running it briefly after each application of solvent will help draw the solvent down along the walls of the assembly. Reinstall the air cleaner and the job is complete.

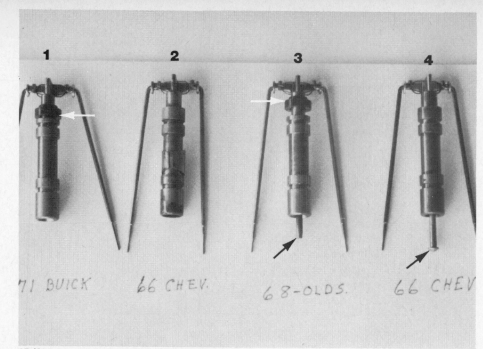

1B/Shown are some power-piston design variances. 1 and 3 have a top seal retainer collar (white arrows). 2 has a spring clip which tensions against the side of the piston well. 4 has no upward restraint. 3 & 4 have a depth-limiting stem protruding from bottom (arrows).

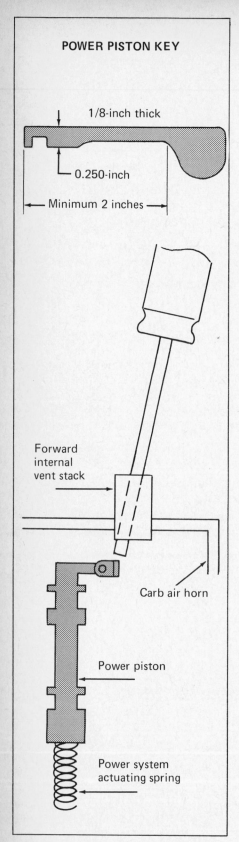

**POWER PISTON KEY**

1/8-inch thick

0.250-inch

Minimum 2 inches

Forward internal vent stack

Carb air horn

Power piston

Power system actuating spring

If the piston happens to stick in the up (rich) position you can push it down with a small screwdriver or use the end of the key as a pusher. If that does not free it, use the key-like tool.

2/Cutaway showing special Doug Roe power-valve piston key hooked to power piston-metering rod cross bar.

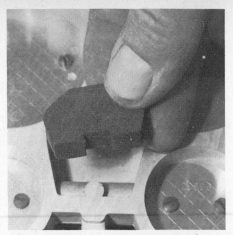

1/Foam pad designed to stop secondary fuel well leakage is only partially effective at best. Epoxy is the real answer here.

1A/Secondary fuel well (large) and primary fuel well (small) plugs and surrounding area being cleaned prior to the application of epoxy.

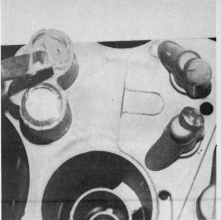

1B/If you get this far into the carburetor, use epoxy to seal the plugs once and for all.

1C/Early model brass plugs sealed with epoxy. Not pretty, but effective.

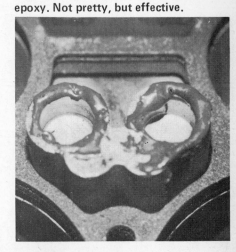

## PREPARATION FOR ASSEMBLY

1. Plugs at the bottom of the secondary and primary fuel wells sometimes leak, causing idle richness. This problem can be cured or prevented by cleaning the surrounding area and sealing them with epoxy. Delco and several parts companies sell a piece of special foam to fit under the secondary fuel wells. This device is only partially effective at best and can cause throttle body warpage if the material is too hard. At best it only seals the secondary-well plug and leaves the smaller primary-well plugs to cause trouble. Epoxy them all and avoid the use of the sealing pad.

2. Hold the throttle body up to a bright light to determine if the throttle blades are positioned correctly in their bores. Do not loosen the screws or attempt to adjust the blades unless there is a very *obvious* alignment problem. Adjustments on these blades are very hard to do correctly, so don't try unless you have an idle overspeed problem from air leakage or a blade hangs up on the bore, etc. If you must—loosen the screws slightly and tap the blade with a small screwdriver or similar object as you exert closing pressure. Tighten the screws before releasing pressure. Some light around the blades is permissible; so if the blades look uniform in fit, leave them alone. RPD four-barrels seat throttle blades against throttle bores, not against an external stop; as is common on some other carburetors.

2/If the throttle blades are badly misaligned, they won't close correctly and large amounts of light will be visible around them. This is not often a problem.

Should the inlet fitting threads strip in the carburetor bowl this handy expansion fitting will save you a ton of money. Push it in the stripped thread area—tighten to expand the sealing rubber and the repair is completed. Buy them from Delco dealers, the Carb Shop or Doug Roe Engineering.

If your Q-jet is a pre-1967 model, epoxy the end fuel plug in place as shown on the right. These plugs have been known to fall out. 1967 and later models with the spun-in plug do not need this precautionary modification.

Delco's stripped fuel-inlet repair nut should be installed with Loctite to ensure a good seal between the nut and the carburetor body. On 1966 Q-jets the fuel passage end plug retainer should be used if the needle and seat modification kit 7036775 has been installed. Or, use the epoxy modification pictured above. Check your Delco dealer for the correct part number to fit your carburetor (7041634 or 7041635).

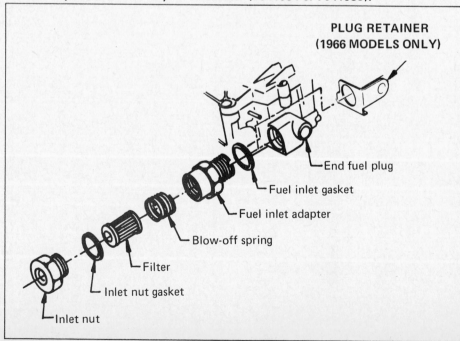

PLUG RETAINER
(1966 MODELS ONLY)

End fuel plug
Fuel inlet gasket
Fuel inlet adapter
Blow-off spring
Filter
Inlet nut gasket
Inlet nut

## ASSEMBLY

1. Turn the bare carburetor fuel bowl upside down and fit the thick throttle-body-to-bowl gasket over the guide/alignment pins. An incorrectly fitted gasket will become immediately apparent because covered screw holes and interferences will prevent further assembly. Air leaks would occur if the carburetor was assembled with the gasket in the position shown in the photo.

2. Place the throttle body on the bowl/gasket assembly and fasten with two (three on some Q-jets) phillips-head screws with a medium-size screwdriver. The bowl is a light alloy so it is possible to strip the threads if the screws are badly over-torqued. If you are strong enough "to go bear hunting with a switch" use some restraint.

3. Place the springs over the idle needles and screw them into the throttle body. Turn the needles in until they seat *lightly* and then back out two full turns.

1/Incorrect placement of throttle-body-to-bowl gasket will cover screw holes and leave large manifold vacuum leaks. It is better to place gasket on the fuel bowl first.

2/Place throttle body on bowl/gasket assembly.

2A/Turn screws tight to seat gasket.

3/Needle installed in throttle body with spring in correct position. Idle-limit caps are no longer used for service. If you cut the caps off, leave them off.

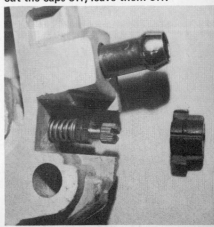

153

4. Let's install the choke components next. Set the carburetor right side up on a fixture and proceed. Slide the fast-idle cam on the shaft of the choke-linkage. The cam lobe and fast-idle steps will be toward the bowl. The fast-idle cam-adjustment tang should be *below* the cam.

5. If your carburetor has its secondary-lockout link on the side of the bowl, set it on its pivot boss and engage the lockout pin on the secondary-throttle shaft.

6. Slide the choke linkage assembly with fast-idle cam along the metal vacuum tube until the end of the shaft is barely inside the hole in the side of the carburetor. To allow the fast-idle cam to pass the secondary lockout, pivot the lockout link counterclockwise for necessary clearance.

7. Hang the intermediate choke lever from the choke-actuating rod and lower into the passage on the side of the carburetor. The link should be suspended in the cavity until the square is centered (visibly) in the window hole (arrow). Push the linkage assembly in toward the carburetor so the end of the large flat-sided shaft tries to enter intermediate dangling lever. Pin the lever against the inside wall of the passage by pushing inward with assembly. By use of the choke-actuating rod, slowly rotate the lever counterclockwise until the shaft lines up with the lever hole and mates. Next slide the linkage until it is flush with its mounting points on the carburetor and fasten with the single screw.

4/Fast-idle cam slides onto shaft of choke linkage as shown by arrow.

5/Where secondary lockout is on side of bowl, slide lever onto pivot so lockout pin on secondary throttle shaft is engaged (arrow).

6/Choke linkage and fast-idle cam being started over metal vacuum tube.

6A/Secondary lockout link pivoted to allow fast-idle cam to slide past.

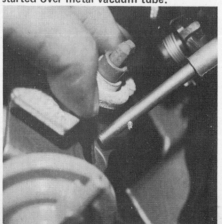

7/Lower intermediate choke lever into passage until flat-sided hole is visible through side hole in carburetor main body (arrow).

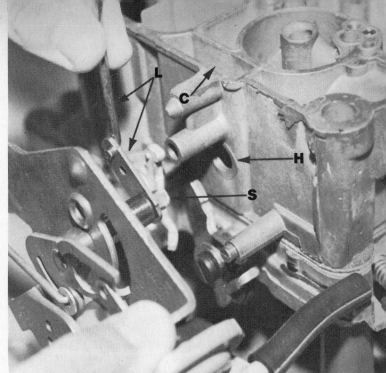

7A/For illustrative purposes intermediate choke rod and lever L are positioned on shaft of a '72 Buick Quadrajet. When actual assembly is done, link and lever will be dangled down in cavity C and shaft S will enter hole H and join it as illustrated.

7B/Rotate lever (arrow) counterclockwise (downward) until shaft slips through slot and choke assembly seats against its mounting pads.

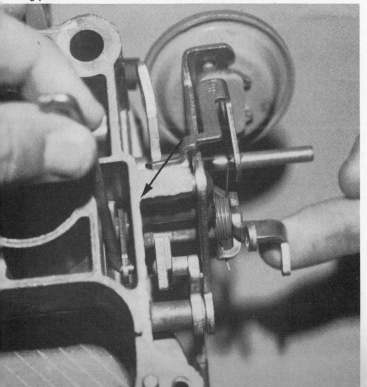

7C/Typical complete choke assembly.

8. Temporarily remove the choke-actuating rod from the intermediate lever by rotating it counterclockwise and lifting it up out of the cavity. Getting it out of the way will make it easier to install the air horn.

9. Connect the vacuum source tube to the vacuum break.

10. Engage the hole in the vacuum-break rod with the end of the air-valve-damper rod. The upper end can be laid over the choke linkage until the air horn is installed.

11. Place the round cork gasket in its well in the hot-idle compensator chamber (only if your carburetor was equipped with HIC).

12. Slide in the bi-metal valve and fasten the cover down with two screws. (Only applies to HIC-equipped carbs).

13. Drop the accelerator-pump check ball into its well and screw in the retaining screw. There is no gasket under the screw.

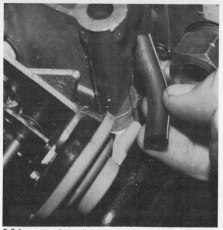

9/Vacuum hose connects to vacuum break.

10/Engaging hole in vacuum-break rod with air-valve-damper rod.

11/New cork gasket in place in hot-idle compensator (HIC).

12/Bimetal in place in HIC cavity. Another HIC location is shown in a drawing on page 109.

13/Partially reassembled float bowl assembly showing: 1 check-ball retaining screw, 2 jets, 3 fuel-inlet valve seat and 4 accelerator-pump spring and 5 power-valve spring.

14. Screw the two primary metering jets into the bottom of the float bowl.

15. Drop the accelerator-pump and power valve springs into their respective wells.

16. Screw the fuel-inlet valve seat (with gasket) into the float-bowl assembly. If your fuel-inlet valve is one of the early-model diaphragm units, drop it into its well (position it carefully) and fasten it down with two screws.

17. Drop the needle into the seat on late-model carburetors. Be sure the float pull clip is still attached to the needle.

18. Hook the float bracket with the needle clip. Insert the float-pivot pin through the two holes in the float and drop the assembly into its slot.

There are two float-pivot-pin versions. The short one (¾" long) can point in either direction when mounting the float. Install the longer one (1¼") with the open end toward the accelerator-pump well. In either case, the plastic stuffer will usually only allow the pivot pin to be installed correctly.

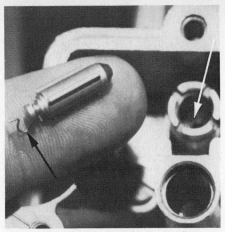

17/Needle with float pull clip in place.

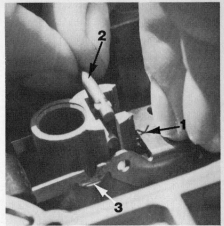

18/Float installation showing 1 inlet needle pull clip on power valve (inboard) side of float housing, 2 first design (short) pivot pin and 3 float-level-adjustment bending point.

**1965-66 needle/seat conversion**—Several manufacturers sell kits to convert the early-model diaphragm valve to a conventional needle and seat unit. I avoid such conversions if possible. The new needle will often alter fuel levels during acceleration and performance maneuvers as compared to the diaphragm system the carburetor was calibrated with. If you should be faced with either converting or throwing away an otherwise good Q-jet bowl, I'd try the conversion. If it drives well this is a small investment as compared to the cost of a bowl or complete carburetor.

Early model diaphragm-type fuel-inlet valve at right compared with needle-and-seat conversion at left.

O-ringed plug is inserted in the fuel-exit well (arrow). O-ring-sealed seat assembly drops in where original diaphragm was located. Small collar sits over it and lines up with the two retaining screw holes and original screws secure the assembly. Inlet needle with pull clip attached drops in the seat housing. Readjust float setting per instructions.

19. Hold the rear of the float down on the fuel-inlet valve and check the float level with the gage furnished in the rebuild kit. If adjustment is required, bend the float arms at the adjustment points next to the power-valve well.

20. Carefully guide the two primary-metering rods into the jets as the power valve is lowered into its well. Do not force the metering rods because *the very thin tips bend easily.*

21. Use a wide screwdriver blade to force the plastic lock/seal into the top of the power-valve well on late models. Some models go with no top seal. If the power piston requires more than light finger pressure to push it down, check the spring to make sure it has guided up into the power valve correctly. Too much spring pressure will not allow manifold vacuum to pull the metering rods down in the jets. This results in very poor fuel economy.

**Float-level gage**—If you have purchased a metering kit or individual pieces you may not have a fuel-level gage. Use a small scale or ruler to measure from a straight edge as illustrated. If you do not have a measurement, try 5/16-inch until you can find out the correct setting for your particular Q-jet model.

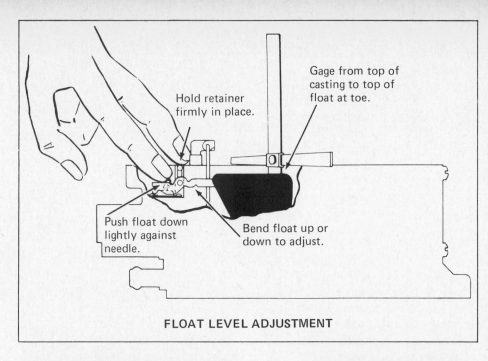

Hold retainer firmly in place.

Gage from top of casting to top of float at toe.

Push float down lightly against needle.

Bend float up or down to adjust.

**FLOAT LEVEL ADJUSTMENT**

19/Check float level using gage supplied in rebuild kit.

20/Be careful here. Don't bend the metering-rod tips. Use caution.

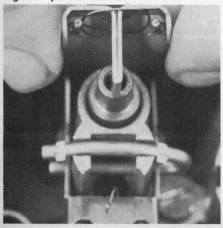

21/Installing plastic seal/lock/retainer for power-piston assembly.

21A/Installed power-valve assembly showing 1 plastic lock (retainer) and 2 metering rods. Be sure small tension spring is in place between metering rods as shown.

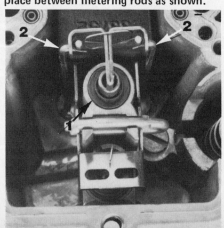

22. Drop the plastic stuffer into position. The top of the stuffer should be no higher than the top of the float bowl. If there is an interference problem, the float-pivot pin may be installed backwards. If you feel uncertain, refer back to disassembly comments.

23. Compare the air-horn gasket to the top of the float-bowl assembly or old gasket and position it in place. A small flap (arrow) built into the gasket goes under the T on the power valve. Slide this flap under the T and then finish positioning the gasket on the carburetor. Failure to do this will not allow proper power.

24. Insert the accelerator-pump rod through its hole in the air horn. If you have not replaced the plastic skirt or the whole pump assembly, spread the skirt outwards with your thumbs. This assures a positive seal against the pump-cylinder wall. As long as the pump cup is not brittle it can be rejuvenated for further service.

25. We are ready to install the air horn. Pick it up as you hold the end of the accelerator pump. Tilt the air horn 90° to engage the air-valve-damper rod.

26. Slowly rotate the air horn down to the carburetor making sure that the four brass tubes go into their respective holes and the accelerator pump goes into its well.

22/Plastic stuffer should be flush with top of bowl if float pivot pin is correctly installed.

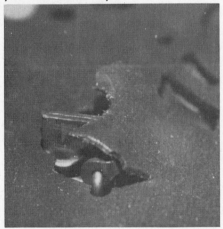

23A/Wrong way to place gasket on bowl assembly. Power valve will never work if you leave it this way.

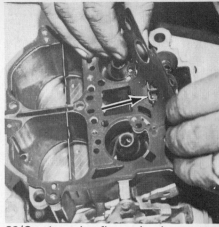

23/Get the gasket flap under the power-piston T.

24/Accelerator pump correctly positioned on air horn. A method of "reconditioning" an old pump skirt is shown. If there are cracks in the plunger it must be replaced.

**More about setting the Q-jet float level—** Here's a neat trick! It is awkward to hold one finger on the float retainer (hinge pin) and another finger pressed just right toward the inlet needle while you make the measurement. If you have the carburetor off the engine, hold one finger against the hinge pin and invert the bowl. The weight of the float will provide pressure against the needle. Measure and adjust as required. This will always provide consistency in float setting.

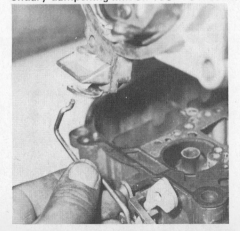

25/Tip air horn 90° to allow installing secondary dampening link on 1967-73 models.

26/Be careful to get the idle and secondary tubes and the accelerator pump cup started into their respective openings. Install the air horn with care.

27. Replace the nine slot-head screws that fasten the air horn to the float-bowl assembly. Don't forget the two hidden screws near the primary nozzles. Start all screws into their holes by hand before tightening from the center out.

28. Fit the bottom end of the accelerator-pump rod into its hole in the throttle linkage.

29. For the carburetors that have a single right-angle bend at the top of the pump rod, rotate the rod into the hole in the pump lever you have selected as best for your application.

30. Push the C clip over the tip of the pump rod and crimp with a pair of needle-nose pliers.

31. If the pump rod has an S bend at the top, you have removed the pump lever during disassembly. Select the pump-rod hole you want and put the lever on the rod. Align the small hole in the lever with the pivot pin and press the pin through the hole with a screwdriver. It will take a little fishing around to get the pin to start through the lever-pivot hole. Sighting through the hole sometimes helps.

Depending on how much pump action is desired, there are two pump-rod holes in the pump lever. The one at the end of the lever is used for normal driving in moderate climates. The inside hole (standard location on some models) gives a larger pump shot for racing or cold climates.

27/Nine air-horn screws include two hidden screws (arrows).

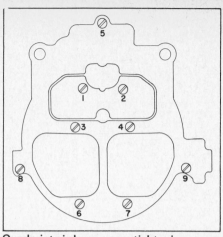

Quadrajet air horn screw tightening sequence

28/Accelerator pump rod has upset tang which mates with keyed hole in throttle lever.

30/Crimp C clip onto accelerator-pump rod after engaging rod into desired hole in pump lever.

**30/Use needle-nose pliers to crimp the clip onto the pump link. Make sure the link cannot be pulled out of the pump arm.**

31/S bend at top of pump rod means you have to locate pump lever so pivot pin can be pushed through it.

31A/Use a screwdriver to push pivot pin through pump lever.

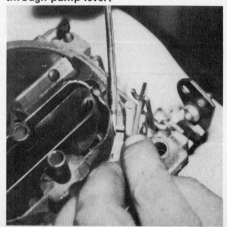

32. Now it is time to insert the choke rod back into the side cavity and connect it to the intermediate hidden lever. Rotate the outside choke lever clockwise and hold it firmly. Grip the top end of the rod and hook it in the lever. Looking down the cavity will help you see how to turn and guide the rod to accomplish this. When you think it is hooked, release the outside lever and move the rod up and down. The outside linkage should respond with movement.

33. Connect the other end of the rod to the choke-blade lever. Some force may be necessary to slip the rod through the hole. Fasten with a C clip similar to the one shown in photo 30A. Some car models used an E clip in this location, but the installation remains the same.

34. Hang the secondary rods from their hanger and drop them into the carburetor. Fasten with a single screw through the top of the hanger.

35. On late-model carburetors with a solenoid, slide the solenoid bracket over the key on the air horn and rotate it down over the locking block. Bend the tangs around the bottom of the block to complete the installation.

32/Choke rod goes into this cavity; is hooked into choke lever by twisting the bent end into a hole. Photo 2B on page 135 illustrates the lever/rod relationship in a cutaway carburetor.

33/Getting choke rod installed is not always easy. Some force may be needed.

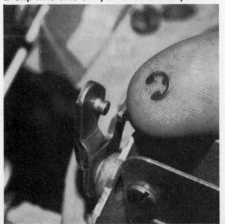

33A/Choke rod is secured in lever with an E clip like this one, or with a C clip.

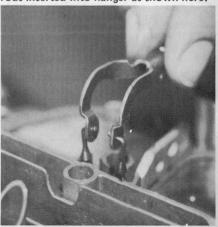

34/Install secondary metering rods with rods inserted into hanger as shown here.

34A/Screw fastens secondary hanger to cam follower.

35/Rotate solenoid bracket into place on keyed boss and fasten with single screw.

**Secondary lockout during choke operation**—There are two types. The one shown below at left in 35A locks the secondary throttle blades closed. At lower right in 35B you see the type which locks the air valve from opening.

35A/Side view of typical Q-jet choke assembly. This model uses a choke link to a lever (arrow) which locks out the secondary throttles when the choke is on.

35B/End view of typical Q-jet assembly. This model has an air-valve lockout linked to the choke (arrow) to prevent secondary throttle operation when the choke is on.

## INSTALLATION

1. Uncover the intake manifold, remove the old gasket and clean the remaining gasket material off the mating surface.

2. Place the new manifold to carburetor gasket over on the manifold using the two studs as a guide.

3. Set the carburetor on the gasket and put in the front mounting bolts. Turn these two bolts until they are slightly more than finger-tight. NOTE: Early models with heat crossover use a stainless-steel shim next to the throttle body. A gasket goes between the shim and the manifold.

4. The nut on the right rear of the carburetor should also be turned slightly more than finger-tight.

5. Hook the throttle-return spring through the eye in the cable-support bracket. Slip the bracket over the left-rear stud and turn down the nut to approximately 10–14 ft. lbs. Tighten the other three mounting bolts to the same torque.

Each installation will have a little different bracket arrangement so install the mounting capscrews and nuts accordingly. Do not over-tighten. You are tightening the two front ones against soft alloy metal and soft gaskets.

6. If the car has an automatic transmission, connect the detent cable with an E clip as shown. Yours may be different.

1/Shop rag is especially helpful as a guard against foreign objects dropping into the manifold unnoticed.

3/Front bolts should be tightened slightly more than finger-tight.

4/Similarly, tighten right rear nut slightly more than finger-tight.

5/Installing throttle-return spring. Note that spring is attached to bracket before securing bracket to stud with a nut.

5A/Now the left rear nut is tightened slightly more than finger-tight. Then tighten the two nuts and the two capscrews to 10–14 ft. lbs.

6/Installing E clip onto automatic-transmission linkage.

7/Socket and ball assembled correctly. Pointed out here by screwdriver.

7. Snap the socket at the end of the throttle cable onto the ball on the throttle lever. Or, if a rod slides through the throttle lever, secure it with the appropriate clip.

8. Attach the PCV vacuum hose.

9. Attach other vacuum hoses per your sketch made during disassembly. Vacuum connections vary so much from model to model that there is no general guide to what connects where. One hose often forgotten is the one to the temperature-control valve in the air cleaner. When you put the cleaner on have a vacuum source for the thermostatic control hose.

10. Connect the thermostat to the choke linkage.

11. Plug the two vacuum hoses and the wire plug into the solenoid on automatic-transmissioned cars (no vacuum hoses on manuals).

12. Use a line wrench to tighten the fuel-line fitting. The best method is to tighten the fuel line modestly—set the air cleaner on the carburetor as a flame arrester in the event of backfire and start the engine. Seconds after it starts and runs, shut it off. Carefully feel around the fuel fitting nut for a leak. Watch your finger closely as a slow-seepage type leak will appear on the flesh and evaporate quickly. If there is seepage, tighten a little more and repeat the check.

13. Now you can secure the air cleaner. Be sure you are in a well-ventilated area or better yet, outdoors. Set the emergency brake, start the engine and let it warm up. Continue checking for any fuel leak.

14. It's time to set the idle.

For the 1971 and later models, set idle according to the tune-up sticker. Earlier models should be set by turning the idle screws clockwise in approximately 1/8-turn increments alternating from side to side until a lean roll (modest roughness) is noted. Back them out approximately 1/8-turn from this point. Be sure the idle speed is close to correct when final mixture setting is attained.

**Note on idle setting**—A drawing on page 109 shows three hot-idle compensator designs. Making idle adjustments requires closing the air entry if the HIC valve is open. Type C can be plugged with a pencil in the vent. Push on the pin of Types A and B to close these valves.

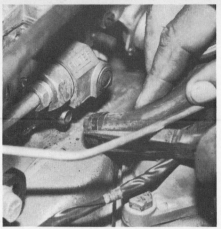

8/Attaching PCV hose to fitting on Q-jet base.

9/This hose controls the temperature valve in the air cleaner. Don't forget it.

10/Hooking up the stat coil linkage may only require sliding link into lever or a clip may install on the link.

11/Two vacuum hoses and a wire plug must be reinstalled to get the CEC solenoid back in service.

**Idle Setting Can Be Dangerous**—At this point I am going to voice a strong word of caution about setting idle. Most cars today are automatic-transmission equipped and final idle settings are made in drive gear. If you are leaning under the hood and open the throttle quickly to "clean it out" or note the engine response, the car may leave the scene with you draped over the fender or grill. Do not take this warning lightly as some very experienced mechanics have been pinned against walls and work benches with badly crushed legs. Emergency brakes will not hold a heavy-throttled engine.

Play it safe—tune it enough in neutral to drive it to an experienced friend with a tachometer and have the final settings made.

If you have a healthy abundance of pride and insist on completing the job, there is a safe way to handle it.

Drive the car's front bumper up against a tree or similar stout object and proceed with the exercise. DO NOT RELY ON CHOCKS OR BLOCKS UNDER THE TIRES.

No matter what you decide—BE CAREFUL!

# Common Sense Adjustments

A number of external adjustments alter carburetor performance and its ability to handle daily driving tasks. We will make no attempt to list specifications and measurements for the hundreds of models that exist. The following Delco drawings show you how to make adjustments. Now we'll explain some important details so you can relate these adjustments to effects on driving and/or maneuvers. Hopefully you will be able to analyze problems and make changes to eliminate any problems you are able to diagnose.

If you reassembled your carburetor without deliberately making changes, things *should* be back to original. If getting the hood down so you can get on with other things is foremost in your mind, then shut it. Drive the car a few days and if a problem exists which might be "adjustment-controllable" seek out the probable cause and fix it. If you need complete specifications on a particular carburetor, refer to the shop manual for that car or truck or get the correct specification from your Delco dealer.

## PUMP ROD ADJUSTMENT

In many cases whether or not you set adjustments to specifications from a manual is unimportant. Simply tailor the pump to fit your needs. If you live in a cold climate, the accelerator-pump requirement will be greater than for those who live in the desert or other warm areas. To obtain ultimate performance and drivability it may be necessary to increase pump capacity in winter and reduce it in summer. This can also apply to metering, choke, idle settings, and even ignition timing. Tune the vehicle to suit *you* in *your* environment.

Start by looking down the primary venturi, using a flashlight to illuminate the area. You should see a stream of fuel from the pump shooter each time the throttle is opened. Do this with the engine off. If this looks OK, then it's a matter of increasing or decreasing capacity. Moving the rod from the outboard hole to the inner

hole increases capacity to give more pump fuel which is helpful for most cold areas. Bending the lever tab upward also increases capacity. Bending it down decreases capacity, which is best for summer heat. Adjustment of the bendable tab on the end of the lever should not exceed 1/32-inch increments for street use. Chapter fourteen comments further on pump alterations.

## IDLE-VENT-VALVE ADJUSTMENT

Vent valves used on pre-1969 models allow vapors to exit the fuel bowl during hot soaks and also let in fine dust. If your car restarts good after short hot soaks, it's either working or you don't need it. Bend the tab and close existing valves when possible. If you have a hot-restart problem and you don't want to go through the adjustment formality, see that the vent valve is off the seat 0.020–0.040-inch when the primary throttle is at curb idle.

## FAST-IDLE ADJUSTMENT

Your fast-idle RPM in neutral with a warm engine should be 1600–2400 depending on make and model. No matter what it is *supposed* to be, one specification cannot be correct for all drivers and geographical locations. You must increase or decrease it to suit yourself. Don't judge it hot—it always sounds fast—that is, 2000 RPM hot will be about 1000 when the engine is cold. Make a subtle change, then evaluate it a few days as you drive.

With the engine warm and shut off, push the throttle open part way, close the choke blade completely and hold it closed as you release the throttle. Start the engine in neutral (brake on) without touching the throttle and check RPM on a tachometer. This is your high-step RPM. Tap the throttle lightly to get it off the high step. This adjustment is critical to engine operation immediately after a cold start. If the engine starts good but runs too slow while still on the high step, increase the warm-engine setting by 200 RPM each change until you are satisfied. If the engine screams with RPM right after start, decrease *warm*-engine RPM setting in similar increments.

## VACUUM-BREAK ADJUSTMENT

The vacuum-break setting strongly influences engine

operation immediately after start and up to the first block distance. If the car spits, coughs, sags and wheezes, bend the vacuum-break rod or tang to give you less choke-blade opening at full diaphragm travel. Increase the blade opening if the car loads up (rich), evidenced by black smoke out the exhaust pipe and extremely slow restarts when the engine stalls.

This—like other things on the carburetor—influences drivability with subtle adjustments. Work in 0.015–0.025-inch increments (a paper clip is approximately 0.035-inch) rather than 1/16 or 1/8-inch increments. You can improvise by measuring with nails, wire, paper clips and common sense when altering to meet your specific needs.

## AIR VALVE DASHPOT ADJUSTMENT

The amount of gap you allow between the link and the end of air-valve-lever slot is only important if you have a sag during highway speed (40–80 MPH) passing type accelerations. The adjustment shown is one of several ways to make the air valve do what *you* want it to.

## CHOKE-UNLOADER ADJUSTMENT

This hot-restart device helps to overcome poor hot starts. Holding the throttle on the floor opens the choke blade a prescribed amount so it does not "choke." The open throttle also lets in much fresh air and the chamber is soon purged of extra fuel and so the engine will start. Make the prescribed WOT unloader adjustment and use it during hot restarts. Remember to hold the throttle open during starting. *Never, never* pump the accelerator when starting a warmed-up engine.

## AIR-VALVE LOCKOUT ADJUSTMENT—SECONDARY-THROTTLE-VALVE LOCKOUT ADJUSTMENT

Be sure this lockout arrangement is adjusted as specified or secondary operation may be hampered.

## CHOKE-COIL ROD ADJUSTMENT

Cool the engine-choke coil so it tries to close the choke blade. This way you can easily see and feel the direction the coil pushes the blade closed. Study your carburetor to reveal whether to lengthen or shorten the rod one rod diameter for increased or decreased choke.

NOTE: PERFORM ADJUSTMENTS IN PROPER SEQUENCE

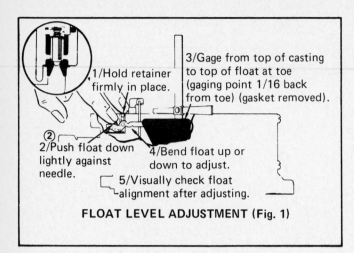

1/Hold retainer firmly in place.

3/Gage from top of casting to top of float at toe (gaging point 1/16 back from toe) (gasket removed).

2/Push float down lightly against needle.

4/Bend float up or down to adjust.

5/Visually check float alignment after adjusting.

**FLOAT LEVEL ADJUSTMENT (Fig. 1)**

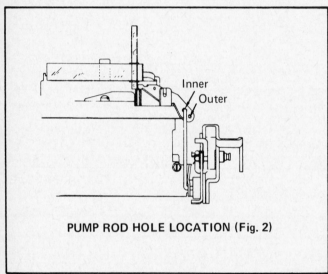

Inner
Outer

**PUMP ROD HOLE LOCATION (Fig. 2)**

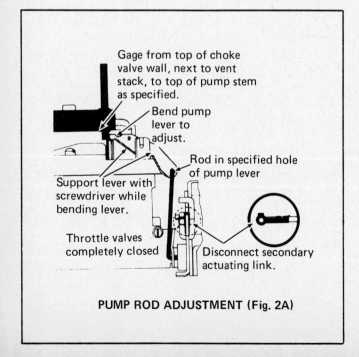

Gage from top of choke valve wall, next to vent stack, to top of pump stem as specified.

Bend pump lever to adjust.

Rod in specified hole of pump lever

Support lever with screwdriver while bending lever.

Throttle valves completely closed.

Disconnect secondary actuating link.

**PUMP ROD ADJUSTMENT (Fig. 2A)**

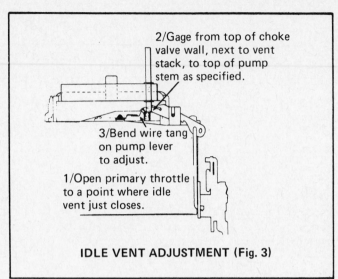

2/Gage from top of choke valve wall, next to vent stack, to top of pump stem as specified.

3/Bend wire tang on pump lever to adjust.

1/Open primary throttle to a point where idle vent just closes.

**IDLE VENT ADJUSTMENT (Fig. 3)**

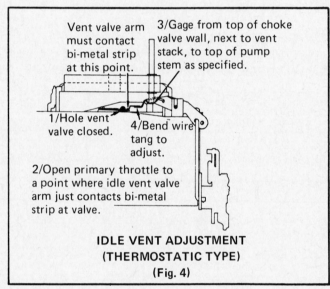

Vent valve arm must contact bi-metal strip at this point.

3/Gage from top of choke valve wall, next to vent stack, to top of pump stem as specified.

1/Hole vent valve closed.

4/Bend wire tang to adjust.

2/Open primary throttle to a point where idle vent valve arm just contacts bi-metal strip at valve.

**IDLE VENT ADJUSTMENT (THERMOSTATIC TYPE) (Fig. 4)**

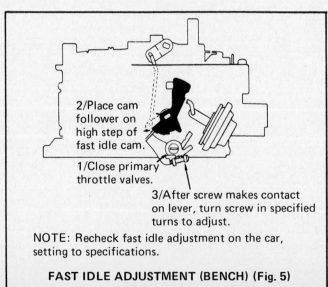

2/Place cam follower on high step of fast idle cam.

1/Close primary throttle valves.

3/After screw makes contact on lever, turn screw in specified turns to adjust.

NOTE: Recheck fast idle adjustment on the car, setting to specifications.

**FAST IDLE ADJUSTMENT (BENCH) (Fig. 5)**

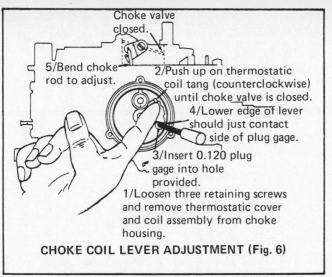

5/Bend choke rod to adjust.

Choke valve closed.

2/Push up on thermostatic coil tang (counterclockwise) until choke valve is closed.

4/Lower edge of lever should just contact side of plug gage.

3/Insert 0.120 plug gage into hole provided.

1/Loosen three retaining screws and remove thermostatic cover and coil assembly from choke housing.

**CHOKE COIL LEVER ADJUSTMENT (Fig. 6)**

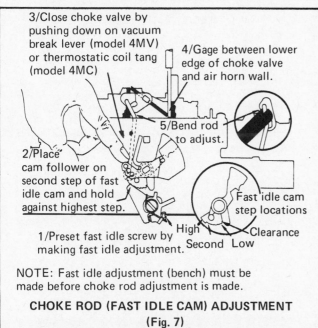

3/Close choke valve by pushing down on vacuum break lever (model 4MV) or thermostatic coil tang (model 4MC)

4/Gage between lower edge of choke valve and air horn wall.

5/Bend rod to adjust.

2/Place cam follower on second step of fast idle cam and hold against highest step.

Fast idle cam step locations

High    Clearance
Second  Low

1/Preset fast idle screw by making fast idle adjustment.

NOTE: Fast idle adjustment (bench) must be made before choke rod adjustment is made.

**CHOKE ROD (FAST IDLE CAM) ADJUSTMENT (Fig. 7)**

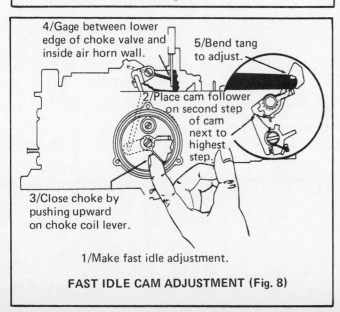

4/Gage between lower edge of choke valve and inside air horn wall.

5/Bend tang to adjust.

2/Place cam follower on second step of cam next to highest step.

3/Close choke by pushing upward on choke coil lever.

1/Make fast idle adjustment.

**FAST IDLE CAM ADJUSTMENT (Fig. 8)**

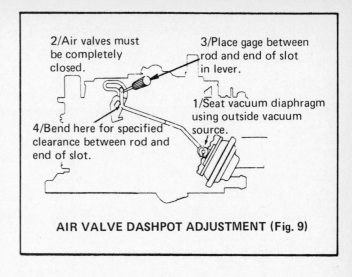

2/Air valves must be completely closed.

3/Place gage between rod and end of slot in lever.

1/Seat vacuum diaphragm using outside vacuum source.

4/Bend here for specified clearance between rod and end of slot.

**AIR VALVE DASHPOT ADJUSTMENT (Fig. 9)**

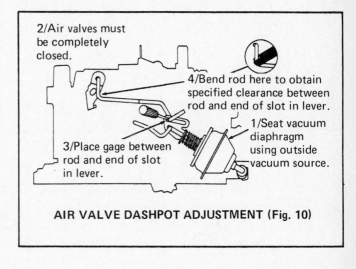

2/Air valves must be completely closed.

4/Bend rod here to obtain specified clearance between rod and end of slot in lever.

3/Place gage between rod and end of slot in lever.

1/Seat vacuum diaphragm using outside vacuum source.

**AIR VALVE DASHPOT ADJUSTMENT (Fig. 10)**

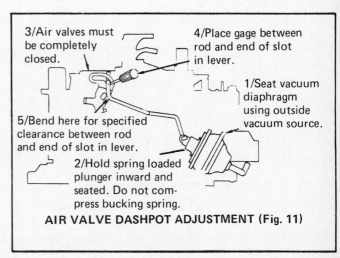

3/Air valves must be completely closed.

4/Place gage between rod and end of slot in lever.

1/Seat vacuum diaphragm using outside vacuum source.

5/Bend here for specified clearance between rod and end of slot in lever.

2/Hold spring loaded plunger inward and seated. Do not compress bucking spring.

**AIR VALVE DASHPOT ADJUSTMENT (Fig. 11)**

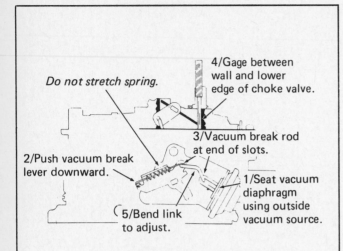

*Do not stretch spring.*

4/Gage between wall and lower edge of choke valve.

3/Vacuum break rod at end of slots.

2/Push vacuum break lever downward.

1/Seat vacuum diaphragm using outside vacuum source.

5/Bend link to adjust.

With choke spring pick-up located in top notch, hold the choke valve toward the closed position. Hold vacuum break diaphragm stem against its seat so vacuum break rod is at the end of slot. The dimension between the lower edge of choke valve and air horn, at choke lever end, should be as specified.

Bend vacuum break rod to adjust.

## VACUUM BREAK ADJUSTMENT
### (Early models without air-valve link to vacuum break)
### (Fig. 12)

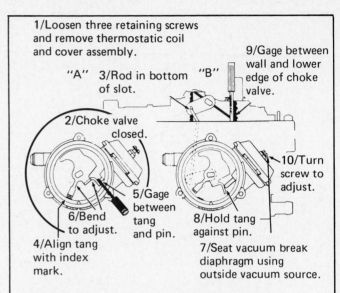

1/Loosen three retaining screws and remove thermostatic coil and cover assembly.

9/Gage between wall and lower edge of choke valve.

"A"  3/Rod in bottom of slot.  "B"

2/Choke valve closed.

5/Gage between tang and pin.

6/Bend to adjust.

4/Align tang with index mark.

10/Turn screw to adjust.

8/Hold tang against pin.

7/Seat vacuum break diaphragm using outside vacuum source.

The vacuum break adjustment must be made as follows: Loosen three retaining screws and remove thermostatic cover and coil assembly.

Step "A"/ With choke valve closed, and choke rod in bottom of the slot in the upper choke lever, align the thermostatic spring pick-up tang directly over the index tab on the inside of the choke housing. After tang is aligned, adjust vacuum break tang to specified dimension between tang and vacuum break pin.

Step "B"/ With vacuum break diaphragm seated, and tang against vacuum break pin, the dimension between wall and lower edge of choke valve should be as specified. Make sure choke rod is in bottom of slot in choke lever when gaging. Turn screw on vacuum break cover to adjust. Install baffle plate and thermostatic cover. Adjust cover to specified mark on choke housing.

## VACUUM BREAK ADJUSTMENT (Fig. 14)

1/Open primary throttle valves and rotate vacuum break lever until tang on lever and adjustment tang both contact end of spring.

2/Close primary throttle valves.

3/With valves closed, the cam follower should have 50-100% contact on low step of cam.

4/Bend tang to adjust.

5/Vacuum break lever must not hit adjustment tang.

## SPLIT CHOKE SPRING ADJUSTMENT (Fig. 13)

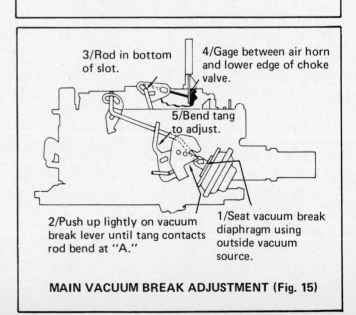

3/Rod in bottom of slot.

4/Gage between air horn and lower edge of choke valve.

5/Bend tang to adjust.

2/Push up lightly on vacuum break lever until tang contacts rod bend at "A."

1/Seat vacuum break diaphragm using outside vacuum source.

## MAIN VACUUM BREAK ADJUSTMENT (Fig. 15)

**NOTE: PERFORM ADJUSTMENTS IN PROPER SEQUENCE**

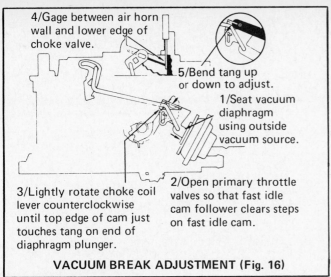

4/Gage between air horn wall and lower edge of choke valve.

5/Bend tang up or down to adjust.

1/Seat vacuum diaphragm using outside vacuum source.

2/Open primary throttle valves so that fast idle cam follower clears steps on fast idle cam.

3/Lightly rotate choke coil lever counterclockwise until top edge of cam just touches tang on end of diaphragm plunger.

**VACUUM BREAK ADJUSTMENT (Fig. 16)**

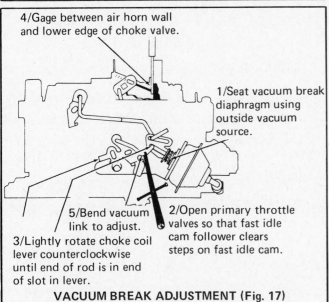

4/Gage between air horn wall and lower edge of choke valve.

1/Seat vacuum break diaphragm using outside vacuum source.

5/Bend vacuum link to adjust.

2/Open primary throttle valves so that fast idle cam follower clears steps on fast idle cam.

3/Lightly rotate choke coil lever counterclockwise until end of rod is in end of slot in lever.

**VACUUM BREAK ADJUSTMENT (Fig. 17)**

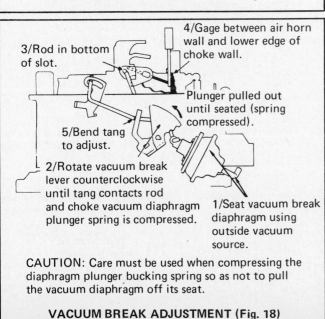

3/Rod in bottom of slot.

4/Gage between air horn wall and lower edge of choke wall.

Plunger pulled out until seated (spring compressed).

5/Bend tang to adjust.

2/Rotate vacuum break lever counterclockwise until tang contacts rod and choke vacuum diaphragm plunger spring is compressed.

1/Seat vacuum break diaphragm using outside vacuum source.

CAUTION: Care must be used when compressing the diaphragm plunger bucking spring so as not to pull the vacuum diaphragm off its seat.

**VACUUM BREAK ADJUSTMENT (Fig. 18)**

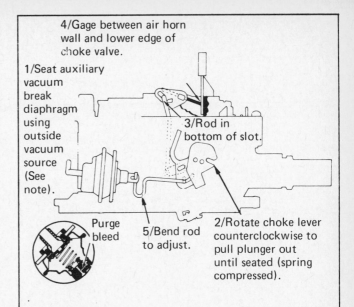

4/Gage between air horn wall and lower edge of choke valve.

1/Seat auxiliary vacuum break diaphragm using outside vacuum source (See note).

3/Rod in bottom of slot.

Purge bleed

5/Bend rod to adjust.

2/Rotate choke lever counterclockwise to pull plunger out until seated (spring compressed).

CAUTION: Care must be used when compressing the diaphragm plunger spring so as not to pull the vacuum diaphragm off its seat.

NOTE: If purge filter is used (see inset), remove vacuum break diaphragm hose and rubber covered filter element from vacuum break tube and, using a small piece of tape, plug the small bleed hole. After adjustment, remove the tape making sure the small bleed hole is open and install the rubber covered filter element over the vacuum break tube.

**AUXILIARY VACUUM BREAK ADJUSTMENT (Fig. 19)**

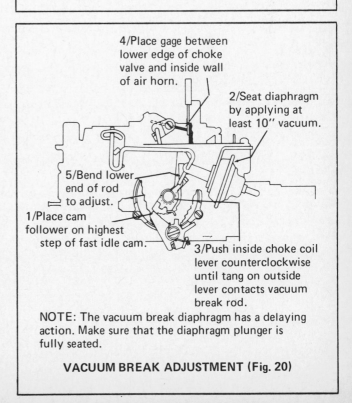

4/Place gage between lower edge of choke valve and inside wall of air horn.

2/Seat diaphragm by applying at least 10" vacuum.

5/Bend lower end of rod to adjust.

1/Place cam follower on highest step of fast idle cam.

3/Push inside choke coil lever counterclockwise until tang on outside lever contacts vacuum break rod.

NOTE: The vacuum break diaphragm has a delaying action. Make sure that the diaphragm plunger is fully seated.

**VACUUM BREAK ADJUSTMENT (Fig. 20)**

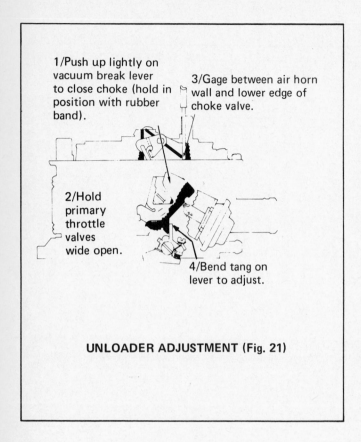

1/Push up lightly on vacuum break lever to close choke (hold in position with rubber band).

3/Gage between air horn wall and lower edge of choke valve.

2/Hold primary throttle valves wide open.

4/Bend tang on lever to adjust.

**UNLOADER ADJUSTMENT (Fig. 21)**

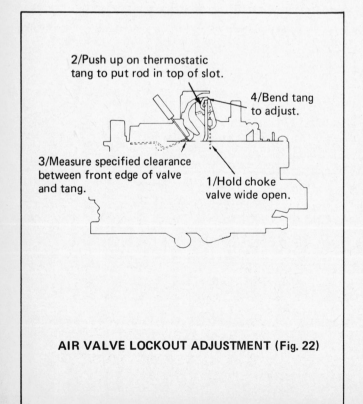

2/Push up on thermostatic tang to put rod in top of slot.

4/Bend tang to adjust.

3/Measure specified clearance between front edge of valve and tang.

1/Hold choke valve wide open.

**AIR VALVE LOCKOUT ADJUSTMENT (Fig. 22)**

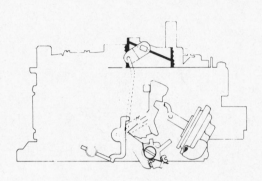

1/Hold choke valve wide open by rotating vacuum break lever towards open choke (*clockwise*).

2/Hold secondary throttle valves slightly open.

3/Measure 0.015 clearance.

4/Bend lever to adjust.

5/Hold choke valve and secondary throttle valves closed.

6/ 0.015 max clearance.

7/Bend pin to adjust secondary lockout lever clearance.

1. Opening clearance
Hold choke valve wide open by rotating vacuum break lever toward open choke (clockwise). With secondary throttle valves held partially open, measure the clearance between lockout pin and toe of lockout lever, as shown.

2. Secondary lockout-pin clearance
With choke valve and secondary throttle valves fully closed, bend lockout pin at point shown to maintain specified clearance between lockout pin and lockout lever.

**SECONDARY THROTTLE VALVE
LOCKOUT ADJUSTMENT
(Fig. 23)**

**NOTE: PERFORM ADJUSTMENTS IN PROPER SEQUENCE**

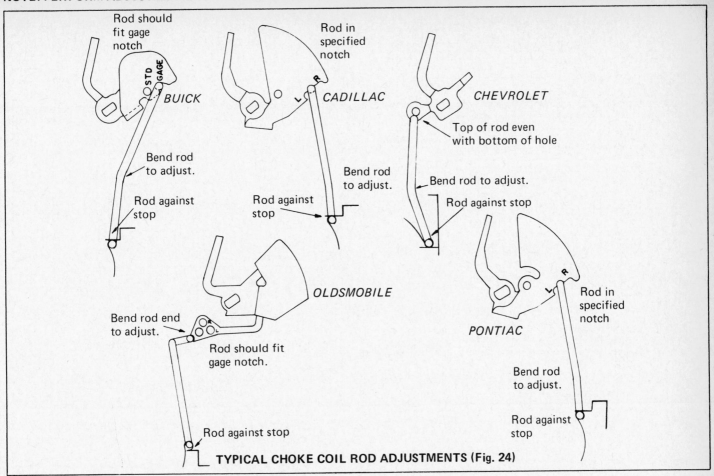

Rod should fit gage notch — **BUICK**

Rod in specified notch — **CADILLAC**

**CHEVROLET** — Top of rod even with bottom of hole

Bend rod to adjust.

Rod against stop

Bend rod to adjust.

Rod against stop

Bend rod to adjust.

Rod against stop

**OLDSMOBILE**

Bend rod end to adjust.

Rod should fit gage notch.

Rod against stop

**PONTIAC** — Rod in specified notch

Bend rod to adjust.

Rod against stop

**TYPICAL CHOKE COIL ROD ADJUSTMENTS (Fig. 24)**

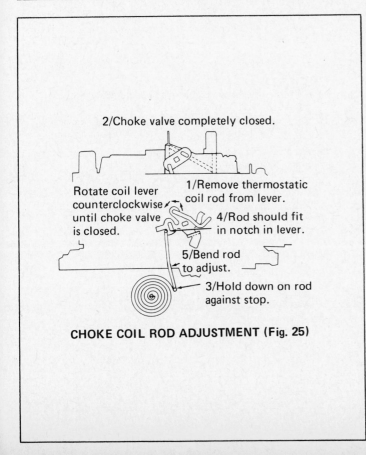

2/Choke valve completely closed.

Rotate coil lever counterclockwise until choke valve is closed.

1/Remove thermostatic coil rod from lever.

4/Rod should fit in notch in lever.

5/Bend rod to adjust.

3/Hold down on rod against stop.

**CHOKE COIL ROD ADJUSTMENT (Fig. 25)**

2/Rotate cover and coil assembly counter-clockwise until choke valve just closes.

3/Align index marks.

1/Place cam follower on highest step of cam.

1. Install choke thermostatic coil, cup baffle and cover assembly with gasket between the choke cover and choke housing.

2. Place the fast idle cam follower on the highest step of the fast idle cam.

3. Rotate choke cover and coil assembly counter-clockwise until the choke valve just closes and the index point on cover aligns with the center index point on the choke housing.

**AUTOMATIC CHOKE COIL ADJUSTMENT (Fig. 26)**

171

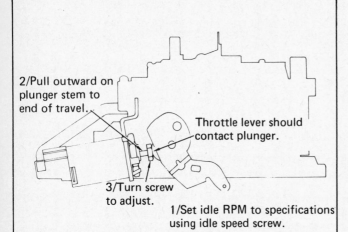

2/Pull outward on plunger stem to end of travel.

Throttle lever should contact plunger.

3/Turn screw to adjust.

1/Set idle RPM to specifications using idle speed screw.

This adjustment is to be made only after: (1) replacement of the solenoid, (2) major overhaul of the carburetor is performed, or (3) the throttle body is removed and replaced.

The following procedure is used to adjust the CEC valve controlled engine speed on a running engine (in "neutral" for manual or in "drive" for automatic transmissions), with air conditioning off, distributor vacuum hose removed and plugged, and fuel tank hose from vapor canister disconnected. Note: For Corvettes, remove fuel tank gas cap instead of hose from canister.

Before proceeding, follow instructions on vehicle tune-up sticker.

1. Adjust curb idle speed to specifications using idle speed screw.

2. Manually extend the CEC valve plunger to contact the throttle lever and pull outward on plunger stem to end of travel.

3. Turn plunger screw to adjust engine speed to car manufacturer's specifications:

|  | Passenger | Trucks |
|---|---|---|
| A/T | 650-D | 650-D |
| M/T* | 850-N | 850-N |
| *350 V8 M/T | 900-N | 900-N |

CAUTION: Do not use the CEC valve to set curb idle speed.

**CEC VALVE ADJUSTMENT (Fig. 27)**

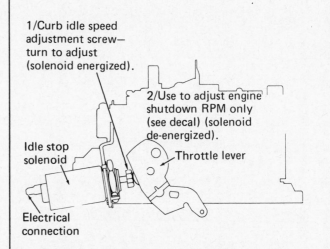

1/Curb idle speed adjustment screw—turn to adjust (solenoid energized).

2/Use to adjust engine shutdown RPM only (see decal) (solenoid de-energized).

Idle stop solenoid

Throttle lever

Electrical connection

The following procedure is used to adjust the idle stop solenoid to control engine speed on a running engine. Follow instructions on vehicle tune-up sticker before proceeding.

1. With engine at normal operating temperature and idle stop solenoid energized (plunger stem extended), adjust plunger screw to obtain specified engine speeds.

Low idle adjustment

2. To set low engine idle speed, with the idle stop solenoid disconnected electrically, adjust idle speed screw on throttle lever to obtain specified engine speed.

**IDLE STOP SOLENOID (Fig. 28)**

# Fuel Supply System

**A**n important part of carburetor performance is the ability of the fuel system to meet engine fuel requirements. Racers often have high-RPM leaning-out problems that can't be helped by metering changes. Just as often, a racer with a super fuel pump will have severe starting problems and over-rich mixtures during low and mid-range running. These problems are attributable to fuel pressure and can be corrected by a sensible approach to fuel-system design.

## FUEL PRESSURES

The Rochester Q-jet likes to see the fuel pressure between 4 and 6 psi, but the carburetor will run adequately at 2 psi and can be run below that for specific applications. Never design a fuel system for a street-driven car with more than six pounds fuel pressure at idle because the fuel level rises with increased pressure, causing idle instability, over-rich low end and starting problems.

## USE A FUEL PRESSURE GAGE

Always monitor fuel pressure on a 0—10 psi gage mounted outside the cockpit of the vehicle. Fuel lines inside the vehicle can be cut during an accident or mechanical failure. If the gage must be inside the cockpit, use an electric unit with wires running from the sender at the pressure source. The gage should be connected to the fuel line as close to the carburetor inlet as possible. Low fuel pressure can be caused by:

1. Pump with insufficient capacity
2. Fuel line too small
3. Restrictive fuel line fittings
4. Too many bends or a crimped line
5. Tank outlet or filter screen restricted
6. Plugged fuel filter
7. Filter installed between the tank and the fuel pump
8. Dirt anywhere in the system
9. Lines too close to a heat source.

## FUEL LINES

Fuel lines must be routed away from heat and firmly secured so that they will not vibrate and fatigue.

Fittings must be

Fuel-pressure pick-up point should be as close to the inlet needle as possible. The Q-jet lends itself to a very sanitary installation, as shown here. The thin casting boss allows only a few threads, so beef up the attachment of the fitting to the carburetor body with epoxy such as Marine-Tex or Devcon F.

Copper, aluminum or steel—Remember copper and aluminum tubing will fatigue much quicker than steel. If the installation is intended for a rather lengthy duration of time, steel line should be used. If copper or aluminum is your choice, mount it so flexing is held to a minimum as these metals work-harden and fail as a result.

An adjustable fuel-pressure regulator in the line between the fuel pump and carburetor ensures against excess pressure causing flooding and hot-starting problems. This one is set at 3 psi (arrow).

Any mechanical fuel pressure gage (directly connected to fuel) should be mounted outside the driver's compartment for safety.

Steel fuel lines with no flexible sections— are what you'll find on GM cars and trucks in nearly all instances. This is the recommendation from Rochester Products Division engineers as they believe flexible line is more subject to failure. A leaky fuel line can be the cause of fire and this should be avoided. If you install a section of hose or other flexible line, remember to check it regularly and replace it whenever the hose "begins to get tired." Also, be sure to use good clamps at each end of the hose to ensure against leaks at these points.

Take-apart type AC pumps are often used for competition cars where the rules require mechanical pumps. This type pump can be rebuilt when required.

nonrestrictive. Some fittings have built-in restrictions. Check every fitting that you use in your fuel-supply system. Make sure that the passage is the same size all the way through the fitting. If it necks down inside, you may be able to open up the passage with a drill.

Avoid 90° (right-angle) fittings wherever possible as these give the most fluid friction and therefore are more restrictive than straight-through or 45° fittings. A right-angle fitting has the same restrictive characteristics as a piece of tubing several feet long.

If a mechanical fuel pump is used on a competition engine, install a 1/2-inch ID line from the tank to the pump. Make sure that the fitting which connects to the fuel tank is as close to the fuel-line ID as is consistent with safety. Fuel-line attachment to the fuel tank is one area commonly overlooked by mechanics just getting started in competition. They will install a large line, but try to feed fuel through a tiny 1/4-inch or smaller diameter opening . . . which does not work. Here is one place where steel, copper or aluminum tubing is a better choice than fuel hose because tubing fittings typically allow a larger through-passage than you can obtain with hoses. This is because a hose fitting has to have enough wall thickness to withstand hose-clamp pressure. Fittings for tubing can have openings which are very close to the actual ID of the tubing itself, thus giving less restriction.

Remember that the important point is to keep the fuel supply to the pump unrestricted. The pump outlet (for an engine-mounted mechanical pump) or regular outlet connection to the carburetor/s can be through 5/16- or 3/8-inch tubing. A fuel pump pushes fuel better than it pulls it. This is why the mechanical pump may be augmented or replaced by a rear-mounted electric pump.

Stock-size (5/16 or 3/8-inch) fuel lines can be used on a street machine, but the pump performance will be compromised by such an installation. Small lines, sharp bends or kinks or right angle fittings will always cause a pressure drop and are more subject to trouble.

Keep the line away from exhaust-system components to avoid excessive heat. Make sure that there is no part of the body that will deflect the exhaust from open headers back into the area where the line is mounted. If a line passes near the exhaust system and there is no other place to route it, insulate the fuel line very thoroughly. The line should be clamped against the chassis or body structure with rubber-lined clamps such as those which are used in aircraft repair shops.

Locate the pressure regulator (if one is used) as close to the carburetor as possible. Lines between the regulator and carburetor/s can be 3/8-inch ID. There are usually two outlets on the regulator. Where two carburetors are used, connect each carburetor to an outlet.

Pressure at the carburetor must be set with the engine idling so that there will be some fuel flow to allow the regulator to function. Use a fuel-pressure gage at the carburetor and adjust the regulator to provide the desired pressure.

When running the car at high ambient temperatures, a cool can should be used just ahead of the regulator (upstream) on the high pressure side so that the fuel will not tend to flash into vapor when it is changed to a lower pressure by the regulator.

Fuel in the pump can be heated when it is pumped from the outlet, through the by-pass and back into the pump inlet. To avoid this problem, which is most apparent in low-demand situations such as idling, some racers install an external by-pass through a 1/16-inch restriction from the pump outlet to the tank. Don't use any by-pass restrictor larger than this or or you could reduce the pump output too much. The by-pass line should dump back into the main tank low enough to enter liquid fuel. If the by-pass fuel is dumped in above the fuel level it picks up air and creates fuel motion, creating vapors. The by-pass idea can also be used with mechanical pump installations. In this case the small-diameter return line is usually connected from the carb inlet to the fuel tank. Some AC fuel filters have a by-pass line connection.

## FUEL FILTERS

A fuel filter should be used between the fuel pump and the carburetor. Never use any kind of filter—other than a simple screen—on the suction side of a fuel pump. This is true regardless of the fuel-pump type. The usual screen at the tank outlet will usually be OK if it is clean.

The filter in the line to the carburetor should be as nonrestrictive as possible. Paper-element filters are excellent for the purpose. If you are concerned about the pressure drop through the filter, "Y" the fuel line to run through two filters so each one supplies a carburetor, or use two filters in parallel in the line to a single carburetor. Mount the filter canister to allow easy replacement of the element.

Sintered-bronze (Morraine) filters found in the inlets of many carburetors are OK for street use if there is little dirt in the gasoline and if the tank itself is clean. If the fuel supplied to the pump is dirty—as may be the case in a dusty area—remove the sintered-bronze filters from the inlets and install an inline filter between the pump and the carburetor.

Pay attention to the fuel filter and replace the paper element or throw away inline filters when the fuel pressure drops. This indicates that the filter has done its work and it is time for a new one.

For all competition use, take the sintered-bronze filters out of the inlets and install an inline filter in the fuel-supply system.

This unit repairs stripped threads in Rochester inlets and also provides a simple way to adapt an in-line filter with minimum plumbing. Available from the Carb Shop or Doug Roe Engineering.

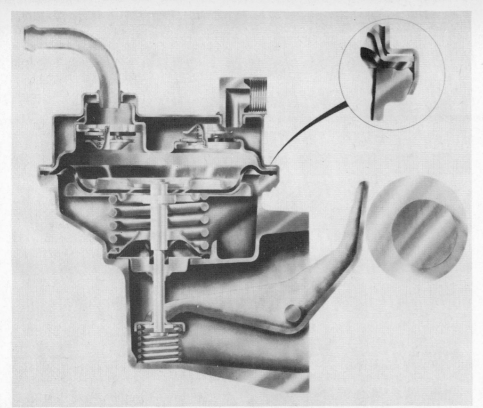

AC "Unitac" mechanical fuel pumps are found on most current GM cars and trucks. This is a non-repairable pump because of the assembly method which crimps the valve and diaphragm housing onto the pump casting. Mechanical fuel pumps such as this operate from an eccentric on the camshaft. Corvair pumps worked off a crankshaft eccentric.

AC EP-11 and EP-12 light-duty electric fuel pumps will supply up to 22 gallons per hour (GPH) at 4.5 to 8 psi fuel pressure. Maximum current draw is 3.5 amperes. These solenoid-operated pumps pulse or cycle at a rate varying with fuel demand.

AC EP-1 and EP-2 heavy-duty electric fuel pumps supply up to 35 GPH at 4.5—8 psi fuel pressure. The pumps can be used in parallel pairs for added capacity. Maximum current requirement is 2.0 amperes per pump. The motors in these diaphragm pumps operate at constant speed, but the diaphragm rests when there is no fuel demand.

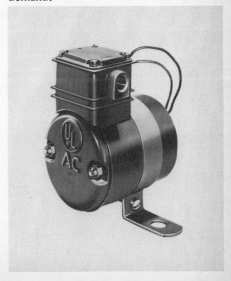

## FUEL PUMPS

There are two common types of fuel pumps: mechanical and electrical. Although the carburetor never "knows" which kind supplies its fuel, let's consider the two types briefly.

**Mechanical**—These engine-mounted pumps are diaphragm types driven by camshaft/crankshaft eccentrics operating a rod or lever. Advantages include low initial cost, simple plumbing and mounting, low noise level and familiarity to the general public because millions have been used over the years. Disadvantages include transferral of engine heat into the fuel, especially when the engine is shut off hot. The mechanical pump sucks fuel through a long line from the tank, further promoting the fuel's tendency to flash into vapor, especially on warm days. Fuel vaporizes more readily with decreased pressures.

**Electrical**—These pumps are not widely used as original equipment except for the in-tank pumps on the Vega, 1969-70 Buick and several foreign cars. The externally- or frame-mounted electrical pump has a high initial cost, is noisier than a mechanical pump (unless special mounting procedures are used) and requires connection to the car's electrical system. It should be mounted near the tank, away from heat and flying rocks, and plumbed to the tank outlet and to the fuel line.

Electrical pumps are typically mounted at the rear—near the tank—as a "pusher" device. Fuel pushed forward to the carburetor has less tendency to flash into vapor as it moves through the line. For competition or hot-weather use the rear-mounted electrical

175

pump provides "vapor-lock insurance." There are several types of electrical fuel pumps: solenoid-operated (AC, Bendix, Lucas), vane-type (AC, Holley) and positive-displacement (gear-type).

When installing an electrical pump, eliminate the mechanical pump if at all possible. It heats fuel and limits pressure to the mechanical pump's output pressure.

Include a safety switch in the circuit so that the pump will not work unless there is oil pressure. For starting before oil pressure develops, run a wire so the starter energizes the pump from the starter solenoid circuit. Once the engine is running, the three-way switch provides voltage to the pump as long as there is oil pressure. A schematic showing the wiring from the switch is included here for reference.

Mount the pump in as cool an area as possible—away from any exhaust-system components. Any fuel pump mounted to the chassis will transmit some noise into the car's body structure. This is no problem on a race car but it can be a source of annoyance on a dual-purpose car driven on the street. The best way to reduce the transmission of pump noise into the body is to mount the pump on rubber-insulated studs. These are mounted in turn on a chassis member or on a stiffened section of the body. Never mount the pump directly to a large flat sheet-metal surface because this will amplify the noise—just the opposite of what you are trying to achieve.

There's good reason to mount the pump near the tank. A pump trying to pull fuel creates a vacuum (low absolute pressure) on the end of a fuel line (especially a long one), which tends to cause the fuel to flash into vapor. This is why professional racers replace the engine-mounted mechanical pump with an electric pump near the fuel tank.

When fuel is pumped forward at high pressure, there is another advantage which may not be quite so obvious. The high pressure more than offsets pressure losses in the lines from friction and acceleration. Pumping the fuel forward at high pressure and then reducing its pressure prior to sending it on to the carburetors ensures that there will always be adequate pressure available at the carburetors. Incidentally, the Holley pressure regulator can be used with other electric or mechanical

Holley high-pressure electric fuel pump pushes fuel from tank at high pressure (9—14 psi) to a regulator (arrow) mounted near carburetor. Adjustable regulator allows setting desired fuel pressure to carburetor. Pressure switch in center (Holley 89R-641A) is essential component for pump installation. It senses engine oil pressure to shut off fuel pump when engine is stopped, regardless of whether ignition is turned off or not.

Typical wiring diagram for an electric fuel pump. AC pressure switch (Chevrolet 3986857) ensures against pump operation unless engine is running. This arrangement does start the pump when the ignition switch is being used to energize the starter solenoid. Starter circuit from ignition switch is not shown in this schematic. Wiring harness shows actual component configuration for the AC PS-9 pressure switch.

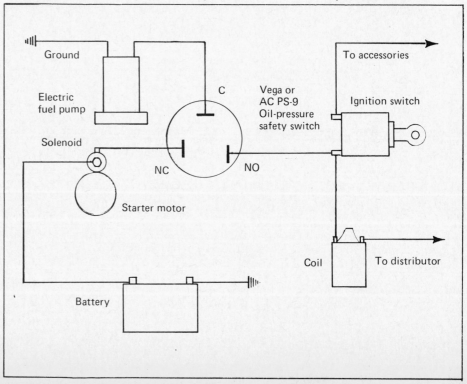

176

pumps.

Keep the pump-inlet line short and large. If a filter is used between the fuel tank and the pump inlet it must be a screen-type filter so that there will be no pressure drop at this point—which could cause fuel vaporization and consequent pump cavitation. It is actually preferable to position the pump so that its inlet is slightly below that tank so that the fuel level will create a "head" or positive pressure at the inlet. On a drag car it helps to place the pump behind the tank so acceleration will tend to push fuel to the pump inlet.

## EVAPORATION CONTROL SYSTEM

See emission chapter fifteen for what you need to know about this important emission-control system as related to the fuel-supply system. If you are tuning a 1972 or later automobile, it has ECS. Some 1971 models also have it.

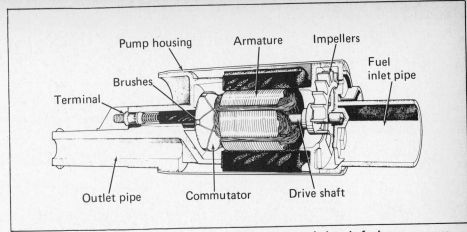

AC's submerged electric fuel pumps, as used on 1969 and later Buick Rivieras and on all Chevrolet Vegas, push the fuel to the carburetor. The design and pump location greatly reduce vapor-lock and hot-start problems. In a hot-soak situation any pressure buildup merely pushes the fuel back into the tank instead of overloading the needle/seat and causing the carburetor to flood.

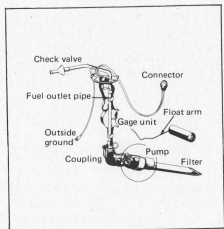

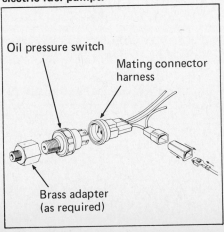

AC PS-9 oil pressure safety switch for electric fuel pumps.

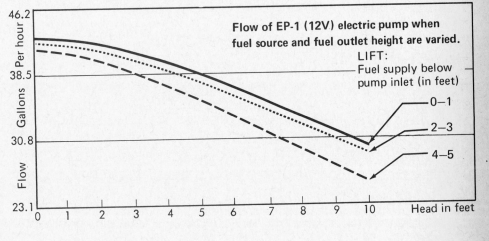

Flow of EP-1 (12V) electric pump when fuel source and fuel outlet height are varied.

LIFT:
Fuel supply below pump inlet (in feet)

0–1
2–3
4–5

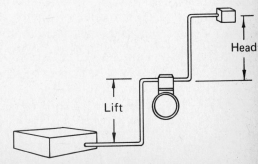

# How to Select & Install Your Carburetor

I am perfectly aware of the widely held belief that "bigger is better." Owners frequently rush out to buy a *bigger* carburetor after their first disillusionment with the performance of their new cars, pickups or vans. Most of these people are not racers, so *a bigger carburetor seldom helps them.* Their engines lose torque—the mainstay of street and highway performance. But, because no baseline performance tests were run before *and* after installing the new larger carburetor, the owner lives with the too-big carburetor. And he'll swear he made the right decision because the car runs "so much better." My own experience has proved that bigger is more often *worse.*

The common cause of no-go performance in today's vehicles is low compression—plus long-overlap camshafts for $NO_x$ reduction. These two minuses are accompanied by a low numerical (high gear) drive-line gearing that locks engine RPM into a range where the cam is least efficient. Highly restricted exhaust systems worsen the problem. In short, our engines are victims of the emission-control era.

First, think about the *combination* of axle ratio, tire diameter and the engine when you build or buy a vehicle for a specific reason: around town or highway driving, trailer towing, carrying a camper, etc. Consider a higher numerical drive-line gearing (lower gear) and stay away from big-diameter rear tires. Pick a camshaft (or engine) to give more power down *below* 4000 RPM (unless you're building a 100% racer, of course!). If you've been buying vehicles for a number of years, one of them probably brings back fond memories. It did all the things you thought important—without emptying your pockets for gas money. It probably had a single- or two-barrel carburetor. What this tells you is: The carburetor/engine/rear-axle ratio/weight/etc. were *matched* to one another. It was a good *combination.* This can be achieved with a four-barrel, but probably not with a large one which allows opening the secondaries at low RPM.

Taxis in big cities often have small carburetors, yet they compete well with big-engine/big-carburetor vehicles going away from the stop lights. These vehicles are special ordered as to carburetion/axle ratio/engine size to do a specific job well. They are not the all-purpose vehicles you get when you walk into the showroom and ask for the banana-yellow one with the red seats. Study the option sheets and order your car to fit your requirements. It's worth the wait!

Suppose you are building/buying a vehicle primarily for the old drive to work routine—mostly around town. Consider an intermediate-size vehicle with a six-cylinder or small V-8 engine if fuel economy is a factor. With most of your driving at 30 to 60 MPH, it is detrimental to gear and carburate for higher speeds. Assuming a tire diameter of 26 inches and an axle ratio of about 4.0:1, the majority of your driving will be at 1600 to 3000 RPM. The Air Flow vs. Engine RPM chart shows a 200 CFM carburetor will supply all the air required for 3000 RPM from a 200 CID engine—at 100% volumetric efficiency. But your engine is not 100% efficient so derate the flow figure by the V.E. (say, 200 CFM x 0.75 = 150). A very *small* carburetor may handle nearly all of your driving requirements.

If drag racing is "your thing" and you are building an all-out engine for operation in the high-RPM range of 6000 or higher, then size the carburetor accordingly to fit the engine size, RPM capability (considering camshaft and gear ratio). Be prepared to sacrifice economy and drivability during any attempt to use this vehicle on the streets or highways.

*Optimize the combination* for the vehicle's intended use. Make every element fit the combination as closely as possible: cam, carburetor, manifold, gear ratio, compression, etc. If the engine displacement is large enough, the Q-jet is a logical choice because the primaries will supply approximately 300 CFM. Today's Q-jet-equipped automobiles will run 60 to 80 MPH on the primaries alone without even opening the secondaries.

Remember the air flow chart shows *maximum* air requirements for an engine. The normal tendency is to select a carburetor size to handle very high RPM, even if the engine will *seldom* be operated at high RPM. My recommendation is to use a gear-ratio/tire-size/RPM/MPH chart or circular slide rule to find the RPM range where most of the intended use of the vehicle will occur. Select the carburetor to serve that RPM range, even if you compromise high-speed capability somewhat.

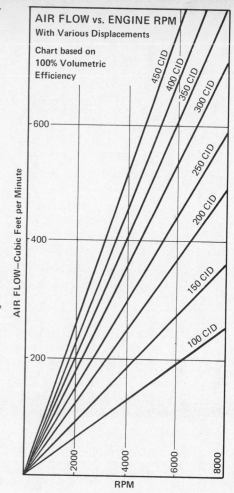

AIR FLOW vs. ENGINE RPM
With Various Displacements
Chart based on 100% Volumetric Efficiency

## ROCHESTER CARBURETOR FLOW RATINGS

| | CFM |
|---|---|
| **Quadrajet** (Air flow at 90° air-valve opening) | |
| 1-7/32-inch venturi (primary) | 800 |
| 1-3/32-inch venturi (primary) | 750 |
| **Model 2G—1-1/4" flange, 1-7/16" throttle bore** | |
| 1-3/32-inch venturi | 278 |
| **Model 2G—1-1/2" flange, 1-11/16" throttle bore** | |
| 1-3/16-inch venturi | 352 |
| 1-1/4-inch venturi | 381 |
| 1-5/16-inch venturi | 423 |
| 1-3/8-inch venturi | 435 |
| **Model 4G** (Throttle-bore—venturi size in inches) | |
| 1-7/16—1-1/8 primary; 1-7/16—1-1/4 secondary | 486 |
| 1-7/16—1-1/8 primary; 1-11/16—1-15/32 secondary | 553 |
| 1-9/16—1-1/8 primary; 1-11/16—1-15/32 secondary | 692 |
| **Monojet** | |
| 1-7/16 in. throttle bore, 1-7/32 in. venturi | 160 |
| 1-11/16 in. throttle bore, 1-5/16 in. venturi | 210 |
| 1-11/16 in. throttle bore, 1-1/2 in. venturi | 250 |

NOTE: 4-barrels rated at 1.5 in. Hg pressure drop.
2-barrels rated at 3.0 in. Hg pressure drop.
1-barrels rated at 3.0 in. Hg pressure drop.

## IMPROVED BREATHING INCREASES VOLUMETRIC EFFICIENCY AND POWER

The stock passenger-car engine is a reasonably efficient air pump—over a fairly wide RPM range—from low RPM to 5,000 RPM or slightly higher. It provides a reasonably flat torque curve over this operating band. Its pumping capabilities can be idealized (optimized) to provide better pumping within a narrower RPM band by improving breathing. By reducing restrictions which cause pressure drop, charge density reaching the cylinder is increased. Improvements are typically made by changes to the carburetor (higher capacity), intake tract (manifold through the ports and valves), exhaust system (headers or free-breathing mufflers) and valve timing (camshaft). Such changes are similar to the things that a pump designer would do to make an efficient pump—he'd work to optimize the performance *at a particular RPM*. But, there is a trade-off. When you make the engine a better pump, the torque peak and the entire torque curve are lifted to a higher RPM band. Lower RPM performance is *reduced* accordingly.

Not all engines can be improved for higher performance—better breathing—without extensive modifications. This is because the designers have purposely optimized the low-RPM performance with restrictions such as small-venturied carburetors, tiny intake manifold passages and ports, small valves actuated by short-duration camshafts with lazy action, low compression, combustion chamber design and restricted exhaust systems. Truck engines, low-performance passenger car engines and industrial engines are good examples. Keep such factors in mind when you are selecting a carburetor, because it is difficult—if not impossible—to upgrade the performance of an engine with such built-in restrictions. As they say, "You can't make a race horse out of a plow horse." If the engine is so restricted that it produces peak power at 4,000 RPM, selecting a carburetor on the basis of

Here are the two primary venturi sizes used in Q-jets. At left is the small 1-3/32-inch one used in the 750 CFM unit; on the right is the bigger 1-7/32-inch one used in the 800 CFM unit. Both retain the super-efficient three-venturi arrangement for excellent metering at low air flows. Flow ratings assume the air valve is positioned at 90° (full open). Photos are slightly larger than actual size.

**Bigger ain't necessarily better!** The larger the carburetor, the higher the air flow must be before the main system begins to feed. If the carburetor is too large, the pump shot will be consumed before the main system starts, assuming the throttle has been slammed full open. This results in a sag or bog so the engine gives less performance than one with a smaller carburetor more closely matched to the engine size.

This 1-3/32-inch venturi Rochester 2GC flows 280 CFM at 3 inches Hg pressure drop. The one shown fits a 1972 307 CID Chevrolet V-8. Millions of these have been sold as original equipment and for replacement since their conception in 1955. Flange size is 1-1/4 inch.

This large Rochester two-barrel flows 435 CFM at 3 inches Hg. Or, if compared to a four-barrel rated at 1.5 inches Hg pressure drop, it flows 307 CFM. 1957 Pontiacs were first to use this size with the 1-1/2-inch flange. They are popular on GM V-8's with large-displacement engines. Many 1974's changed to Q-jets to obtain finer mixture control to meet emission requirements and improve drivability.

feeding that same engine at 6,000 to 7,000 RPM is not wise. It's unlikely that the engine will ever run at those speeds—at least, not without extensive modifications. And, a too-large carburetor will definitely worsen the performance which was previously available. It is extremely rare when your transportation vehicle will perform its duties (performance, economy and drivability) better with a replacement carburetor. With a bigger one it's even less likely.

### CHOOSING CARBURETOR CAPACITY

Every carburetor ever sold as original equipment or as an aftermarket unit is a compromise unit in one way or another. A couple of RPD's competitors' carburetors feature huge flow capacity and mechanical secondaries. There is no doubt of their performance capabilities for all-out go-power.

As stated in the introduction to the book, Rochester leads the field in volume sold. They build units that are leaders or strong contenders in all areas except full-race applications. All of their carburetors have been designed to serve the majority of applications with good drivability and economy while maintaining good performance potential.

If you want RPD two-barrel units to serve in the latter capacity it takes more than one per engine because their largest one is rated at 435 CFM at 2.0-inches Hg. The Quadrajet with its very efficient primary serves well in drivability and performance and has huge secondaries to provide power potential with 750 and 800 CFM at 1.5-inch Hg pressure drop (at 90° air-valve opening).

The larger the carburetor, the higher the air flow must be before the main systems begin to feed. If the carburetor is too large, the pump shot will be consumed before the main systems start. This results in a sag or bog. This is why double-pumper four-barrel carburetors are offered by Holley. Rochester's largest carburetor, the Quadrajet, provides a supplement of fuel to handle this transition period when fuel is needed to cover sudden throttle openings. On most units you will notice two holes above or below the leading edge of the secondary air valves. Those holes are fuel feeds that respond to high depressions that exist under the air valve when the huge secondary throttle blades are opened. During the first few degrees of air-valve opening the holes are uncovered and they respond with a spray

of fuel that fulfills the demand momentarily until the nozzle has time to respond. This is RPD's way of handling this requirement as opposed to a second pump. Most enthusiasts and mechanics are not even aware that this secondary "pump" system exists on the Q-jet.

### UNPACKING YOUR CARBURETOR

If you are buying your carburetor in a store, unpack it before you pay for it. Look at the *inside* of the box as you take out each piece. Note whether one side of the box shows evidence of being smashed or damaged from movement of the carburetor. If such is obvious, check that side of the carb very carefully. Look it over slowly and thoroughly to make sure that there has not been any shipping damage. Don't make a hurried examination. Although Rochester packaging engineers constantly improve the packaging, their best efforts are often in vain when the freight companies mishandle the shipments. Be aware that it is quite possible for the *outside* of a carton to look perfectly intact—yet the contents may be damaged.

Nature provided the desert tortoise with packaging which has lasted this one 80 years. I doubt if RPD can equal that, but their current packaging efforts are certainly the best this author has seen.

Warranty for Rochester Products Carburetors and Emission Controls— See page 135 for details on the warranty which is applicable to these items.

It has become a real challenge at RPD to get the carburetor to the user in the condition it left their inspection facility. The carburetor is placed on a specially designed molded base piece and then capped with a durable air-horn cover and vacuum-sealed in tough plastic. This package is then placed in a strong carton. Not indestructible but certainly well engineered. Most of the carburetors arrive in good condition because of this carefully engineered packaging.

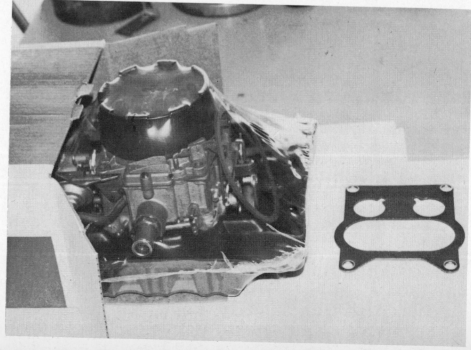

Give the carburetor a visual inspection, then hold the carburetor in your hand as you operate the throttle lever to make sure that the throttles open fully and close without binding or sticking. A bent throttle lever can occur. When it does, it is possible for the throttles to remain partly open (not return fully to idle or fast-idle position) when the lever is actuated.

If you have ordered your carburetor through the mail, give it the same inspection immediately upon its arrival. Don't just open the box and look inside to make sure that it is the correct carburetor—although that is important, as sometimes the wrong carburetor may be shipped—check it over thoroughly as just mentioned.

Should there be a problem, proceed carefully if you try to fix it yourself. If it is a simple thing—such as a bent throttle lever—you may be able to remedy the problem yourself easier than returning the unit. Be aware many levers are peened on shafts and it's important they remain on tight. If shipping damage or attempted repair loosens them, damage exists that will give you future problems.

**Shaft peening is illustrated on this air-valve lever.**

**Carburetor selection** may be limited by the type of manifold available for your engine. Smaller carburetors can nearly always be installed with adapter flanges. Larger carburetors should not (in most cases!) be mounted on a manifold designed for a smaller carburetor *unless* the manifold can be modified to eliminate restrictions at the carburetor-to-manifold flange.

If the casting has been smashed, it is obvious that the part will have to be replaced. How this is accomplished depends on how the carburetor was shipped to you (parcel post, motor freight, United Parcel Service, etc.). In some instances, the settlement of damage claims will be between yourself and the freight company (referred to as "the carrier"), and in other instances, the firm that sold you the carburetor will have you return it for replacement. The main thing to remember is that you will save yourself a lot of time and trouble if you will find out what the supplier requires *before you send or take the carburetor back to your dealer for replacement.*

Order the carburetor *early.* It may not be obvious to you, but one of the first things to remember in building your engine is to get the correct parts together and then get the car assembled and tested before setting impossible dates for completion and first competition attempts. Many cars do not show up at major events. The reason is often because the builder has been over-optimistic about the amount of time required to obtain all of the parts and then to assemble and test the combination. And, the owner loses his entry fee—as well as not getting to race!

With the advent of stock classes in various aspects of racing, getting the proper model carburetor may mean you make it through "tech" inspection as opposed to sitting on the sidelines.

Late-model cars are certified with specific model carburetors and must remain in that dress, if they are to remain legal to emission certification.

Dealers cannot possibly stock every model carburetor released in the past few years. Give your parts man an opportunity to order your pieces in from a warehouse location. You'll be glad you did. Be sure to get the other pieces which you need at the same time, including fuel lines, extra gaskets and jets, air cleaner, tubing wrench, new fittings and a fuel filter.

### BEFORE INSTALLING YOUR CARBURETOR

Check the screws which attach the top of the carburetor. Also check the screws which hold the throttle body. Gaskets will compress after they have been installed and it is important to check the

**Open throttles extend below the throttle body. Handle the carburetor with care when it is off the manifold or out of its shipping container. Avoid letting your friends play with the carburetor unnecessarily! Damage inevitably occurs when some unknowing person holds the throttles open—and sets the carburetor down hard onto a workbench or other surface. Chances are good you won't be able to run the carburetor until you've purchased a new throttle body—at your expense.**

screws as described so that there will not be any fuel or air leakage after the carburetor is installed.

If you happen to take the carb apart to look at its insides, put it back together immediately upon finishing your inspection so that the gaskets will not shrink into an unusable condition.

A lot of magazine articles have indicated that it is a good idea to take the throttle butterflies off of the shafts or to loosen the screws so that the throttles can be centered in the throttle bores. DON'T DO IT. The throttle levers contact adjustable stops and the carb is checked for the correct air flow before it leaves the factory. If you hold the carb up to a light, you will see clearances around the throttles . . . that's the way it is supposed to be! You can end up creating more trouble for yourself than you can imagine if you happen to break off one of the screws in the throttle shaft. These are staked at the factory so that they will not turn. A quick visual inspection of the carburetor will indicate that the screws have been staked.

In general, you should plan to bolt the carburetor on and run it before making any changes. Forget all the rumors and bench-racing discussion that you have been participating in lately. Start with what the Rochester engineers have found to work successfully. They produce hundreds of thousands of carburetors every year—for all kinds of applications—and chances are awfully good that the carburetor will be very close to correct when it is bolted onto your engine.

For alterations not spelled out in factory manuals stick to this book for authoritative information. A great deal of research has gone into this. Should a real special problem arise not covered in this volume, write a brief, precise letter to Doug Roe Engineering and they will try to assist. Enclose a stamped self-addressed envelope.

## REMOVING & INSTALLING YOUR CARBURETOR

1. Remove the air cleaner, carefully detaching any lines to the cleaner and marking them with masking tape so that they can be reassembled correctly.
2. Remove existing carburetor as follows (not all items apply to every application, of course):
a. Remove steel fuel line fitting carefully

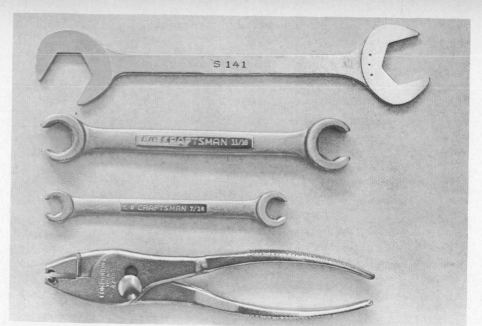

Essential tools for the carburetor technician. Pliers remove/install Corbin clamps on fuel/water hoses. They are made by several firms. Tubing wrenches in center are Sears Craftsman. Popular 1/2—9/16-inch size is not shown but should be in your tool-box. Top wrench is a Mac S-141 for fuel-inlet nuts. You really need one of those!

In-line filters are efficient and less restrictive than the small bronze or paper filters supplied as standard equipment in many carburetors. This filter is in the fuel line to a Rochester Q-jet feeding a turbocharger on author Doug Roe's competition off-road buggy.

because it is reused in most instances. A fuel-line wrench which contacts four of the nut flats should be used if you can obtain one because the fuel-line nuts are usually quite soft and tend to round off instead of turning. When this happens you will probably have to resort to the handy Vise-Grip pliers to get the nut out. Then cut off the fuel line very close to the end, install a new nut and reflare the end of the tubing. And, going through this drill of cutting and re-flaring requires a tubing cutter and a flaring tool. These may be rented from your near-by auto-parts house in some instances. Getting the line off the engine to allow this work will probably require disconnecting the fuel line at the fuel pump. If the nut at the carburetor rounded off, you can be sure that the one on the fuel pump will do the same. So, if you can, go buy that tubing wrench before taking the one off at the fuel pump or you may have to create an entire new fuel line to get the carburetor installed.

In some instances cutting off several inches of the fuel line and replacing that portion with a quality neoprene fuel line may be the answer. Be sure to obtain the required fittings before depending on this alternative. Do not use clear plastic materials. A quality line will be marked for fuel use and will have a fiber reinforcement. Use good hose clamps at each end. Flexible hose is not recommended between the pump and carburetor by GM. All GM cars have steel tubing between the fuel pump and carburetor.

b. Remove throttle linkage and automatic-transmission controls from throttle lever. Take off throttle-return spring and note its anchor points for correct reassembly.

c. Remove PCV hose if it is attached to carburetor.

d. Remove distributor spark hose/s, labelling these so vacuum and timed-spark hoses are kept separate for correct reinstallation.

e. Remove fresh-air hose if one is used and label it. Remove any other hoses such as gulp-valve hose (for air-injection-equipped cars) and connections to power brakes, etc. Label these.

f. Detach choke linkage at the carburetor, noting whether it pulls or pushes to close the choke plate.

g. Remove the nuts and/or bolts and lock-washers attaching carburetor to manifold, taking care to avoid dropping into the car-buretor. Covering the top of the carburetor is a good idea at this point. If you drop any

**Tubing vs. hose sizes**—Tubing is typically measured by its outside diameter; hose is measured by its inside diameter. An easy way to remember this is that hose will fit over tubing of the same listed size: 3/8-inch hose slips over 3/8-inch tubing. Pipe is also measured by inside diameter.

**Steel fuel lines with no flexible sections**— are what you'll find on GM cars and trucks in nearly all instances. This is the recommendation from Rochester Products Division engineers as they believe flexible line is more subject to failure. A leaky fuel line can be the cause of fire and this should be avoided. If you install a section of hose or other flexible line, remember to check it regularly and replace it whenever the hose "begins to get tired." Also, be sure to use good clamps at each end of the hose to ensure against leaks at these points.

**Avoid gasket stacks or packs**—These compress unevenly so that the carburetor base warps to bind the throttle shaft. This can cause one of the corners of the throttle base (mounting flange) to break off. Always use the correct gaskets.

No gasket sealer is required if the carburetor base and manifold flange are clean and flat. Do not stack gaskets together to get clearance for lever operation during special applications. Do it right or you could end up buying an entire new carburetor body or throttle base. Because these parts are not ordinarily stocked by even the largest dealers, there could be a long wait for pieces to reassemble your carburetor.

nut, lockwasher, cotter key, clip, piece of linkage or whatever, stop. Find and recover it *before* proceeding. Observe and note the position of any brackets held on the engine by the carburetor-attachment hardware. Lay these aside so that they will not fall into the manifold when the carburetor is pulled off. Before you lift the carburetor off of the manifold, check carefully to make sure that there is nothing loose which can fall into the manifold . . . such as a nut, bolt, fitting, cotter key or whatever.

h. Carefully pull the carburetor off of the manifold. If it sticks, tap it gently at each side with a rubber or plastic mallet.

i. Clean the intake-manifold mounting surface carefully, taking care to keep pieces of the gasket from getting into the manifold.

3. Install new carburetor as follows (not all steps apply to every application):

a. Install any new studs which are required. Use double-nut technique to install studs. Use Loctite on studs if available. Install manifold-flange gasket. If a metal heat shield is used to protect the carburetor base from exhaust gases in the heat-riser passage, place the thick (0.080-inch) manifold-flange gasket under steel heat shield. No gasket is used between the shield and the carburetor. Pay attention to the gasket combination used by the automobile manufacturer. Duplicate it as nearly as possible.

b. Some items used on the original carburetor may have to be transferred to the new one. These may include linkage ball connections, or lever extensions.

c. There may be some items on the new carburetor which must be removed to adapt it to a particular application.

d. One of the easiest ways to prevent manifold leaks is to check the carburetor base gasket. Hold it against the manifold flange to make sure there is plenty of clamping area to hold the gasket in position. Look to see whether any openings do not match up. When you are satisfied the gasket is correct for the manifold, hold the gasket against the carburetor base to see if it fits the throttle bores, mounting holes and general shape of the base.

e. Flush fuel line. Pull primary wire from coil and insulate its end with a piece of tape so it won't spark. Hold a can under the open end of the fuel line and catch the fuel as the starter is used to crank the engine. Connect fuel line to carburetor or fuel-line assembly before securing the carburetor. It allows you to start the fuel-line

Pay attention to what you are doing. This brand-new Q-jet was installed by a know-it-all expert do-it-yourselfer who didn't take time to examine the carburetor he took off. He overlooked the heat shield which is typical for Chevrolets through 1969 models. But, when he brought the carburetor back just as it is shown here, the carburetor was to blame. Of course! Needless to say, the power piston and bore required a careful cleaning to make the unit operable.

Surface of heat shield 3884576 at left marked TOP goes against carburetor base. Center gasket 3884574 fits between heat shield and manifold. Arrows indicate where gasket and shield must align or exhaust leaks will occur. This arrangement was used on Chevrolets through 1969. Gasket/spacer 3998912 at right used on 1970 and later Chevys has phenolic bushings around stud holes.

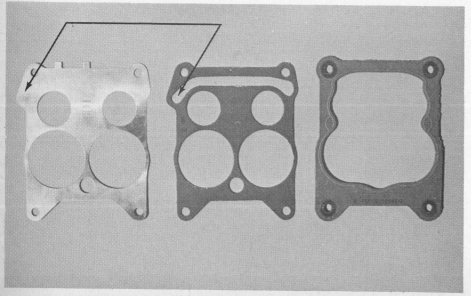

nut and run it in a few threads with your fingers. This prevents cross threading and a fight to get the nut started.

f. Place carburetor on manifold and add any brackets held in place by the attachment hardware—including throttle-linkage and solenoid brackets if such are used. Install nuts and/or bolts and tighten them gently against the carburetor flange. Then cross-tighten the bolts/nuts alternately to a maximum of 14 ft-lbs torque. Overtightening can cause a warped throttle body which will bind the throttle shafts so throttle action is stiff. This can cause poor closing and thusly erratic idle. In the worst case, overtightening may snap off a corner of the carburetor base.

g. Connect throttle and transmission linkage. Connect throttle-return spring. Operate linkage to ensure nothing binds as the throttle is fully opened and then closed. Make sure that the throttle returns all the way to idle without sticking or binding with choke held wide open.

Make sure that the levers don't hit anything as the throttle linkage is actuated. Make this check by having a friend actuate the linkage with the foot pedal as you observe the operation of the linkage. Look in the carburetor to see for yourself whether the throttle plates are 100% open in both the primary and secondary barrels. Don't try to make this check by operating the throttle with your hand at the carburetor or by looking at what the levers *appear* to be doing. Slack or play in the linkage may allow full throttle when moved at the carburetor, but give less than full throttle when the foot pedal is depressed. If the foot pedal movement does not give full throttle, examine the linkage to see what minor changes or adjustments may be needed in the linkage connecting to the carburetor. Some linkages allow enough adjustment to get the throttle blades wide open before the pedal rests against the floor. The two should happen simultaneously or excessive loads will be exerted on the carburetor throttle lever and eventually it will loosen from the shaft. Adjust wide open to correlate with the pedal being floorboarded. Do not change the linkage or levers on the carburetor itself because simpler adjustments are usually provided on the vehicle cable housing or linkage rods.

Don't think that this checking the full throttle is only for amateurs. Experts have been tripped up on this point . . . whether they care to admit it or not. Check both the pump and the throttle levers to make sure that nothing is in the way.

h. Install choke linkage for divorced choke. Hold throttle lever partially open and operate choke manually to make sure that it operates freely.

i. Reconnect appropriate hoses to carburetor. If additional fittings should be required, these can usually be transferred from the old carburetor.

j. Remove air-cleaner stud from old carburetor and install it in the new one, or use one provided. Install air cleaner and make sure it does not hit any portion of the linkage.

4. Crank the engine or turn on the electric fuel pump to fill the carburetor with fuel. Check the fuel-inlet fitting/s for leaks and tighten fittings to eliminate any leaks.

5. Depress the accelerator pedal to floor and allow to return to normal. This charges the manifold with fuel, sets the choke and fast idle. With foot off accelerator pedal, crank engine until it fires, repeating the previous two steps as required. Assembly quality eliminates most problems but if flooding should occur, stop to correct the cause. First try tapping the carburetor lightly with the handle of a screwdriver. If this does not stop the flooding, remove the fuel line and take off the carburetor top. Check the float setting. If it is OK, the chances are good that there is dirt under the needle. The needle and seat are separate items, so pull the needle out of the seat and blow out the seat with compressed air from both sides.

---

**Effect of air leakage on idle speed**—Any air leak into the intake system affects idle speed and quality. Air leaks can be caused by worn valve guides, worn throttle-shaft bores in the carburetor, holes in any hose connected to intake-manifold vacuum, non-sealing gaskets where the manifold joins the cylinder head or carburetor. Leaks in any component attached to the manifold—such as vacuum motors, power-brake accumulators, etc., can also cause problems.

---

Look for the air-valve stop tab on your carburetor (black arrow). When against its stop, the air-valve blades may be either fully opened (90° position) or partially closed (only 70° in some models). The air-valve opening may relate to what the designers found necessary to obtain desired distribution in a particular manifold/engine combination. Or, it could relate to the maximum opening required by low-volumetric-efficiency smog-type engines which could not use any more capacity. The mixture distributor tab on the underside of the air valve is another "fix" for improved distribution in a particular application. For most performance uses it is essential to get the air valve fully open so the maximum carburetor capacity will be available. Failure to do this can cost you a loss of as much as 170 CFM on the biggest Q-jet.

Typical choke thermostat to carburetor linkage connection is shown here.

6. Fast-idle setting is usually adequate for most vehicles. If a particular application requires adjustment, use the following procedure: with engine running, advance throttle and place fast-idle cam such that speed screw contacts top step of cam. On this step, set at approximately 1600 RPM.

Remember if your vehicle has T.C.S. (Transmission Controlled Spark advance) your adjustment of fast-idle speed is being made with no vacuum advance in neutral. In the morning you will have vacuum advance and the speed may be too fast (objectionable). Make subtle changes to get it where it suits you.

## GET ACQUAINTED WITH IT!

Once you have installed your new carburetor, live with the new combination for a few days before beginning your tuning efforts. Naturally, you will want to set the idle correctly (to exact specifications if your car has emission-control devices), check the ignition timing with a timing light and make sure the spark plugs, points, cap and wires are all in tip-top shape.

By driving the car, you will learn what you have. If you are drag racing, get your times consistent by working on your driving technique *before* you start tuning. There are enough variables without adding inconsistent driving.

Many details related to tuning are provided in a later chapter on the carburetor and performance.

## BLUEPRINTING ROCHESTER CARBURETORS

"Blueprinting" has become an important word in the enthusiast's vocabulary. The word, as applied to an engine, means bringing all of the parts of a high-production engine into the correct relationship by matching and mating, balancing and remachining as required to get the combination into a near-perfect state suitable for performance applications. Blueprinting is often done on brand-new engines right out of the crate or just off the showroom floor. This attention to detail can provide impressive performance improvement as an immediate payoff for the time and care invested in the project.

"Blueprinting" is now applied to the precise preparation and rebuilding of any component in the engine or chassis. In racing application this is an accepted thing.

Southern California drag racer Jim Mehl does a thorough job of checking hoses and connecting linkage after making changes to his Q-jet.

With emission regulations as they are today, beware of altering production-built units for street use. Tearing a new carburetor apart and attacking it with drills, trick linkages and so forth is like overhauling a new watch or micrometer when you first get it . . . something you probably wouldn't consider doing. There would be too much chance of ruining it— and voiding the guarantee—before you could use it.

Carburetors are made very precisely because they have to accomplish an extremely accurate job of metering fuel and air into the engine. *They have to be correct before they are shipped.* Every single carburetor which will be used as original or replacement equipment on a production automobile must pass critical computer-controlled tests which check whether that individual carburetor performs to tight specifications which ensure that it will provide correct performance (including low emissions) when it is installed. The unit has to pass the qualification tests before it can be shipped. These requirements are laid down by law.

Production techniques used to manufacture these carburetors have been developed over years of making carburetors to be right when an automobile manufacturer bolts one onto an engine.

The author is fully aware that numerous magazine articles are printed every year—pointing out "fixes" to be made to new or used carburetors for more performance. Almost without exception, these have been tried by competent test personnel in field tests—and discarded as being unworkable or not generally applicable. Great care in selecting, installing and tuning your carburetor is the best "trick" you can pull out of your bag or tool box to get the utmost in performance and satisfaction.

If alteration and/or blueprinting is required for your application, study the tuning chapter carefully for many helpful hints.

**Fuel supply system**—Read the last chapter for important details on fuel lines, fuel filters and fuel pumps.

# The Carburetor—Manifold Relationship

## BASIC DESCRIPTION

The intake manifold provides a mounting for the carburetor and connecting passages between the carburetor throats and the cylinder-head ports. The manifold has to divide the air and fuel equally among the cylinders at all speeds and loads. Each cylinder should receive the same amount of the air/fuel mixture with the same A/F ratio as all other cylinders to allow minimum emissions and for maximum power production.

As part of the design, the passages should have approximately equal length, cross-sectional area and geometric arrangement. For various reasons, this cannot always be done, so flow-equalizing features are sometimes used to make unequal-length passages perform as if they were nearly equal.

Because the manifold has to distribute both gas and liquid, the designer has to consider the fact that there is nearly always some liquid moving around on the walls and floor of the manifold. Controlling this liquid fuel by various devices such as sumps, ribs and dams is extremely important in ensuring equal A/F ratios for all cylinders. As an aid to vaporization and as part of controlling liquid fuel which has not been vaporized (or which has dropped out of the mixture due to an increase in pressure), heating of the manifold is also essential in all except all-out racing manifolds. To avoid gravity influences which could cause uneven distribution of the liquid fuel, the manifold floor and carburetor base are typically parallel to the ground—instead of being parallel to the crankshaft centerline which may be angled for drive-train alignment.

Small carburetor venturis aid air/fuel velocity through the carburetor and into the manifold and generally aid vaporization and hence distribution. High-speed air flow through the

Top and side views of the low-rise big-port aluminum manifold 3977609 supplied on the 450 HP 454. Here you see the direct comparison with the low-rise small-port cast-iron manifold used with a Q-jet on 390 HP and lower performance 454's. These are cross-H two-plane designs. Both have a sheet-metal shield directly under the heat passage to avoid exhaust heating oil which is splashed under the manifold.

carburetor and manifold tends to maintain turbulence and both the velocity and the turbulence maintain fuel droplets in suspension, thereby tending to equalize cylinder-to-cylinder A/F ratios. If the mixture is allowed to slow, as will occur wherever a passage size increases, fuel may separate from the air stream to deposit on the manifold surfaces. This causes variations in the A/F supplied to the cylinder.

The design of the passages should be such that the mixture flow is sufficiently fast to maintain good low and mid-range throttle response without reducing volumetric efficiency at high RPM. And, the internal design of the passages must carry the air/fuel mixture without forcing the fuel to separate from the air on the way to the cylinder head. When a mixture of fuel and air is forced to turn or bend, the air can turn more quickly than the fuel and the two may separate. When fuel separates from the air, a cylinder being fed with the mixture after the bend will receive a different A/F ratio than if the fuel had stayed with the mixture stream.

Pulsing within the manifold must also be accommodated. This phenomenon, also called back-flow or reversion by some, occurs twice during a four-cycle sequence. A pulse of residual exhaust gas enters the intake manifold during the valve-overlap period when both the intake and exhaust valves are off their seats. Another pulse toward the direction of the carburetor occurs when the intake valve closes. Either of these pulses may hinder flow in the manifold. In the most severe cases, the pulsing travels through the carburetor toward the atmosphere, and appears as a fuel cloud standing above the carburetor inlet. This is called "standoff."

It should be noted that the carburetor meters fuel into the air stream *regardless of the direction of the stream*—down through the carburetor—or up through it from the manifold. The phenomenon occurs in some degree in most manifolds and in a marked degree in others—as we will discuss later in this section.

In many performance applications A/F ratio differences between cylinders are remedied by using unequal jet sizes. Richer (larger) jets are used in a section of the carburetor feeding lean cylinders and leaner (smaller) jets are used for those carburetor barrels feeding rich cylinders. Although un-

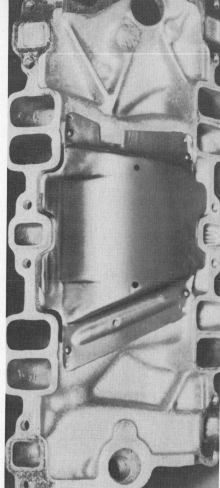

Underside of Chevrolet small-block high-rise manifold shows heat shield to prevent oil from being heated by contact with the exhaust heat passage.

equal jetting—commonly called cross- or stagger-jetting—is a partially effective remedy, in such situations *it is the manifold that is at fault*.

Manifolds also provide for mounting other accessory devices, sometimes serve as tappet-chamber covers and may provide coolant passages or attachments.

### BASIC MANIFOLD TYPES

There are a number of different manifold types: single-plane, two-plane, high-rise single- or two-plane, individual or isolated-runner (IR) and IR with plenum chamber (Tunnel Ram).

**Single-plane manifolds**—Roger Huntington, in a *Car Life* article, "Intake Manifolding," said, "The very simplest possible intake-manifold layout would be a single chamber that feeds to the valve ports on one side and draws from one or more carburetor

venturis on the other side. This is called a 'common chamber' or 'runner' (*single-plane*) manifold. Common-chamber manifolds have been designed for all types of engines—in-line sixes and eights, fours, V8's. It's the easiest and cheapest way to do the job. In fact, most current in-line four and six-cylinder engines use this type of manifold."

In the previously mentioned article, Huntington also wrote, "But we run into problems as we increase the number of cylinders. With eight cylinders there is a suction stroke starting every 90° of crankshaft rotation. They overlap. This means that one cylinder will tend to rob air-fuel mixture from the one immediately following it in the firing order, if they are located close to each other on the block. With a conventional V8 engine there are alternate-firing cylinders that are actually adjacent—either 5-7 on the left in AMC, Chrysler, GM firing order or 5-6 or 7-8 in the two Ford firing orders."

Two-plane (Cross-H) manifolds have been shown to be more throttle- and torque-responsive at low- and mid-range than most—but not all—single-plane designs. As of 1970-71, new single-plane manifolds were introduced with high mixture stream speed, good fundamental air/fuel distribution and adequate inlet flow at high RPM to permit strong running at both low and high engine speeds. Edelbrock's *Tarantula*, and *Torker* and Weiand's *X-terminator* and *X-celerator* are examples.

Dual-port-type manifolds may appear at first glance to be two-plane designs, but they are two single-plane manifolds. A small- and a large-passage manifold are stacked one above the other in a single casting. The network of small passages connected to the primary side of the carburetor provides high-speed mixture flow for good distribution and throttle response at low- and mid-range RPM. Larger passages connected to the secondary portion of the carburetor supply the extra capacity required for high-RPM operation. It should be noted that experimental dual-port, *two-plane* manifolds have been developed by the automobile manufacturers. Development work on the concept has been done by Ford, AMC, International Harvester, Holley Carburetor and the Ethyl Corporation. Most of this work has been oriented toward emission reduction. Volvo produced a manifold in the mid-sixties that used a similar idea and the current manifold on the Mazda rotary (Wankel) engines uses the same general idea.

Carburetor selection is thoroughly covered in chapter ten, but it should be noted here that smaller carburetors can be used effectively with single-plane manifolds because the common chamber damps out most pulsing (which reduces flow capability). A small carburetor tends to quicken mixture flow and thereby improves throttle response at the low- and mid-range.

**Two-plane manifolds**—These are simply two single-plane manifolds arranged so that each is fed from one half (one side) of a two- or four-barrel carburetor—and each of the halves feeds one half of the engine. A cross-H manifold for a V8 is a typical example which will be familiar to most readers. Each half of the carburetor is isolated from the other—and from the other half of the manifold—by a plenum divider. The manifold passages are arranged so that successive cylinders in the firing order draw first from one plane—then the other. In a 1-8-4-3-6-5-7-2 firing order, cylinders 1, 4, 6, 7 draw from one manifold plane and one half of the carburetor. 8, 3, 5, 2 are supplied from the other plane and the other side of the carburetor.

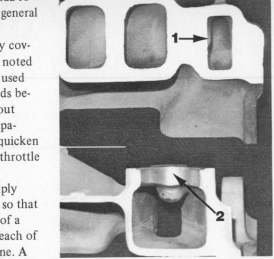
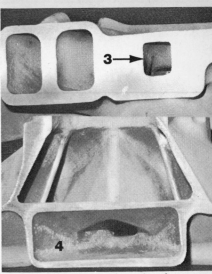

Cross-sections from a Weiand manifold illustrate how manifold provides other functions. Built-in water passage 1 connecting two cylinder heads to central water outlet 2. Exhaust crossover passage 3 from cylinder heads connects to heat-riser section 4 under carburetor.

Late O.E.M. Q-jet on Edelbrock TM-20 Tarantula (oval port, big-block Chevy). Carburetor mounting flange is designed for standard-base Holley, but will accept spread-bore units when fitted with conventional S.B./S.F. (spread-bore/standard-flange) carb adapter (Trans-Dapt, Eelco, etc.). Rear wheel HP normally up 18-31 (corrected) over that obtained with stock cast-iron manifold and same Q-jet. This is a single-plane manifold.

**Detecting manifold leaks**—An instant indication of a manifold leak is a rougher-than-normal idle with the throttle at the factory-set curb-idle position. There may be an obvious hissing or whistling sound. First check all hoses which should be attached to the carburetor base or to the intake manifold. Make sure that no hoses are cracked or broken. Take special care to check the underside or hidden portions of the hose where a leak might not be obvious with just a glance. Also make sure that all manifold vacuum ports have plugs installed where required. Many mechanics use a small squirt can filled with gasoline to help locate leaks in manifold gasketing. When gasoline enters the manifold through a leak, the idle speed increases noticeably. Care must always be used when using gasoline because of the ever-present fire hazard.

# V-8 MANIFOLD TYPES

**CROSS-H**

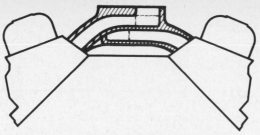

Cross-H or two-level manifold feeds half of cylinders from one side of carburetor — other half from other side of carburetor. Two sides of manifold are not connected.

**SINGLE-PLANE**

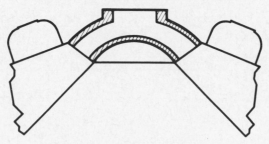

Single-plane manifold has all cylinder intake ports connected to a common chamber fed by the carburetor.

**PLENUM-RAM**

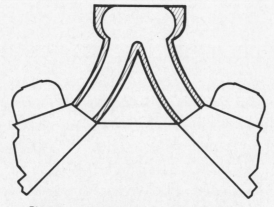

Plenum-ram manifold has a plenum chamber between passages to intake ports and carburetor/s.

**ISOLATED-RUNNER**

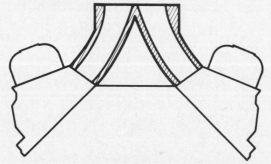

Isolated-runner (IR) manifold uses an individual throttle bore of a carburetor for each cylinder. There is no interconnection between the intake ports or throttle bores.

Ford Cobrajet 428 CID engine used a Q-jet as factory-stock equipment.

Classic cross-H manifold design was chosen by Ford engineers for their 427 CID single-overhead-cam drag-racing engine. Ford Motor Company photo.

Cross-H manifold design is clearly shown on this 1973 Pontiac Super-Duty 455 CID engine. Left side of Q-jet supplies lower plane of manifold (black arrows). Right side of carb supplies upper plane.

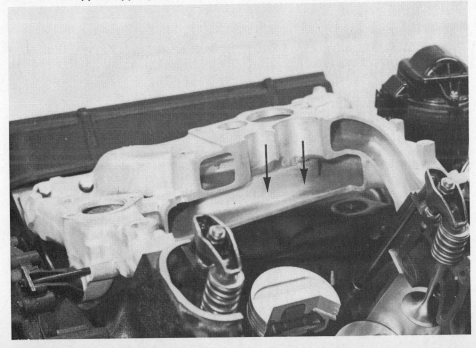

Because there is less air mass to activate each time there is an inlet pulse, throttle response is crisper and mid-range torque is improved—as compared to a conventional single-plane manifold.

The division of the manifold into two sections may cause flow restriction at high RPM because only one half of the carburetor flow capacity and manifold volume are available for any intake stroke. Thus, the divider is sometimes reduced in height—or removed completely—to make more of the carburetor and manifold capacity available. This kills bottom-end performance while top-RPM capability is being increased because the mixture speed is reduced at low RPM due to the volume increase on each side or section.

On engines built for high-RPM operation, divider removal can sometimes be a tuning plus. For street/track applications where the camshaft and other engine pieces are chosen to build torque into an engine, the divider should be left in place. The super-tuner can remove small amounts of the divider until he reaches the optimum divider height for a particular engine-parts combination and application.

**High-rise or high-riser?** Optional high-rise high-performance manifolds have been offered by some of the automobile manufacturers and by several of the aftermarket manifold makers, too. High-rise should not be confused with high-riser, which merely refers to spacing up the carburetor base so that there is a longer riser available to straighten flow before it enters the manifold. The main function of a high-riser is to improve distribution by eliminating directional effects caused by a partially opened throttle, as described in the engine requirements chapter. A high-rise manifold, on the other hand, aligns the cylinder-head port angle with that of the manifold passage or runner. In the case of a V8 engine, this usually means that the entire network of manifold runners is raised—with the carburetor mounting.

Both high-rise and high-riser designs may be used with either single-plane or two-plane manifolds.

**Independent-runner manifolds**—Independent- or isolated-runner (IR) manifolds are race-only devices which use one carburetor throat and one manifold runner per cylinder-head inlet port. Each of these runner/carburetor-throat arrangements is totally

Two Chevrolet big-block aluminum high-rise manifolds for four-barrel carburetors clearly show how cutting away the plenum divider changes a manifold. The lower manifold with the cutaway divider has been changed to a single-plane open-plenum design. The top manifold with divider intact is a two-plane cross-H design. An adapter is required to use the Q-jet on either of these large-port manifolds.

Weiand high-rise single-carburetor manifold shows how positioning carburetor base can provide a straight path for mixture flow to the intake ports.

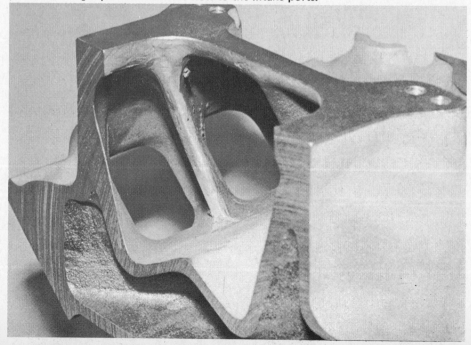

This circle-track racer has a Q-jet on the factory high-rise manifold. An adapter takes care of the difference in throttle-bore and bolt-hole spacings. Use of center divider helps keep torque up to ensure good acceleration off the corners.

isolated from its neighboring cylinders. Carburetors used in IR setups should have complete fuel-metering functions for each of the carburetor barrels because there is no way that the cylinders can share functions. The Quadrajet does not lend itself to this type of manifold because of the small primary bores. An important benefit of an IR-type manifold is that it allows tuning to take advantage of ram effect. Shorter tuned lengths cause the torque peak to occur at higher RPM.

Carburetors used in IR systems must be much larger than would be needed for other manifold types because each cylinder is being fed by just one carburetor throat or barrel. Another reason for needing a larger carburetor is that the severe pulsing which exists in such systems tends to reduce carburetor flow capacity in the RPM range where standoff becomes severe. In any IR system, some method of standoff containment should be used. The usual method is a stack long enough to contain the standoff atop the carburetor inlet.

Two views of Chevrolet high-rise manifold which is supplied as a stock item on the Z-28 and LT-1 engines. The two distinct levels below the flange identify this as a cross-H design. Ribs in manifold floor are distribution devices to direct mixture and to aid in controlling liquid fuel when vaporization is not perfect. Details on modifying the center divider and using a spacer for increasing high-RPM power are covered in H. P. Books' *How to Hotrod Small-Block Chevys*.

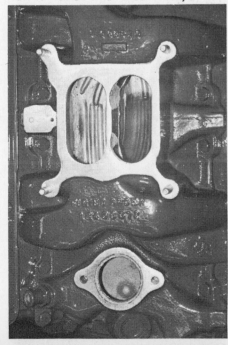

**Plenum-ram manifold**—This "almost-IR" manifold uses a plenum chamber between the carburetor base and manifold runners. This chamber helps to dissipate the strong pulsing so that less of it enters the carburetor to disrupt flow. And, just as important, it allows the cylinders to share the carburetor flow capacity. In the typical dual-quad plenum manifold, three or four cylinders will be drawing mixture from the plenum which is being fed by eight carburetor bores.

The sharing of carburetor capacity allows the Tunnel-Ram/Hi-Ram approach to manifolding to be "very forgiving" in terms of the carburetor capacity which is used, so long as the carburetors are not too large for the RPM range which is to be used. In the case of the typical Pro-Stock drag racer, 6,000 to 8,500 RPM is usual. With a large-displacement engine, there is plenty of air flow at 6,000 RPM to start the main system. The Quadrajet has never been popular on this type setup but the great flow response of these units could widen the useable RPM range with no sacrifice at top end.

## MULTIPLE CARBURETOR MANIFOLDS

Until about 1967, two or more carburetors were considered essential for a modified or high-performance engine. No self-respecting automotive enthusiast would consider building an engine with less than two carburetors unless the rules required it. But that situation has been changing rapidly over the past few years and the present-day trend is definitely toward the use of a single four-barrel carburetor for *all* street—and many competition—applications. The availability of sophisticated four-barrel carburetors with small primaries and large, progressively operated secondaries has allowed the construction of single-carburetor manifolds which provide excellent performance and low emissions, too.

The only application remaining for multiple carburetion—at least as we once knew it—is on drag cars and boats. And, it also fits into the kind of extended high-RPM operation which is encountered at the Bonneville National Speed Trials. Multiple carburetion is used for this type of use to reduce the pressure drop across the carburetors to an absolute minimum so that there will be the least possible HP loss from this restriction. Of course, low-RPM operation is not a con-

Here's a Q-jet with a velocity stack on a Weiand high-ram manifold. This type installation definitely requires more pump capacity. It might be aided by retargeting pump nozzles to point down toward throttles instead of hitting the booster venturis. We used an adapter here to get it all fitted together.

sideration for such uses, so the extra venturi area or carburetor flow capacity does not create any problems.

A great many engines have been sold with two- and three-carburetor manifolds as original stock equipment, but by 1972 all of the manufacturers except Chrysler had converted their high-performance engines to single-four-barrel configurations.

Should you be the proud owner of a car with 2 x 4's or 3 x 2's, by all means avoid buying new mechanical linkages to open the throttles simultaneously. A lot of

words are scattered throughout this book about the need for adequate mixture velocity and we have told you how to obtain it through selecting the correct carburetor capacity for the engine size and RPM. Simultaneous opening of multiple throttles (except for individual runner manifolds) goes completely against all of these recommendations, making the car hard to drive because the velocity through the carburetors at low and medium speeds is drastically reduced—as compared to the more desirable progressively operated secondaries.

Details of the 1957 Oldsmobile J-2 engine carburetion. Top photo shows cutaway air cleaner housing which connected to an oil-bath air cleaner at each side of the engine. White arrow indicates vacuum diaphragm which opened the front and rear carburetors within about one second from the time the center throttle was opened to 75% where it contacted a vacuum switch (black arrow). A vacuum reservoir provided vacuum for throttle actuation because low manifold vacuum existed when the carbs were signalled to open.

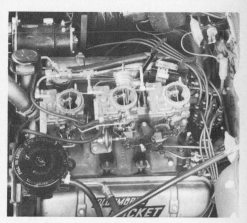

1966 Olds 4-4-2 had three two's. Center is a 2GC (with choke), end ones are 2G's. This had a mechanical linkage (no vacuum diaphragms!) which started to open the end carbs when the center carb was 65% open.

348 CID Chevrolet 3 x 2 used vacuum actuation of end carbs. Single diaphragm received vacuum from reservoir when vacuum switch (arrow) was contacted by 75%-opened throttle on center carb.

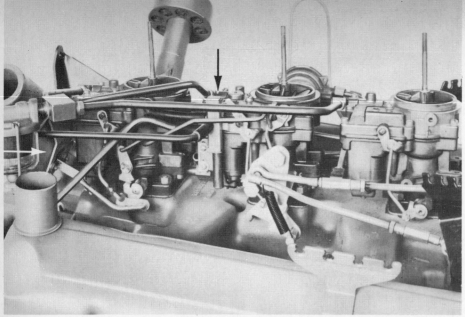

1958 Pontiac Tri-Power carburetion used single diaphragm (white arrow) to open end carb when 75%-opened throttle contacted vacuum switch (black arrow). System included vacuum reservoir.

Vic Edelbrock, President of Edelbrock Equipment Co., with his 1974 Scorpion intake manifold with a Q-jet mounted for the benefit of super-stock drag racers. Air gap between cam cover and underside of runners helps keep mixture cool and dense. Manifold works best above 6250 RPM with high-port modified heads and radical cams.

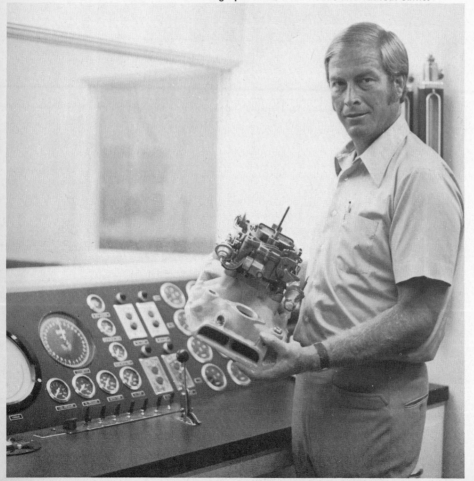

## MINICAR MANIFOLDS

Most of the stock (original-equipment) manifolds for small engines are single-plane designs. But, the aftermarket manifold makers have taken diverse approaches in their designs, using modifications of the single-plane approach and two-plane types, too. Most of the units available as this book went to press in 1973 were for the Pinto and Vega, but a few manifolds for imports—notably Datsun and Toyota—were also beginning to appear.

**Pulsations in intake manifold affect main-jet size**—Main-jet size depends in some degree on the number of cylinders fed by the venturi because fuel flow is affected by air-flow pulsations in the manifold. Where there are only one or two cylinders fed by a single venturi with no balance chamber or plenum connecting the cylinder/s with the other cylinders, pulsing can be extreme at wide-open throttle. Such pulsing requires using a smaller main jet because fuel is pulled out of the discharge nozzle in both directions of air flow—into the carburetor—and out of the carburetor as pulsing passes the discharge nozzle toward the entry.

**Distribution problems?** Getting a manifold to provide the desired distribution to all cylinders can be a problem. Remember that air/fuel separation in the manifold deposits fuel on the inlet manifold's plenum floor. Delivery of this fuel to the cylinders can be improved with slots and dams. To cure a rich-running cylinder, epoxy glue a length of 1/8-inch diameter aluminum rod across the floor of the entry to the manifold runner feeding that cylinder. A lean cylinder may be helped toward normalcy by grinding 1/8-inch wide by 1/8-inch deep fuel collection slots in the plenum floor and part way into the manifold runner.

Additional A/F mixture corrections may require the use of distribution tabs on the booster venturis. The photos you see in this chapter show some of the things we've had to do to plenum floors to get good distribution in several of our Tarantula series.

Jim McFarland
Edelbrock Equipment Co.

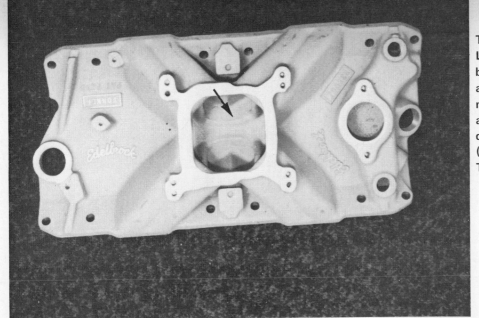

Tarantula Torker for small-block Chevrolet. Longitudinal mixture distribution dams bridge the branch dividers of cylinders 1—3 and 5—7, 2—4 and 6—8. This manifold has negative taper in its branches and incorporates deliberate port mismatch at the cylinder heads in an effort to reduce reversion (stand-off) pressure as much as possible. This is a single-plane manifold.

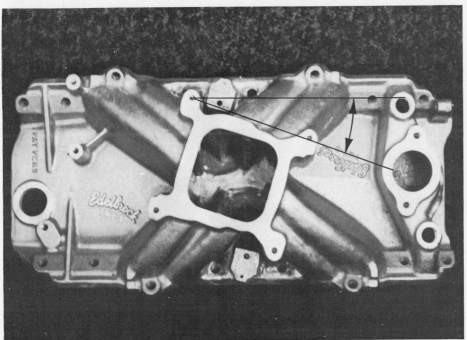

Tarantula for Mark IV Chevrolet V8 engines (both oval- and rectangular-port cylinder heads) incorporates rotated carburetor pad for improvement in cylinder-to-cylinder mixture distribution. Air/fuel dams are also placed in plenum floor in same fashion as small-block Chevrolet V8 Torker Tarantula. This design has shown the most significant reduction in emissions (HC, CO, $NO_x$) of all Edelbrock manifolding (HC=17.9%, CO=43.0%, $NO_x$=20.9% below best obtained with O.E.M. two-plane design and same carburetion).

1—Experimental Pontiac Tarantula designed for Rochester Q-jet base dimensions. Torque-type design incorporating tall, narrow branches. Fuel dam across floor of entry to cylinder number 6 (arrow) was required to prevent 3—6 firing sequence fuel buildup and overenrichment of cylinder number 6. This is a fiberglass prototype. CO concentrations are used to read distribution, a function of exhaust emission levels and a preliminary indication of what can be expected when the manifold is cast and subjected to lengthy emissions testing.

# The Carburetor & Engine Variables

**IDENTICAL ENGINES IDLING AT 700 RPM**

INITIAL TIMING 8° BTC
Relatively little air flow past carburetor throttle plate which is in nearly closed position.

INITIAL TIMING 4° BTC
Larger throttle plate opening is needed to allow more air into the fuel to achieve 700 RPM.

"It must be the carburetor." "You need your carburetor rebuilt, fella!" "We probably should get a different carburetor." How many times have you heard friends—or even yourself—making these or similar statements? Well, there are a number of things affecting the ability of the carburetor to perform as you think it should on your engine: spark timing, valve timing, temperature, density, compression ratio and exhaust back pressure. None of these are as easy to "change" as the carburetor, but unless changes are made in these other vital areas, there may not be any real reason to change your carburetor in an attempt to cover up what appear to you to be engine faults. Get acquainted with all of the factors affecting the engine—and carburetor—before making unnecessary changes or griping unnecessarily about the carburetor's performance.

## SPARK TIMING

Late-model cars have labels which specify engine idle speed and spark advance settings. These have been carefully worked out by the factory engineers to ensure that the engine combination in that vehicle will have minimum emissions within specified limits. The trend is to use a retarded spark at idle and during the use of the intermediate gears. Some smog systems which lock out vacuum advance in the intermediate gears have a temperature-sensitive valve which brings straight spark advance into operation if the engine starts to overheat and during cold operation prior to the time that the engine water temperature reaches 80°F.

New legislation put into effect in early 1973 will not let the manufacturer activate and deactivate emission-reducing hardware with temperature and time-control devices. This is to provide better emission reduction during all types of driving rather than intermittently during a specific driving schedule, namely the driving schedule legislated into law.

Most carburetors provide ports for timed spark and straight manifold vacuum. In general, a retarded spark reduces oxides of nitrogen to a minimum by keeping peak pressures and temperatures at lower values than would be obtained with an advanced spark setting. It also reduces hydrocarbon emissions. This is good for reducing emissions—but bad for economy, drivability and heating of the coolant. Fuel is being burned in the engine, but its energy is largely wasted as a heat flow into the cylinder walls and as excess heat in the exhaust manifolds—which also aggravates the problem of heating in the engine compartment. The cooling system has to work harder. In essence, fuel is still burning as it passes the exhaust valve. Thermal efficiency of the engine is less because there is more wasted energy.

The use of a retarded spark requires richer jetting in the idle and main systems to get off-idle performance and drivability. A tight rope is being walked here because the mixture must not be allowed to go lean or higher combustion temperatures and a greater quantity of oxides of nitrogen will be produced. Also if mixtures are richened too far in the search for drivability, CO emissions increase.

Because retarding the spark hurts efficiency, the throttle plate must be opened further at idle to get enough mixture in for the engine to continue running. This fact must be considered by the carburetor designer in positioning the off-idle slot. It must also be considered because high temperatures at idle due to the retarded spark and high idle-speed settings definitely promote dieseling or after-run.

Dieseling is caused primarily by the greater throttle opening, but it is aggravated by the higher average temperatures in the combustion chambers. Higher temperatures tend to cause any deposits to glow so that self-ignition occurs. On a car equipped with

---

**Dieseling**—this is the tendency of an engine to continue running in an irregular and rough fashion after the ignition has been switched off. This problem is aggravated (1) by anything which remains hot enough to ignite air/fuel mixtures—such as any sharp edges in the combustion-chamber area and (2) by the fast-idle settings used to meet emission requirements.

---

an anti-dieseling solenoid, turning the ignition on causes the solenoid to move the throttle to its idle setting. When the ignition is turned off, the solenoid retracts the idle setting about 50 RPM slower than a normal idle setting (a more closed idle position for the throttle) so that the engine has less tendency to run on. If it can't ingest enough mixture to support the engine's continuing to run, the dieseling is usually stopped effectively.

There is also a Combined Emission Control (CEC) solenoid on 1971 and later GM cars. This is not an anti-dieseling solenoid. It operates on a different principle for accomplishing a different approach to the emission-reduction problem, as described in the emissions chapter.

On carburetors equipped with timed spark advance (no advance at closed throttle) a port in the throttle bore is exposed to vacuum as the throttle plate moves past the port—usually slightly off-idle. The spark-advance vacuum versus air-flow calibration is closely held in the production of the carburetor because this has a dramatic effect on HC and $NO_X$ emissions. The distributor advance, once held to be so important for economy, has now become an essential link in the emission-reduction chain.

## VALVE TIMING

Valve timing is the one item with the greatest effect on an engine's idling and low-speed performance. Adding valve overlap and lift with a racing cam allows the engine to breathe better at high RPM, but it worsens manifold vacuum at idle and low speeds—creating distribution and vaporization problems. These are obvious because the engine becomes hard to start, idles very roughly—or not at all—has a bad flat spot coming off idle and has very poor pulling power (torque) at low RPM. This is especially true when a racing cam is teamed with a lean, emission-type carburetor.

Because manifold vacuum is reduced, the signal available to pull mixture through the idle system is also reduced, causing a leaning of the mixture. Also, the throttle will have to be opened further than usual to get enough mixture into the engine for idling. This can place the off-idle slot/port in the wrong relationship to the throttle so that there is insufficient off-idle fuel to carry the engine until main-system flow begins.

The idle-stop solenoid is an anti-diesel device. Curb idle is set with the solenoid energized. This holds the throttle further open than it will be when the ignition is turned off. When the ignition is turned off, the throttles close to reduce the amount of air entering the engine, thereby reducing the tendency to diesel.

The distributor vacuum port (arrow) or timed-spark port is exposed to manifold vacuum as the primary throttle blades are opened. The shape and location of the spark port varies between carburetor models. Some do not have them. When the primary throttles close, the port is no longer exposed to manifold vacuum so there is no vacuum advance at idle.

The CEC solenoid (1) has two positions and controls distributor advance vacuum from the timed-spark port in the carburetor. When the solenoid is energized in reverse or the high forward gears, vacuum is available for the distributor. This also advances the position of the throttle blades from curb-idle to a more open position used to control HC emissions during deceleration. We have included more details about this system in the Emissions Chapter 15. Electrical connections are in plug 2, hoses 3 and 4 connect vacuum from carburetor and distributor.

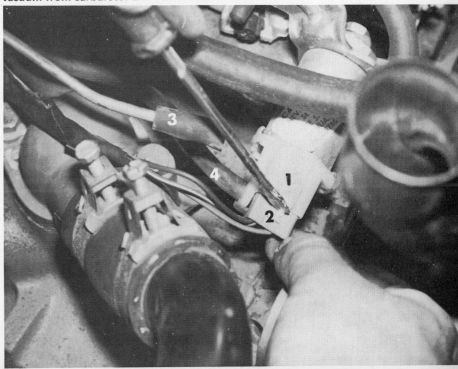

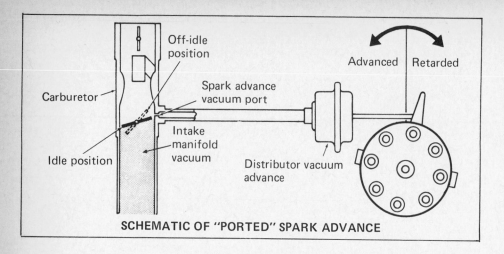

**SCHEMATIC OF "PORTED" SPARK ADVANCE**

And, the idle mixture has to be made richer to offset the poor vaporization and distribution problems. Part of the poor distribution and poor drivability problem stems from the overlap period. When both the exhaust and intake valves are open simultaneously, some of the exhaust gas which is still in the cylinder at higher-than-atmospheric pressure pushes into the intake manifold to dilute the incoming charge. This charge dilution effectively lowers the combustion pressure. This is especially true up to that RPM where the overlap time interval becomes short enough so that the reverse pulsing becomes insignificant.

Starting with 1971 model vehicles, camshaft overlap—that period of time when both intake *and* exhaust valves are open—has taken on new meaning to engine designers. An extensive study of camshaft designs revealed a reduction of $NO_x$ emissions was provided by many high-performance camshaft designs. This is because the mixture dilution (intake mixture diluted by residual exhaust) caused by the valve overlap causes inefficiencies at the low emission-test RPM—cooling the chamber and giving lower $NO_x$. The sad part to this story is that many engines are being released with cam timings not conducive to good torque and low-end power—with induction and exhaust systems and gearing which "pin" you in that range. This is one more of the unfortunate facts which cause today's domestic engines to fall short of our needs and expectations.

Reverse pulsing through the venturis at wide-open throttle can cause added fuel flow so that a richer mixture is created. Once the main system starts, the discharge nozzle will deliver fuel in response to air flow in either direction. More information on valve timing, reverse pulsing and so forth is contained in the ram-tuning section of the following chapter. Read it carefully!

When the manifold vacuum is reduced (pressure increased toward atmospheric), the power valve may start to operate, or at least to open and close as the manifold vacuum varies wildly. This is not a valid reason to remove the power assembly, but it may require altering the spring which will help the power valve close at the lowest vacuum (highest pressure) which occurs during idling.

In extreme cases, a "wild" racing cam may magnify these problems to the point where the car becomes undrivable for anything except a competition event. This is especially true if the carburetion capacity has been increased to match the deep-breathing characteristics of the cam.

There have been many cases where a racing cam was installed at the same time that the carburetor was changed. If the mechanic did not understand what was occurring in the engine—and many do not—the carburetor was often blamed for poor idling and the bad flat spot off idle when the real culprit was the racing camshaft! Detailed information on tuning an engine with a racing camshaft is contained in the performance-tuning chapter.

## TEMPERATURE

Temperature greatly affects carburetion. It affects mixture ratio because air becomes less dense as temperature increases. A density change reduction reduces engine power capability, even though main-jet corrections of approximately one main jet size smaller for every 50°F. ambient temperature increase can be made to keep mixture ratio correct.

Maximum power production requires that the inlet charge be as cold as possible. For this reason, racing engines are usually designed or assembled so that there is no exhaust-gas or jacket-water heating of the intake manifold. Stock-type passenger-car and truck engines, on the other hand, use a heated intake manifold because the warmer mixture—although not ideal for maximum power—helps drivability. A warm-air inlet to the carburetor and/or an exhaust hot spot or water-heated intake manifold greatly aid vaporization. Good vaporization ensures that the mixture will be more evenly distributed to the cylinders because fuel vapors move much more easily than liquid fuel.

Icing occurs most frequently at 40°F. and relative humidity of 90% or higher. It is usually an idle problem with ice forming between the throttle plate and bore. It usually occurs when the car has been run a short distance and stopped with the engine idling. Ice builds up around the throttle plate and shuts off the mixture flow into the engine so that the engine stops. Once the engine stalls, no vaporization is occurring, so the ice promptly melts and the engine can be restarted. This may occur several times until the engine is warm enough to keep the carburetor body heated to the point where vaporization does not cause the icing.

We have also seen cases of "turnpike icing" in the venturi itself when running at a relatively constant speed for a long period of time under the previously stated ideal icing conditions. In this case, ice build-up chokes down the venturi size so that the engine runs slower and slower. Vaporization of fuel removes vast quantities of heat from the surrounding parts of the carburetor, hence there is a greater tendency for this phenomenon to occur where vaporization is best, namely, in small venturis. Icing is no longer a major problem because most cars are factory-equipped with exhaust-manifold stoves to warm the air which is initially supplied to the air cleaner inlet. Thermostatic flapper valves (on some engines) shut off the hot-air flow when underhood temperatures reach a certain level.

On some high-performance engines, a vacuum diaphragm opens the air cleaner to a hood scoop or other cold-air source at low vacuums and heavy loads.

At the other end of the thermometer we find a phenomenon called percolation. It usually occurs when the engine is stopped during hot weather or after the engine has been run long enough to be fully "warmed up." Engineers call this a "hot soak." In this situation, no cooling air is being blown over the engine by the fan or vehicle motion. Heat stored in the engine block and exhaust manifolds is radiated and conducted directly into the carburetor, fuel lines and fuel pump.

Fuel in the main system between the fuel bowl and main discharge nozzle can boil or "percolate" and the vapor bubbles push or lift liquid fuel out of the main system into the venturi. The action is quite similar to that which you have seen in a percolator-type coffee pot. The fuel falls onto the throttle plate and trickles into the manifold. Excess vapors from the fuel bowl—and from the bubbles escaping from the main well—are heavier than air and also drift down into the manifold. This makes the engine very difficult to start and a long cranking period is usually required. In severe cases, enough fuel collects in the manifold so that it runs into cylinders with open intake valves. The fuel washes the oil off of the cylinder walls and rings and severe engine damage can result.

Percolation is aggravated by fuel boiling in the fuel pump and the fuel line to the carburetor because this can create fuel-supply pressure as high as 11 to 15 psi—sufficient to force the inlet valve needle off its seat. Fuel vapor and liquid fuel are forced into the bowl so that the level is raised. This makes it that much easier for the vapor bubbles to lift fuel to the spillover point.

Solving the percolation problem requires several solutions. The main system is designed so that vapor bubbles lifting fuel toward the discharge nozzle tend to break before they can push fuel out of the nozzle. Fuel levels are carefully established to provide as much lift as can be tolerated. In some instances, the fuel will be made to pass through an enlarged section at the top of the main well or standpipe to discourage vapor-induced spillover.

Gaskets and insulating spacers are used between the manifold and

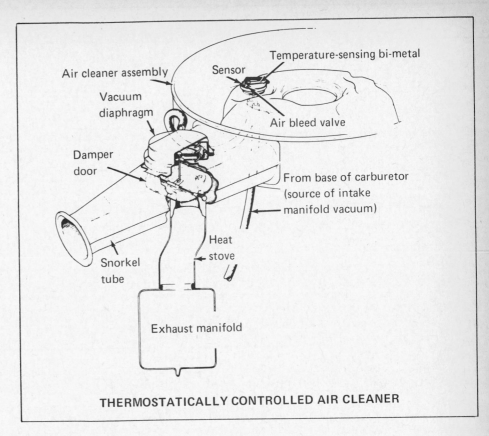

**THERMOSTATICALLY CONTROLLED AIR CLEANER**

Heat from an exhaust-manifold stove is piped through the flex duct (arrow) to the air cleaner's inlet snorkel. Carburetor air is regulated to approximately 100°F. by a thermostatic valve (arrow) in the air cleaner which allows manifold vacuum to open the flap in the snorkel so cooler underhood air can enter or at least mix with the hot air from the exhaust-manifold stove.

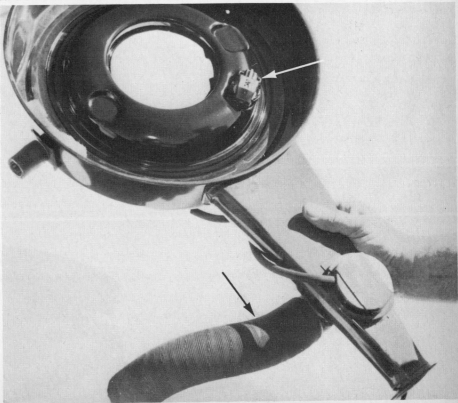

carburetor and between the carburetor base (throttle body) and fuel bowl. In some instances, an aluminum heat deflector or shield is used to keep some of the engine heat away from the carburetor.

As mentioned in the distribution section of the engine requirements chapter, an abnormally warm day in winter when fuel typically has a high vapor pressure will cause hot-starting problems due to percolation.

Another way that percolation and hot-starting problems are reduced is through the use of internal bleeds in the fuel pump and/or a vapor-return line on the carburetor ahead of the inlet valve. When the bleed and/or return line are used, any pressure buildup in the fuel line escapes harmlessly into the fuel tank or fuel-supply line.

Another bad effect caused by high temperature is boiling of the fuel in the fuel line between the pump and the carburetor. Fuel can even boil in the fuel pump itself. When the fuel pump and line are filled with hot fuel, the pump can only supply a mixture of vapor and liquid fuel to the carburetor. Because very little liquid fuel is being delivered during an acceleration after a hot soak, the fuel level drops, causing leaning. In fact, the bowl may be nearly emptied, partially exposing the jets. When the jets are partially exposed, the carburetor cannot meter a correct air/fuel mixture because the jets are designed to work with liquid fuel—not a combination of liquid and vapor.

This condition is called vapor lock. Rochester designers combat this problem by isolating the fuel-metering

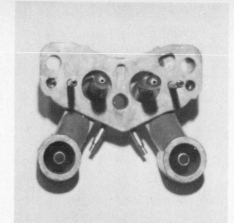

The two-barrel cluster tends to keep some of the metering components cooler. It includes the emulsion tubes, idle tubes, main and idle air bleeds and the main nozzles. A thick gasket is often used between the cluster and the carburetor body so as to insulate the cluster from some of the heat retained in the carburetor.

There's nothing very mysterious about a fuel-cooling can (cool can). Fuel passing through the coiled line is cooled by ice or dry ice and alcohol in the can. Fuel temperature is lowered to ensure that carburetor/s receive liquid fuel. Cold fuel under pressure is not likely to flash into vapor when it enters the carburetor and sees only atmospheric pressure.

components from hot metal areas with good gasket materials used as heat barriers. In the case of the 2G, 4G and H series carburetors the idle tubes, idle-fuel channels, nozzles and main fuel aspirator channels are contained in a *cluster* unit. These are held in place on a gasket by retention screws. The Monojet and Quadrajet have a thick insulating gasket between the throttle body and fuel bowl to prevent excessive heat transfer to the metering orifices and channels. In some cases the throttle-body-to-manifold gasket is extra thick so less heat will be transferred from the engine into the carburetor.

In extreme cases, fuel lines may have to be rerouted to keep them away from areas of extreme heat, such as the exhaust system. If the lines cannot be relocated, it is usually possible to insulate them. This is especially important for high-performance on racing vehicles. Cool cans also help, as do high-performance electric fuel pumps located to *push* fuel to the carburetor.

Formula racing cars are often seen with fuel radiators or coolers.

## DENSITY

In the engine requirements chapter we related volumetric efficiency of the engine to the density of the air/fuel mixture which its cylinders receive. We showed that the higher the density, the higher the HP.

**Pressure bleeds**—GM, Ford and AMC cars typically include a bleed in the fuel pump to allow pressure to bleed back toward the tank during a hot soak or whenever the engine is stopped. Some of the cars also have a fuel-return line leading from the carburetor fuel inlet or the filter to the tank. The bleed and/or return line reduce the percolation tendency. Chrysler products do not usually have either a bleed or a return line and as a result they suffer dramatically from hard starting due to percolation. Pressure in the fuel line to the carburetor may reach as high as 18 psi in these cars.

It is easy to calculate relative air density from barometer and temperature readings, but many tuners prefer the K & D Air Density Meter as a guide for selecting the correct main jet according to air-density changes. Air density changes from hour to hour—day to day— and most certainly from one week to the next. The meter is available from K & D Accessories, P.O. Box 276-R, Longview, WA 98632.

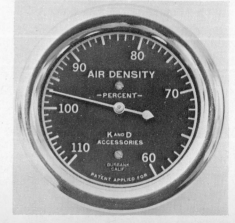

The density of the mixture depends on atmospheric pressure which varies with altitude, temperature and weather conditions. And, the mixture density is also affected by the intake-system layout. Density is increased when the carburetor is supplied with cool air and when the intake manifold is not heated. Density is reduced if the inlet air is heated or if the air/fuel mixture delivered by the carburetor is heated as it enters the manifold. Further reductions in density occur as the mixture picks up heat from the manifold and cylinder-head passages, hot valves, cylinder walls and piston heads.

As with almost all other engine variables, there are tradeoffs. Warming the mixture reduces its density—but also improves distribution, especially at part throttle.

Although density has an effect on carburetor capacity, its greatest effect is on mixture. Because the *major* density changes which occur are due to altitude changes, let's consider the effects of driving from dockside at Los Angeles to the 5000-foot altitude of Denver. The flow and pressure-difference expressions look like this:

$$Q \sim A \sqrt{\frac{\Delta_p}{\gamma}} \quad \text{or} \quad \Delta_p \sim \left(\frac{Q}{A}\right)^2 \times \gamma$$

where

$Q$ = volume flow

$\gamma$ = density

$\Delta_p$ = pressure difference

$A$ = carburetor-venturi area

Regardless of density, the volume taken in remains the same at a constant RPM. But, from sea level to 5000 feet, the density reduces by 14%. The pressure difference $\Delta_p$ also reduces by 14%, giving a reduced pressure drop so that the carburetor acts as if it were *larger*. But, because A is part of a squared function, the apparent increase in A due to density is $\sqrt{14\%} = 3.7\%$ increase. Thus, if the main system nozzle started flowing at 1000 RPM at sea level, the nozzle would start at 1037

RPM at the 5000-foot altitude. The change in carburetor capacity (air flow) is really quite minor and can be neglected as being of any importance when choosing a carburetor for use at higher altitude.

But, the change which occurs in mixture cannot be neglected. Although the density of the air which is entering the carburetor is reduced, the fuel density remains the same and A/F increases with altitude, as shown in the following equation.

$$A/F \sim \frac{Area_v \times \sqrt{\Delta_p \times \gamma_a}}{Area_{mj} \times \sqrt{\Delta_p \times \gamma_f}}$$

where

$Area_{mj}$ = main-jet area

$Area_v$ = venturi area

$\Delta_p$ = pressure difference

$\gamma_f$ = fuel density

$\gamma_a$ = air density

The standard rule of thumb is to reduce the jet size by one percent flow for each 1000-foot altitude increase. Because there is 1.5 to 2.0% fuel-flow change between jet sizes ranging from 0.040 to 0.080 inch, it would require approximately 0.002 inch smaller jets to keep equivalent air/fuel ratios at the 5000-foot altitude.

## COMPRESSION RATIO

High compression improves engine performance by increasing the burning rate of the air/fuel mixture so that peak pressures and peak torque can approach the maximum of which the engine is capable. High compression also increases emissions of hydrocarbons and oxides of nitrogen.

Until about 1970, high-compression engines with up to 11:1 c.r. were available in high-performance cars. By 1971, manufacturers were reducing compression ratios and by 1972, most cars had no more than 8.5 or 8:1 c.r. By reducing compression, the burning rate of the air/fuel mixture is slowed so that peak pressures which enhance the formation of $NO_x$ cannot be reached. Reduced compression also increases the amount of heat which is transferred into the cylinder walls because more burning is still going on as the piston is descending, thereby raising

the exhaust temperature.

Lowering the compression reduces HC emissions by reducing the surface-to-volume ratio of the combustion chamber. The greater the surface-to-volume ratio, the more surface cooling which occurs, thereby increasing HC concentration.

The use of low compression increases the air/fuel requirement at idle because the greater amount of residual exhaust gas remains in the clearance volume and combustion chamber when the intake valve opens, causing excessive dilution of the mixture. This can cause off-idle drivability problems. In effect, low compression provides a degree of exhaust-gas recirculation (EGR) without plumbing or hardware as described in the emissions chapter.

Raising or lowering the compression ratio of an engine does not normally affect the main-system fuel requirements, so jet changes are not usually required when such modifications are made. Raising compression may require slightly less ignition advance in some cases.

Compression ratio and octane requirement are closely related. As compression is increased, octane must also be increased (a higher-octane fuel used) to avoid detonation and preignition—often referred to as "knock." Similarly, lowering an engine's compression reduces its octane requirement.

After World War II, engines were designed and produced with high-compression designs to obtain higher efficiencies. By 1969, some had been available with 11 to 1 compression! Then the need for reduced emissions began to be approached on the basis of getting ready for the future.

First, the auto makers asked the fuel companies to start "getting the lead out" in preparation for the coming emission equipment which will not be able to tolerate lead in the exhaust.

Engineers are working to create engines which will operate on lead-free gasoline. Tetraethyl lead (Ethyl compound) is one of the most commonly used anti-knock additives employed as an octane-increaser in gasolines. Lead compounds are being studied for their effects on plant life, the atmosphere and humans—and there is a serious move to rid these from exhaust on an ecological basis. But, this is not the real reason that engineers want to eliminate lead from fuel, as previously mentioned. The expensive catalyst used in catalytic mufflers—expected to be standard parts by 1975-76—is literally destroyed when contaminated with lead and lead by-products. Thus, the automotive engineers would like to ensure that there will be little likelihood of the users being able to buy gasoline which could wreck the catalytic mufflers when they are used. These and other emission devices are discussed in the emission chapter.

In 1971, engines with lower compression ratios became standard items in all lines. And, by 1972, most engines were being made with only 8 to 1 c.r. The gasoline refineries are turning out a lower octane product and the trend is clearly to still lower octanes. Where 100+ octane gasolines were commonplace in the late sixties, 90-octane is expected to be the standard by 1975 unless a new breakthrough in emission hardware is discovered.

## EXHAUST BACK PRESSURE

Exhaust back pressure has very little effect on carburetion at low speeds. For high performance, low back pressure is desired. The effects of exhaust restriction increase with the square of the RPM. If the RPM doubles, back pressure will increase four times, etc.

## VAPOR LOCK AND GASOLINE

High performance and vapor-lock conditions should never be allowed to integrate. Use of the correct fuels can prevent this. The fuel blend you use is often not of your own choosing and yet can cause a good engine to run very sour. I think it's important to pass comment on this subject.

The term *vapor lock* refers to a leaning-out condition caused by vaporous fuel. Carburetor metering orifices are designed to pass required amounts of "solid" or liquid fuel. They will not pass enough highly vaporized fuel to run an engine properly. A vapor-lock condition is evidenced by anything from a slight sag to a complete stall.

Much time is spent each year by auto manufacturers making each model and engine combination meet certain performance standards. Vapor-lock tests run in the heat of the Arizona desert determine correct fuel-inlet needle-seat size, fuel-line size and routing, fuel-pump requirements, metering, etc. So you will bet-

ter understand vapor lock and how to combat it, explanations of test procedures and fuel characteristics follow.

## VOLATILITY

The words *volatility* and *vapor pressure* encompass explanation of the vapor characteristics of gasoline which sometimes cause annoying disturbances in vehicle operation. Hopefully, these explanations will assist you in future problem diagnosis.

In domestic and most foreign-car internal-combustion engines, gasoline is metered in liquid form through a carburetor where it is mixed with air and vaporized before being routed to the cylinders. *Volatility*—the tendency to evaporate or change from a liquid to a gaseous state—is important. Motor fuels must have this characteristic. A gasoline that does not vaporize readily may cause hard starting and poor warm-up. Unequal distribution of fuel to cylinders, poor economy, high emissions and poor performance are also by-products of inadequately vaporized mixtures.

Too volatile a gasoline may boil in fuel lines, fuel pumps and/or carburetors, causing a decrease in fuel flow to the engine. This results in rough operation, poor performance and even stalling—*vapor lock*. Selection of the right volatility requirements for the application and ambient conditions can minimize these problems.

There is no conceivable way in the operation of today's automobile to avoid occasional seasonal or geographical problems. During winter or high-altitude cold areas, a gasoline of volatility suitable for satisfactory starting and running will be very susceptible to vapor lock during a sudden warm spell or if the car is driven from a mountain area to the hot desert. Flagstaff, Arizona has 7000-foot altitude and bitter-cold winters. Phoenix, less than 150 miles away, is below 1500-foot altitude and often has 70°–90°F. winter weather. Fuel blended by petroleum companies for one area or season will not work well in opposite environments.

Volatility limits are established in terms of vapor pressure and distillation-test results. Fuel companies blend more than one type gasoline as well as seasonal grades for each type. These grades differ with respect to the vaporization tendencies.

## VAPOR-PRESSURE MEASUREMENTS

The Reid Vapor Pressure (RVP) chart shown here will aid you in capturing the concept of seasonal-fuel variations around the nation. If you live in cold country, the winter RVP numbers will be high. This can "give you fits" during spring warm spells.

*Vapor pressure* is defined as the force exerted on the walls of a sealed container by the vaporized portion of a liquid at a certain temperature. On the other hand, it is the force which must be exerted on the liquid to prevent it from vaporizing further. Vapor pressure increases with temperature for any given gasoline. The RVP is measured at a temperature of 100°F. in the presence of a volume of air four times that of the liquid gasoline. It is expressed as pounds per square inch absolute.

The larger the RVP number, the more volatile the fuel.

## DISTILLATION

The main body or total hydrocarbon content of gasoline is measured in terms of volatility in a test called distillation. This test determines temperature at which certain portions of the fuel are evaporated under closely defined conditions.

In summary, it can be said these distillation and vapor-pressure characteristics define and control starting, warm-up, carburetor icing, acceleration, vapor lock, crankcase dilution, and to some extent fuel economy. Fuel companies understand the needs of the motorist and make considerable effort to meet them year around—across the nation.

As with any major task there are compromises and when a vehicle is driven from mountain cold to desert heat or Indian summer hits a winter area, such conditions exceed the latitudes of protection and problems result. The unaware motorist will most likely blame the carburetor.

With this understanding of gasoline, let's take a brief look at tests which auto manufacturers perform to determine performance and drivability levels of their vehicles.

## VAPOR-LOCK TESTS

In addition to all the laboratory tests run on components and total vehicles, there are the seasonal and

# REID VAPOR PRESSURE

## 9 RVP FUEL

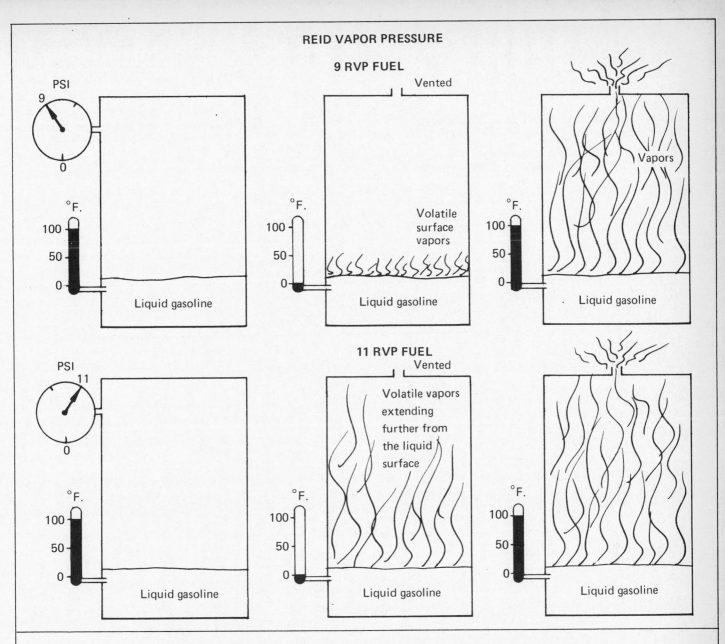

## 11 RVP FUEL

## TYPICAL RVP VARIANCES OF FUEL AROUND THE UNITED STATES

| City or Area | Summer | Winter |  | Summer | Winter |  | Summer | Winter |
|---|---|---|---|---|---|---|---|---|
| Albuquerque | 7.3 | 10.0 | Des Moines | 8.8 | 11.9 | New York | 9.4 | 13.3 |
| Amarillo | 7.0 | 10.8 | Detroit | 9.6 | 13.0 | Northwest Pennsylvania | 9.2 | 13.2 |
| Atlanta | 9.1 | 12.2 | El Paso | 6.8 | 9.8 | Oklahoma City | 8.6 | 11.5 |
| Bakersfield | 8.2 | 10.3 | Fargo | 9.0 | 11.2 | Omaha | 9.5 | 12.2 |
| Baltimore | 9.0 | 12.6 | Hartford | 9.6 | 13.2 | Philadelphia | 9.1 | 12.9 |
| Billings | 8.3 | 11.8 | Houston | 8.8 | 11.4 | Phoenix | 7.4 | 10.6 |
| Birmingham | 8.9 | 11.5 | Indianapolis | 9.1 | 12.5 | Pittsburgh | 9.1 | 13.1 |
| Boston | 9.2 | 13.1 | Jackson | 8.9 | 11.7 | Portland, Oregon | 9.4 | 11.6 |
| Buffalo | 9.7 | 12.2 | Jacksonville | 9.0 | 10.9 | Richmond | 9.5 | 11.9 |
| Casper | 7.6 | 10.6 | Kansas City | 8.7 | 12.5 | Salt Lake City | 7.5 | 11.9 |
| Central Michigan | 9.5 | 12.7 | Little Rock | 8.7 | 11.8 | San Antonio | 8.3 | 10.8 |
| Charlotte | 9.2 | 11.6 | Los Angeles | 9.1 | 10.7 | San Francisco | 9.2 | 11.2 |
| Chicago | 9.0 | 12.7 | Louisville | 8.6 | 12.3 | Seattle | 9.0 | 11.9 |
| Cincinnati | 9.3 | 12.5 | Memphis | 8.2 | 11.4 | Shreveport | 8.8 | 11.2 |
| Cleveland | 9.9 | 12.9 | Midland, Texas | 7.4 | 9.8 | Spokane | 8.6 | 11.5 |
| Corpus Christi | 8.4 | 10.8 | Milwaukee | 9.4 | 12.7 | St. Louis | 8.9 | 12.5 |
| Dallas/Fort Worth | 8.0 | 10.6 | Minneapolis/St. Paul | 9.1 | 12.2 | Toledo | 9.4 | 13.4 |
| Davenport | 9.3 | 12.3 | Newark | 9.1 | 13.1 | Tulsa | 9.0 | 12.2 |
| Denver | 7.7 | 10.3 | New Orleans | 9.0 | 11.3 | Wichita | 8.2 | 11.9 |

geographical test bogies (criteria) each GM vehicle model must meet before production release. A base run is made first with seven-pound fuel (RVP) in a qualified* production car. A base run consists of three WOT accelerations from a standing start, to one mile. ET's are taken at 1/4, 1/2 and 1 mile on each run. The times from each segment (1/4, 1/2, 1) are averaged and these average times represent the capability of the vehicle or a 100% performance factor. This is the base-line bogy.**

After obtaining their base-line bogy on a given vehicle, the special tester will prepare for the test fuel runs. The manufacturers require all vehicles to run a minimum of 90% of their potential performance with 10 RVP at 100°F. To meet the hot weather requirements these tests are performed during the sizzling months of June, July and August in the desert.

An ideal test condition would be a 100°F. day with no clouds. A steady sun load is also part of the weather requirement. With this, a test driver drains the fuel tank completely and installs 10 RVP fuel.

With all test fuels the accelerations follow two simulated hot-engine conditions: (1) 15-minute hot soak, (2) five-minute hot idle.

After temperatures are stabilized throughout the vehicle by a prescribed warm-up it is parked in the

soak shed for a 15-minute hot soak. This special shed has a garage-type door in each end for easy entry and exit. It has no roof so the car hot soaks with full sun load and no cooling benefit from possible breezes. When this simulated 15-minute "coffee stop" is ended the engine is started. It is run a few seconds and put into gear as the shed door is opened electrically. The accelerator is pushed WOT simultaneous with starting the watches. Times for the 1/4, 1/2 and 1 mile are recorded with no added accessories operating (such as air conditioning). If the time to the 1/4 is identical to the base run, the performance is 100% (possible but not likely). Let's say there was some vapor lock between the 1/4 and 1/2 markers. This will reflect a longer time. Even when the driver doesn't note surge-sags, etc. longer times will reveal minor vapor lock. Have you mysteriously lost fractions of seconds at a strip and wondered why? I hope you haven't spent a lot of money trying to get the quick time back when a new batch of low-RVP fuel would have done it. Or, if the engine is being run in a competition application, a fuel-cooling can or fuel radiator.

Getting back to the loss of performance from the 1/4 to the 1/2. Let's say base time was 40 seconds and the test fuel time was 43.6.

$$\frac{40.0}{43.6} = 91.7\%$$

That would be considered acceptable with a 91.7% performance. Had the test fuel run time been 45.8 seconds, the performance factor would have been 87.03%. Unacceptable.

This latter vehicle would go into a development program until all portions of the acceleration were 90% or better.

Each vehicle has to meet the same bogy following stabilization and a five-minute idle. Most cars exceed the 90% requirement before they are released to the public. Many will be up around 98%.

I pointed this out because an original-equipment manufacturer goes to a lot of trouble to make vehicles drivable year around. Be aware that fuel can cause many problems that will be blamed on the carburetor. Also remember when you install off-brand components from high-volume marketers, they will not be thoroughly tested. If you are a competitor, settling for anything less than 100% in every run will likely cause you to slip out of contention.

Incidentally, the use of air conditioning greatly aggravates vapor-lock tendencies by markedly increasing the under-hood and fuel temperatures. This is of no importance to the racer, but worth remembering for ordinary driving.

---

*Built and tuned exactly to the production blueprints.
**"Bogy" means a reference point or specification.

# The Carburetor & Performance

**T**his chapter will introduce you to the various happenings and conditions which affect the ultimate performance which can be obtained from any engine and carburetor combination.

It is difficult to separate this chapter's contents from those of the High-Performance Tuning chapter which follows because most of the items and details are completely interrelated. It is tough to talk about one without discussing the other.

Study this chapter thoroughly *before* you proceed to the nuts and bolts instructions in chapter fourteen.

## CARBURETOR RESTRICTION

Carburetors are tested at a given pressure drop at Wide-Open Throttle (WOT) to obtain a CFM rating indicative of flow capacity. One- and two-barrel carburetors are tested at 3.0-inch Hg pressure drop. Four-barrel carburetors are tested at 1.5-inch Hg pressure drop.

When you want to make comparisons, use these formulas:

$$\text{Equivalent flow at 1.5 in. Hg} = \frac{\text{CFM at 3.0 in. Hg}}{1.414}$$

$$\text{Equivalent flow at 3.0 in. Hg} = \text{CFM at 1.5 in. Hg} \times 1.414$$

The one- and two-barrel rating was adopted because low-performance engines typically showed WOT manifold-vacuum reading of 3.0-inches Hg. When four-barrel carburetors and high-performance engines became commonplace they were rated at 1.5-inches Hg pressure drop because of *two* reasons. First, this rating was close to the WOT manifold vacuum being seen in these engines. Secondly, although we've never seen this mentioned anywhere else, most of the carburetor-testing equipment had been designed for smaller carburetors. The pump capacity on this expensive test equipment was not adequate

NHRA Super-Stock Chevy with Q-jet on Edelbrock Tarantula manifold. Arrows indicate water connections for Stahl Associates warm-up water circulation from push vehicle. Note in-line fuel filter. Cool can is used but not shown in this photo.

to provide 3.0-inches Hg pressure drop through larger carburetors. 1.5-inch Hg was about the limit of the pump capacity. Hence the 1.5-inch rating came about as a "happy accident."

For maximum output, it is essential to have the carburetor as large as possible—*consistent with the required operating (driving) range.* The driving range must be considered as discussed in the chapter on selection and installation.

Using a lot of flow capacity—more than calculations indicate necessary for the engine—can reduce inlet-system restriction and increase volumetric efficiency at WOT and very high RPM. Such carburetion arrangements can compete with fuel injection in terms of performance because the restrictions are minor. However, it should be noted that the dual-quad installations typically used by professional drag racers are not capable of providing usable low- or mid-range performance. These engines are typically operated in a very narrow range of 6000 to 8500 RPM or so. Such installations have not usually used Rochester carburetors.

## AIR CLEANERS

Because every engine needs an air cleaner to keep down expensive cylinder wear which is caused by dust, it makes sense to use one which will not restrict the

carburetor's air-flow capabilities. By avoiding restrictions in the air cleaner, you allow your engine to develop full power.

The only time that an engine might possibly be run without an air cleaner would be on drag-car or drag-boat engines being operated where there is no dust in the air or pit—a very unlikely situation, to say the least. Even then, when an air cleaner is removed from the carburetor, the air-cleaner base should be retained because its shape ensures an efficient entry path for the incoming air and helps to keep the incoming air from being heated by the engine.

The accompanying table shows the results of tests which dispel some of the common fallacies about air cleaners and their capabilities. In general, a tall, open-element air cleaner provides the least restriction. It also increases air-inlet noise. Note that some of the air cleaners will allow full air-flow capability. These are the air cleaners which should be used by racers, even if their use requires adding a hood "bump." An air cleaner which gives full flow capability to the carburetor provides impressive top-end power improvements, as compared to one which restricts flow. For instance, the use of two high-performance Chevrolet air cleaners stacked together—instead of one open-

Dual snorkels or inlets are usually one of the trademarks of a high-performance vehicle as it comes from the factory. One above is a 1972 Pontiac GTO. Chevrolet Z-28 high-performance air cleaner is one of the least restrictive single-element units available today.

element cleaner—improved a 1969 Trans Am Camaro's lap times at Donnybrook, Minnesota by one full second.

It is very important to check the clearance between the upper lid of the air cleaner and the top of the carburetor's pitot or vent tubes. Air cleaner elements vary as much as 1/8 inch in height due to production tolerances. The shorter elements can place the lid too close to the pitot tubes so that correct bowl reference pressures are not developed. Whether the pitot tubes are angled or flat on top, there should always be at least 3/8-inch minimum clearance between the tip of the pitot tube and the underside of the air-cleaner lid.

You have noticed the long "snorkel" intakes on modern air cleaners. These are put there to reduce intake noise—not to improve performance. High-HP engines nearly always have two snorkels for more air and perhaps for "more image," too. For competition, the snorkel can be removed where it joins the cleaner housing. Additional holes can be cut into the cleaner housing to approximate an open-element configuration to improve breathing. Or, it is sometimes possible to expose more of the element surface by inverting the cleaner top. As shown in the accompanying table, an open-element design is least restrictive.

When it comes time to race, use a clean filter element. Keep the air cleaner base on the carburetor if you possibly can, even if you have removed the air cleaner

cover and element. But, be sure to secure the base so that it cannot vibrate off to strike the fan, radiator or distributor.

Because the carburetor is internally balanced, that is, the vents are located in the air horn area, no jet change is usually required when the air cleaner is removed.

## VELOCITY STACKS

Velocity stacks are often seen on racing engines. These can improve cylinder filling (charging) to a certain extent, depending on a great many other factors. For instance, when velocity stacks are used on carburetors on an isolated-runner manifold, the stacks may form part of a tuned length for the air column.

Velocity stacks also provide a straightening effect to the entering air. And, they can contain fuel standoff—which is typical with isolated-runner-design manifolds. Remember that the velocity stacks need space above them to allow air to enter smoothly. Mounting a hood or air-box structure too close to the top of the stacks will reduce air flow into the carburetor. Two inches should be considered a bare minimum clearance between the top of a velocity stack and any structure over it.

When an air cleaner is used on a carburetor equipped with stacks, keep the two-inch recommended clearance between the top of the stacks and the underside of the air-cleaner lid. This may require using two open-element air cleaners "glued" together with Silastic RTV or other sealant—and a longer stud between the carburetor top and the cleaner lid.

Velocity stacks have not been popular with Quadrajet installations to date. This can change as Quadrajets become more popular in performance applications.

## DISTRIBUTION CHECKING

Distribution checking to determine whether all cylinders are receiving an equal mixture becomes ex-

---

### AIR CLEANER COMPARISON

| Cleaner Type | WOT Air Flow (CFM) |
|---|---|
| None | 713 |
| Chevrolet 396 closed-element with single snorkel | 480 |
| Chevrolet 396 closed-element with single snorkel cut off at housing | 515 |
| Same as above, but with two elements | 690 |
| Chevrolet high-performance open-element unit | 675 |
| Same as above, but with two elements | 713 |
| 14-inch diameter open-element accessory-type air cleaner | 675 |
| Chevrolet truck-type element (tall) used with accessory-type base & lid | 713 |
| Foam-type cleaner (domed flat-funnel type) | 675 |

NOTE: All data obtained with same carburetor. New, clean paper elements used in all cases except the foam type, which also was new.

Holley Tests
October 1971

tremely important when a manifold or carburetor change is made. Even though a previous carburetor/manifold combination may have provided nearly perfect distribution you cannot take the chance that one or more cylinders will be running lean. Although several ways of checking distribution are detailed in the distribution section of the engine requirements chapter, the man at the race track usually has only one way to do it. He must rely on the appearance and color of the plug electrodes and porcelains. He can also observe the color of the piston tops with an inspection light. These can provide a lot of valuable information about what is happening in the engine.

An accompanying plug-color chart explains some of the things to look for. Because the part of most interest is the base of the porcelain—which is "buried" in the plug shell—a magnifier-type illuminated viewer such as AC's should be an early purchase for your tool box.

Before continuing, we need to mention that checking plug color gives only a rough idea of what is occurring in the way of mixture ratio and distribution.

From the list of items which cause plug-appearance variations (later in this section), it is easy to see that using plug color to check distribution will only be helpful when the engine is in good condition. Engine condition can be checked quickly with a compression gage to make sure that all cylinders provide equal compression at cranking speed. Or, for a more accurate check, a leak-down test can be used to compare cylinder condition. Details on building a low-cost leak-down tester were provided in *Car Craft* magazine's September 1973 issue, page 58.

Remember that new plugs take time to "color"—as many as three or four drag-strip runs may be needed to "color" new plugs. Plug color is only meaningful when the engine is declutched and "cut clean" at the end of a high-speed full-throttle, high-gear run. If you allow the engine to slow with the engine still running, plug appearance will be meaningless. Plug readings can be made after full-throttle runs on a chassis dyno with the transmission in an intermediate gear so that the dyno is not overspeeded—but road tests require high gear to load the engine correctly. Similarly, plug checks can be made where the

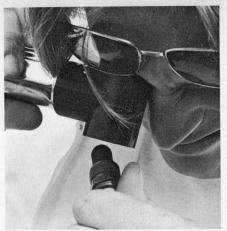

Color and appearance of plug porcelain can be checked with an illuminated viewer with built-in magnifying glass. One of these should be in every tuner's tool kit. Engine must be "cut clean" by turning off the ignition and declutching or getting into neutral at the conclusion of a wide-open-throttle full-power pass in top gear. If this is not done, plug-color readings will be meaningless.

## PLUG COLOR CHART

**Rich**—Sooty or wet plug bases, dark exhaust valves.

**Correct**—Light-brown (tan) color on porcelains, exhaust valves red-brown clay color. Plug bases slightly sooty (leaves a slight soot mark when plug base is turned against palm of hand). New plugs start to color at base of porcelain inside shell and this can only be seen with an illuminated magnifying viewer for plug checking. For drag engines with wedge (quench-type) combustion chambers, speed and elapsed time become more important than the plug color. Plugs may remain bone-white with best speeds/times. A mixture which gives the light-brown color may be too rich for these engines.

**Lean**—Plug base ash-grey. Glazed-brown appearance of porcelains may also indicate too-hot plug heat range. Exhaust valves whitish color.

NOTE: Piston-top color observed with an inspection light through the plug hole can be a quicker and sometimes more positive indicator of mixture than plug appearance. Careful tuners look at *all* indicators to take advantage of every possible clue to how the engine is working.

engine has been running at full-throttle against full load applied by an engine dyno. It is much easier to get good plug readings on the dyno because full power can be applied and the engine cut clean without difficulty. Plugs can be read quickly because you can get to them easier than in the usual car or boat installation. However, don't think that plug heat range and carburetor jetting established on an engine dyno will be absolutely right for the same engine installed in your racing boat or chassis. Airflow conditions past the carburetor can easily change the requirements—perhaps so unevenly that different cylinders will need different changes.

It would be nice if every plug removed from an engine looked like the others from the same engine—in color and condition—but this is seldom ever achieved! Color and other differences indicate combustion-chamber temperatures and/or air/fuel ratios are not the same in every cylinder—or that related engine components need attention. The problem is greatly complicated in engines where there is a great difference between the cylinders in turbulence and efficiency. The big-block Chevrolet is a notable example as described in the H. P. Books *How to Hotrod Big-Block Chevys.*

If differences exist in the firing end of the plugs when you examine them, the cause may be due to one or more factors: unequal distribution of the air/fuel mixture, unequal valve timing (due to incorrect lash or a worn cam), poor oil control (rings or excess clearance). Problems within the ignition system which can also lead to plugs not reading the same or misfiring include: loose point plate, arcing in the distributor cap, defective rotor, cap or plug wires/connectors, cross fire between plug wires, defective primary wire or even a resistor which opens intermittently. Pay special attention to cleanliness of the entire ignition system including the inside and outside of the distributor cap and the outside of the coil tower. Also, clean the inside of the coil and cap cable receptacles. Any dirt or grease here can allow some or all of the spark energy to leak away.

If the cylinders have equal compression and the valves are lashed correctly, a difference in plug appearance may indicate that there is a mixture-distribution problem. It is sometimes possible to remedy this with main-jet changes. For instance, if one or more plugs show a lean condition, install larger main jets in the throttle bore/s feeding those cylinders. Should all of the plugs appear to show a rich condition, install smaller main jets in the throttle bore/s feeding those cylinders. The real problem occurs when several cylinders fed from the *same* throttle bore show different mixture conditions—some lean with some correct—or some rich with some correct—or perhaps a combination of all three conditions! This shows a manifold fault which cannot be corrected with jet changes. Correcting such conditions requires manifold rework which is beyond the scope of this book.

## COLD AIR & DENSITY

Density of the mixture has been thoroughly discussed in the engine-requirements chapter where we showed that higher density inlet air improves the engine's power capability proportionately to the density increase. Improving density with the use of cool inlet air and a cool inlet system was also mentioned in the previous chapter on engine variables. So, let's examine the practical aspects. What can you do to keep the density "up" to get the best HP from your engine?

First, the under-hood temperature is not ideal for HP production because— even on a reasonably cool day—air reaching the carburetor inlet has been warmed by passing through the radiator and over the hot engine components. Under-hood temperatures soar to 175°F. or higher when the engine is turned off and the car is standing in the sun. An engine ingesting warm air will be down on power by more than you might imagine. Assume that the outside (ambient) air temperature is 70°F. and the under-hood temperature is 150°F. Use the following equation:

$$\gamma_{oa} = \frac{460 + t_{uh}}{460 + t_{oa}} \times \gamma_{uha}$$

$$\gamma_{oa} = \frac{460 + 150}{460 + 70} \times \gamma_{uha}$$

$$\gamma_{oa} = 1.15 \, \gamma_{uha}$$

where

$\gamma_{oa}$ = outside air density

$\gamma_{uha}$ = under-hood air density

$t_{oa}$ = outside air temperature

$t_{uh}$ = under-hood air temperature

In the example here, outside-air density is 115% of the under-hood-air density—or 15% greater. The available HP with outside air will be 115% of that where the carburetor is fed underhood air. If the engine produces 300 HP with 150°F. air-inlet temperature, it can be expected to produce 345 HP with 70°F. air-inlet temperature, provided the carburetor meters the correct amount of fuel.

Cold air gives more improvement than ram air because 1% HP increase is gained for about 5°F. drop in temperature. This assumes the mixture is adjusted to compensate for the density change and that there is no detonation or other problems. Using outside air instead of under-hood air is climate-limited because too-cold temperatures may lead to carburetor icing.

If air scoops are being used to duct air to the carburetor, do not connect the hose or scoop directly to the carburetor. Instead, connect the scoop to a cold-air box or to the air-cleaner housing to avoid creating turbulence of the incoming air as it enters the carburetor air horn. Be sure to retain the air-cleaner element as a diffuser to reduce the turbulence of the air as it enters the carburetor. Otherwise, a difficult-to-diagnose high-speed miss can be expected to occur. More details on cold-air boxes are provided in the ram air section later in this chapter.

If you plan to use a cold-air kit which picks up cold air at the front of the car, be prepared to change the air-cleaner filter element at regular intervals—perhaps as often as once a week in dustier areas. If you leave the filter out of the system, plan on new rings or a rebore job in the near future because your engine will quickly wear out. You'll be better off ducting cold air from the cowl just ahead of the windshield. This is a high-pressure area which will ensure a supply of cool outside air to the carburetor. That area still gets airborne dust, but it is several feet off the ground—away from the grimy grit encountered at road level. If you want to keep your car "looking stock," using fresh air from the cowl is another way to get performance without making the car look like a racer.

Cars with stock hoods and stock or near-stock-height manifolds may be equipped with fresh-air ducting to the air cleaner from the plenum chamber across

Constructing your own cold-air package is not difficult. Flexible ducting can be mated to a reworked air-cleaner housing at the carburetor end. The kitchen-utensil department of most any large store will yield a big funnel which can be fitted in place of an existing headlight housing.

---

**Increased density means larger jets**—No matter how a density increase is obtained— whether by increased atmospheric pressure, a cold manifold or a cooler inlet-air temperature—the increase must be accompanied by the use of larger main jets. The size increase (in area) is directly proportional to the square root of the density increase (in percent).

---

the cowl just ahead of the base of the windshield by using parts from GM cars such as some of the high-performance Chevrolets.

Or, you may prefer to use one of the fresh-air hoods which have been offered on some models. These typically mate a scoop structure on the hood with the air-cleaner tray on the carburetor/s. Thus, the under-hood air is effectively prevented from entering the carburetor.

If a tall manifold such as an IR or plenum-ram type is being used, the hood will have to be cut for clearance and a scoop added to cover the carburetor/s. Whether the scoop is open

Smokey Yunick with a very smooth Camaro racing machine. Carburetor on prototype Smoke Ram manifold is fed cool air under pressure from a source ahead of the radiator. Note special duct under hood. It mates with an air-cleaner base housing a tall truck-type element which is not restrictive.

at the front or back depends on the air flow over the car. The optimum entry for air into the scoop may have to be determined by testing. An optimum entry will provide an air supply which is above atmospheric pressure and is non-turbulent. In general, the area of the scoop opening should be approximately 12% larger than the area of the carburetor venturis. The roof of the scoop should be positioned 1-1/2 inch above the carburetor inlet. Any more clearance may create detrimental turbulence—any less will restrict air flow into the carburetor.

Like the fresh-air hoods, scoop-equipped hoods can be mated with a tray under the carburetor/s to ensure that warm under-hood air cannot enter the carburetors to reduce inlet-air density. Consider using air-cleaner elements as diffusers as has already been mentioned.

Drag racers have been doing a great deal of experimentation on scoops and fresh-air supplies from the cowl area ahead of the windshield. The trend in 1972-73 has been to reduce the entry area or to gradually diminish the cross-sectional area of the scoop or air box from the plenum. This increases the velocity of the air being supplied to the carburetors and may also increase turbulence of the air so that it becomes harder to turn the air smoothly to enter the carburetor.

Homemade cowl-induction system for a Camaro TransAm racer. This one connects air cleaner into plenum chamber usually found at base of windshield — a good high-pressure area to use. The air cleaner element must be used to ensure smoothing the entering air stream, thereby eliminating turbulence which could disrupt metering.

Drag racers will always keep the hood open and avoid running the engine between events so that the engine compartment stays as cool as possible. It is helpful to spray water onto the radiator to help cool that component. This ensures that the engine water temperature will be usable and that the radiator will not heat incoming air any more than is absolutely necessary.

Although "seat-of-the-pants" feel may indicate stronger performance from a cold engine, the real fact of the matter is that the engine coolant temperature should be around 180°F. or so to allow minimum friction inside the engine. The idea is to keep the engine oil and water temperature at operating levels while taking care to keep down the inlet-air temperature.

While we are talking about density improvement by using cold inlet air we should remember that a heated manifold reduces density. The exhaust heat or jacket water to the manifold should be blocked off to create a "cold" manifold when performance is being sought after. There are various ways of accomplishing this. In some instances there will be intake-manifold gaskets available which close off the heat openings. Or, a piece of stainless steel or tin can metal can be cut to block off the opening when slipped between the gasket and manifold. Many competition manifolds have no heat riser and therefore the manifolds are "cold" to start with. Some car makers offer shields which fit under the manifold to prevent heating the oil with the exhaust

Corvette fresh-air hood opens at base of windshield in high-pressure area. Hood structure mates with air-cleaner assembly.

Camaro fresh-air hood takes in high-pressure air at base of windshield. Duct built into hood structure mates with foam-rubber gasket atop air-cleaner assembly. Note how hood opening overhangs the cowl in this installation.

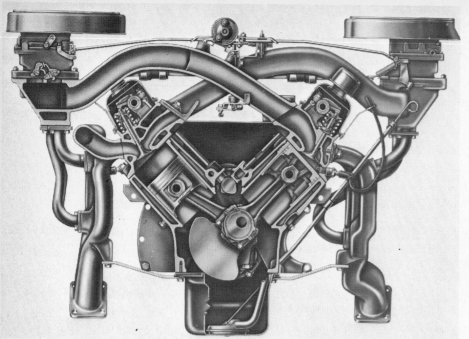

1963 Chrysler 413 CID 300J Ram Induction engine had eight equal-length intake ducts. Each four-cylinder set was fed by one carburetor and an equalizing tube connected the two sets. Duct lengths were selected to give best torque improvement at 2800 RPM— about 10% better than conventional manifolding. The strong torque increase provided a noticeable acceleration improvement over a 1500-RPM range . . . from 50 to 80 MPH. This engine is considered to be a classic example of ram-tuning use in a production car.

heat cross-over passage. These shields also prevent the hot oil from heating a cold manifold—or at least reduce that tendency. This can be a cheap way to gain approximately 10 HP on a small-block V-8 engine.

If the manifold is water-heated it is a simple matter to reroute the hoses so the water does not flow through the manifold.

## RAM TUNING

Ram tuning can give better cylinder filling—and thus improved volumetric efficiency—in a narrow speed range by taking advantage of a combination of engine-design features:

Intake and/or exhaust-system passage or pipe lengths

Valve timing

Velocity of intake and exhaust gases.

Although ram tuning improves torque at one point or narrow RPM band, the improvement tends to be "peaky" so that the power falls off sharply on either side of the peak. It is generally understood that ram tuning is a *resonance* phenomenon. Resonance is

sought at a tuned peak with the knowledge that *the power gained at that point will be offset by a corresponding loss at other speeds.*

Ram tuning can be used to add mid-range torque—as was done on some of the Chrysler six-cylinder and V-8 engines in the early 1960's. These gains are obtained at the expense of top-end power. Or—more usual for high-performance engines—low- and mid-range torque may be sacrificed to take advantage of top-end improvements.

Individual intake-manifold passages for each cylinder (ram or tuned length) can be measured from the intake valve to a carburetor inlet if there is one venturi per cylinder. Or, the length may be measured from the intake valve to the entry to a plenum chamber fed by one or more carburetors. This is a function of manifold design. *The longer the passage length, the lower the RPM at which peak torque will occur.* Although equations have been written to describe where ram-tuning effects will occur, most of these have been oversimplifications

which leave out the effects of manifold-passage size and the sizes of intake ports and valves. Making any of these larger raises the RPM at which best filling occurs. This explains why best drivability and street performance is obtained with small-port manifolds and heads.

Valve timing greatly affects the RPM capabilities of an engine. Cylinder filling can be aided at high RPM by holding the intake valve open past BC (Bottom Center). Consider what happens. At low speeds, holding the valve open past BC allows part of the intake charge to be blown back into the intake manifold as the piston rises on its compression stroke. This reduces manifold vacuum and drastically affects idle and off-idle air/fuel mixture requirements, as described in the engine-variables chapter.

As RPM is increased, faster piston movement creates a greater pressure drop across the carburetor. Air enters the carburetor with higher velocity, giving greater acceleration—and momentum—to the mixture travelling toward the valve.

Thus, as the piston approaches BC on the intake stroke, cylinder pressure is rising toward that at the intake port. And, pressure at the intake port is being increased by air-column momentum in the intake-manifold passage supplying it. Thus, filling improves with RPM until friction losses in the manifold exceed the gain obtained from delayed valve closing.

The past two paragraphs are true, regardless of whether ram tuning is being used or not. Now let's consider what is happening in the manifold passage as the valve is opened and closed. When the piston starts down on its intake stroke, a rarefaction or negative-pressure pulse is reflected to the carburetor inlet. As this pulse leaves the carburetor, atmospheric pressure rushes in as a positive-pressure pulse. When the passage length is optimum for the RPM at which peak torque is sought, the positive-pressure pulse arrives near the time when the valve is closing, assisting in the last part of cylinder filling. Note the interrelationships of passage length, mixture velocity, valve timing and RPM. It's a complex process, to say the least!

The mixture attains a velocity of up to 300 feet per second or more (depending on RPM) as it travels through the port during the intake stroke. Because the mixture has mass (weight), it also has momentum which is useful for aiding cylinder filling when the intake valve is held open after BC (sometimes to 100° past BC) while the piston is rising on the compression stroke. Manifold design enters in at this point because an isolated-runner (IR) system allows much longer delay in closing the intake valve than the usual single- or two-plane manifolds. This is because the IR system supplies only one cylinder per carburetor venturi. There is no other cylinder to consider from the standpoint of mixture dilution, sharing (robbing) or adverse pulsing caused by another cylinder.

When the intake valve shuts, incoming mixture piles up or stagnates at the valve backside, reflecting a compression (positive-pressure) pulse or wave toward the carburetor inlet. As this wave leaves the carburetor inlet, it is followed by a negative-pressure pulse which allows atmospheric pressure to reenter the manifold as a positive-pressure pulse . . . back to the valve. This bouncing or reflective phenomenon repeats several times until the inlet valve again

opens. Pressure at the carburetor inlet varies from positive to negative as the wave bounces back and forth in the inlet passage, but pressure tends to remain positive at the backside of the closed valve . . . except when the valve is open.

At certain engine speeds, the reflection or resonance phenomenon will tend to be in phase (in synchronism) with the intake valve opening and closing so that the positive pulses will tend to "ram" the mixture into the cylinder for improved filling.

Although the pulsations are in phase only at certain speeds (yes, there can be multiple peaks!), the mixture column in the intake passage provides some ram effect at all speeds due to its own inertia.

Best filling is obtained with the intake-system and exhaust-system resonances in phase at the same RPM.

On the exhaust side, exhaust gas enters the pipe at a pressure of 80 psi or higher because the exhaust valve is opened before BC—before the conclusion of the power stroke—while there is still pressure in the cylinder. Thus, the exhaust stroke gets a "head start" so the piston does not have to work so hard in pushing out the exhaust. The exhaust "pulse" starts a pressure wave travelling at the speed of sound (in hot gas) to the end of the system. From the end of the system, a rarefaction or low/negative-pressure pulse is reflected back to the exhaust valve at the same speed. Tuned systems are constructed with lengths to allow this pulse to arrive during the overlap period to ensure complete emptying of the cylinder. The idea here is to reduce exhaust residuals in the clearance volume. This reduces charge dilution and provides more volume for A/F mixture—hence greater volumetric efficiency.

Tuned-exhaust effects on the main-jet requirement vary. If the exhaust causes stronger intake-system pulsing, a smaller main jet could probably be used. Or, if the effect is to lessen pulsing, a larger jet might be required. Predicting these effects is extremely difficult, so tuners tackle each situation on an individual cut-and-try basis.

**Summary of ram tuning**—The peaky results obtained with ram tuning can seriously reduce engine flexibility and this point must not be overlooked. It is all too easy to get over-excited about the spectacular results

obtained from racing motorcycle engines, so let's look at that area briefly before leaving the subject of ram tuning.

Motorcycle engines are typically built as single-cylinder units, each with its own carburetor and exhaust pipe. This allows the designer to take advantage of two very important features: (1) no mixture-distribution problems, and (2) pressure phenomena in the intake and exhaust systems can be relied on for ramming or "supercharging" the cylinder at a desired RPM—usually quite high.

Thus, very-high-output bike engines obtain HP outputs of up to 2.9 HP per cubic inch! However, racing motorcycles with engines of this ilk have 8- to 11-speed gearboxes to allow using the very narrow RPM band in which power is produced. Some of these engines will not produce noticeable power below 6000 RPM!

### RAM AIR? How fast IS your car?

The forward motion of your car cannot be used to induce enough air pressure into the carburetor to provide any appreciable benefits. The idea of adding a mild form of supercharging at low cost by connecting a hose to the carburetor from a scoop on the hood or front of the car is certainly an appealing one. But, unless the vehicle is moving *very fast*, the idea is not as good as using a cold air box without ram effect so the filtered air can flow out of the box feeding the carburetor air inlet. This helps to avoid any disturbance of the carburetor venting. Any ram-air arrangement must pressurize all parts of the carburetor, including the float bowl, to ensure against disturbing the inter-

---

**More about ram tuning**—Complete details on ram-tuned intake systems are in Philip H. Smith's book, *The Scientific Design of Exhaust & Intake Systems.* Perhaps one of the best single articles written on the subject was by Roger Huntington in the July 1960 *Hotrod* magazine, "That Crazy Manifold." July and August 1964 *Hotrod* magazines had two further articles by Dr. Gordon H. Blair, PhD. All are worth reading. However, the use of a dynamometer to measure the real performance of a specially constructed "tuned" system—either intake or exhaust—is absolutely essential.

---

nal pressure balance of the unit. Ram air can provide minor horsepower increases at very high speeds: +1.2% at 100 MPH, +2.7% at 150 MPH and +4.8% at 200 MPH. You'll note ram-air intakes on professional drag race cars and on Formula 5000 single-seater race cars. A careful observer will note small entries to these ram devices, sometimes not much larger than the total venturi area of the carburetors/injectors being used. A lot of engineering is required to get these ram-air arrangements to work just right. Air straighteners are often used in the entry to smooth the air flow so there is less turbulence. Air cleaners may be used on the carburetor/s in some cases to provide smoothing of the air flow just before it makes a 90° turn into the carburetor air horn.

## HEADER EFFECTS vs. JETTING

Headers will usually reduce exhaust back pressure so that the engine's volumetric efficiency is increased (it breathes easier!). The main effects of headers will be seen at wide-open throttle and high RPM. No jet changes will be needed in most instances if the carburetor has been sized correctly for the engine displacement/RPM. The carburetor will automatically compensate for any increased air flow by increasing the fuel flow in the correct proportion. The mixture ratio will not be affected.

The ram-tuning section details how a tuned exhaust system which alters the pulsing seen by the carburetor may make main-jet changes necessary to compensate for any increase/reduction in intake-system pulsing.

**Don't mix jets**—If you tune several cars with different types of Rochester carburetors, remember that different styles of jets are used. Look at the charts on pages 52 and 243. Monojet and Quadrajet carburetors have jets with a square shoulder under the head of the jet. Jets for the **B**, **2G**, **4G** and **H** have a tapered shoulder under the head of the jet.

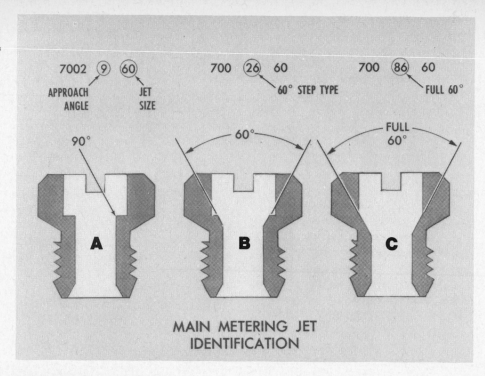

MAIN METERING JET IDENTIFICATION

## MODEL B, 2G, 4G & H MAIN METERING JETS

Three types or groups of main metering jets are used on these models to obtain accurate calibration. Any replacement jet must be of the same major group as the original equipment jet to avoid changing the flow calibration. The group in which each jet belongs can be readily identified by visual examination of the approach angle (nearest to surface with stamped jet number). For example, it is possible to have a jet from each of these groups A, B, C with the number 60 stamped on the jet face: A—7002960; B—702660; C—7008660.

### GROUP A MAIN JETS
#### 90° SQUARE APPROACH (ZINC PLATE)

| Part | Stamped On Jet | Part | Stamped On Jet |
|---|---|---|---|
| 7002943 | 43 | 7002956 | 56 |
| 7002945 | 45 | 7002957 | 57 |
| 7002946 | 46 | 7002958 | 58 |
| 7002947 | 47 | 7002959 | 59 |
| 7002948 | 48 | 7002960 | 60 |
| 7002949 | 49 | 7002961 | 61 |
| 7002950 | 50 | 7002963 | 63 |
| 7002951 | 51 | 7002964 | 64 |
| 7002952 | 52 | 7002965 | 65 |
| 7002953 | 53 | 7002966 | 66 |
| 7002954 | 54 | 7002967 | 67 |
| 7002955 | 55 | 7002969 | 69 |

### GROUP B MAIN JETS
#### 60° APPROACH (SHORT TAPER)

| Part | Stamped On Jet | Part | Stamped On Jet |
|---|---|---|---|
| 7002639 | 39 | 7002651 | 51 |
| 7002640 | 40 | 7002652 | 52 |
| 7002641 | 41 | 7002653 | 53 |
| 7002642 | 42 | 7002654 | 54 |
| 7002643 | 43 | 7002655 | 55 |
| 7002644 | 44 | 7002656 | 56 |
| 7002645 | 45 | 7002657 | 57 |
| 7002646 | 46 | 7002658 | 58 |
| 7002647 | 47 | 7002659 | 59 |
| 7002648 | 48 | 7002660 | 60 |
| 7002649 | 49 | 7028554* | 54 |
| 7002650 | 50 | 7028559* | 59 |

### GROUP C MAIN JETS
#### 60° APPROACH (LONG TAPER)

| Part | Stamped On Jet | Part | Stamped On Jet |
|---|---|---|---|
| 7008660 | 60 | 7008679 | 79 |
| 7008661 | 61 | 7008680 | 80 |
| 7008662 | 62 | 7008681 | 81 |
| 7008663 | 63 | 7008682 | 82 |
| 7008664 | 64 | 7008683 | 83 |
| 7008665 | 65 | 7008684 | 84 |
| 7008666 | 66 | 7008685 | 85 |
| 7008667 | 67 | 7008686 | 86 |
| 7008668 | 68 | 7008687 | 87 |
| 7008669 | 69 | 7008688 | 88 |
| 7008670 | 70 | 7028660* | 60 |
| 7008671 | 71 | 7028661* | 61 |
| 7008672 | 72 | 7028662* | 62 |
| 7008673 | 73 | 7028663* | 63 |
| 7008674 | 74 | 7028665* | 65 |
| 7008675 | 75 | 7028669* | 69 |
| 7008676 | 76 | 7028670* | 70 |
| 7008677 | 77 | 7028673* | 73 |
| 7008678 | 78 | 7028686* | 86 |

*Stainless Steel

## FIRE — A Hazard You Can Minimize

Although the following cautions and warnings concerning fuel and its related hazards may seem too brief, it could not be emphasized any stronger if the entire chapter was devoted to the possible dangers.

Gasoline is one of the most powerful substances known. Pound for pound gasoline has three times as much energy as TNT and ten times the energy of dynamite.

Gasoline vapors can be extremely volatile especially when combined with the circumstances surrounding carburetor work. A hot engine, a float bowl of fuel spewing vapors in the confines of an engine compartment, and maybe a cigarette-smoking friend trying to lend a hand spell disaster, both to the persons involved and possibly to their prized vehicle. A dropped trouble light can break and ignite the vapors, too.

The element of fire is ever present on the race track and in the pits. The heat of competition lends itself to situations which from a safety standpoint are something less than desirable. Number 1, let's practice safety. Number 2, be prepared with an adequate extinguisher. Many lives have been saved and cars salvaged because of the fire extinguishers available on the market today.

Care in installation and observation of what is happening when fuel pressure is applied to the carburetor (watching for flooding, etc.) can go a long way towards eliminating fire. Where a car or a carburetor is being worked on in a closed area containing a flame—such as the pilot light or burner of a water heater or furnace—fire danger is extreme. It just takes one spill to generate sufficient vapor to be carried across the floor to the flame. Then the trouble begins in the form of an explosion or a fire or both. Remember that a carburetor removed from a car contains gasoline and this should be drained before carrying the carburetor inside to work on it. This is especially true when you work in an area which has any kind of open flame.

"Seventy-five percent of all auto fires directly result from leaving off the air cleaner," say M. L. Wikre and D. C. Whiting of the Los Altos, California Fire Department. "Fuel spews out of the carburetor onto the manifold—or standoff collects on the underside of the hood—then a backfire ignites the fuel. The owner usually tells us that he left the air cleaner off because he had been tuning the car." These fire experts point out that the air cleaner prevents a backfire from igniting any stray fuel. The air cleaner itself does not support combustion very well. It's interesting to note that flame arrestors are required on the carburetors of all marine inboard gasoline engines. If a fire starts in the air cleaner, a lot of smoke can be expected, but not much burns if the element is the usual paper type. Nearly all of the auto fires not caused by leaving the air cleaner off seem to be caused by a bad fuel line or connection between the fuel pump and carburetor.

If a fire extinguisher is carried in the car, the fire can usually be put out quickly and with little damage. A 2-1/2-pound ABC dry-chemical extinguisher covers all three classes of fire found in cars. A—upholstery and the interior are common burnables, B refers to gasoline/oil fires, C is an electrical fire.

These extinguishers are small, lightweight and virtually foolproof. Volume for volume, they have much better extinguishing capabilities than carbon-dioxide ($CO_2$). Also, powder retards the fire from flashing up again at hot spots and wiring.

Drawbacks of the dry-chemical extinguisher are that the powder goes *everywhere*, making a mess which must be cleaned up afterward. And, it can go into the engine, especially if there is no air cleaner. If a lot of powder has to be directed into the carburetor air inlet, some will actually get into one or more cylinders through an open intake valve. The piston will compress this into a cake which may prevent the engine from turning when you attempt to restart. Even if you can crank the engine over, the powder is abrasive. So, if very much gets in the engine, pull the cylinder heads and clean out the powder before running the engine again.

$CO_2$ extinguishers are preferred by many because they cannot damage the engine and there is no after-mess to clean up. This is their greatest *plus*. But, because $CO_2$ fights fire by replacing the oxygen, a much larger extinguisher (than dry chemical) is required to match the power of a dry-chemical unit. The fire experts we talked with recommended a 50-pound $CO_2$ unit! In open areas, especially when the wind is blowing, $CO_2$ dissipates very quickly and sometimes will not put out a burning fuel line or wire.

If you have a $CO_2$ and a dry-chemical extinguisher on hand, always use the $CO_2$ first. If that gets the fire out, there's no mess to clean up. For that reason, a large $CO_2$ extinguisher is the thing to have in your garage—*if you can afford it.* $CO_2$ extinguishers require more upkeep than a dry-chemical type and they should be professionally checked and recharged every year.

If you have a fire under the hood, don't throw the hood open because hot gases and flames will rush out. Open the hood just a crack and shoot the extinguishing media in. Better yet, shoot it in from under the engine.

Check everything before restarting the engine or you could start a worse fire—just when your extinguisher is all used up. Check all of the ignition wires and fuel lines to see whether they have burned or melted. Check the fuel line connections to make sure that they have not loosened.

Professional racers are turning to on-board extinguisher systems which pipe pressurized FE1301 Freon to all critical areas. While the first cost of these may seem expensive, the added margin of safety which these give the driver who may have to exit from a flaming vehicle certainly offsets their seemingly high first cost. FE1301 extinguishes about three times better than $CO_2$. The Freon dissipates afterward with no mess and no harm to the engine or other components.

*Blame accidents on whomever you choose, but safety around an auto is mostly up to you.*

*Doug Roe*

215

# High-Performance Tuning

## INTRODUCTION

Did you pick up this book and turn to this performance section after scanning the index? If so, you have some things in common with many car enthusiasts. Many of us scan the front cover and the index page listing the contents of a magazine or book, then buy it or put it back on the rack depending on what it offers. Take this one home because it offers you knowledge in carburetion obtainable in no other volume.

Building a car with new ideas or a completely new approach is a big challenge. The ultimate of satisfaction in this world of automobiles is to take a vehicle, an engine or perhaps a carburetor that has had thumbs turned down on it by many and proceed to make it so great—make it perform so well—that the skeptics can only go off mumbling to themselves with their second-place trophies and memories of last seeing your rear bumper stickers.

This chapter deals primarily with high-performance tips on the Rochester Q-jet, with added information on the 4GC and 2GC. Because the Q-jet is so versatile in its adaptability from one engine to another, it does not take volumes of explanation to get you running at peak performance with nearly any displacement engine you might choose.

Rochester carburetors have put a great number of trophies on the mantel for a few with imagination and the determination to stick with their convictions. RPD, as a company, doesn't splash the pages of hot-rod magazines with big performance ads. They do not sponsor race cars or performance programs. That is their policy by their own choosing, and they just happen to be the world's largest carburetor manufacturer.

RPD is the leader in supplying units for new-car production and without a doubt is the leader in new design ideas. In the area of emissions they are constantly striving to put more exacting mixtures of fuel and air to the cylinder by improved carburetion.

Author Roe with trophies won during the 1969 Annual Hill Climb at Clifton, Arizona with the 1960 Corvair. After winning overall at several other Arizona Hill Climbs from 1965 through 1969, the car/driver combination was referred to as the Arizona Hill Climb Champion. We purposely retained the criticized swing axle-rear suspension. It made winning triple the fun. It's hard to describe the smug pleasure derived from running competitively with many $10,000 to $20,000 cars. The Corvair did well in A Sports Racing class at SCCA events.

Rochester's Quadrajet logo stands proud.

## PERFORMANCE TUNING

Where carburetor calibration changes are discussed in the following sections of this chapter, these are intended FOR COMPETITION ONLY. This means for drag race, off-road, road race and marine applications—anything where the vehicle is NOT TO BE USED ON STREETS OR HIGHWAYS.

Most new RPD carburetors for bolt-on replacement use are calibrated to provide original-equipment specification performance or better when installed without changes. Any modification to the calibrations will probably cause the carburetor/engine combination to be unable to meet exhaust-emission standards/regulations.

John Lingenfelter was holding both ends of NHRA's Super Stock NA Automatic record as we took this book to press late in 1973: 114.79 MPH, 11.97 seconds ET. There's more performance to be had from its Q-jet-plus-Edelbrock-manifold carburetion system.

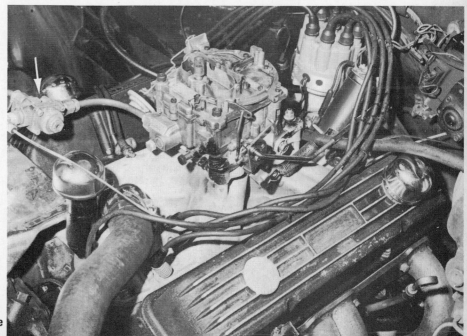

This Quadrajet is adapted to an Edelbrock manifold which sits amid numerous other pieces of performance hardware. Arrow indicates 1/8-inch line to fuel-pressure gage. Note Holley pressure regulator (white arrow).

There is a good chance you have a Rochester carburetor or will have one in the near future. The Q-jet is used almost exclusively on GM V8-powered cars and occasionally is found on other makes such as the 1970 Ford 429-cubic-inch-displacement Cobrajet. RPD's two-barrel and single-barrel carburetors also are well represented under the hoods of various makes and models.

If you are harboring any ideas of taking the family bus down to the strip or have plans for a more ambitious program of racing endeavor, try a Q-jet or another Rochester model. The following recommendations will help you if you do.

If you haven't done so already, go back and read chapter thirteen straight through *before* you proceed any further. The details in the previous chapter must be understood for this tuning chapter to be fully meaningful.

## Q-JET: YOUR GOALS

This carburetor is very versatile as we mentioned before and the tricks you must know can be covered on a few pages—a big plus for the Q-jet. If you have the will to win, study these pages—gather a handful of tools for making subtle changes—and go blow the doors off many of your competitors.

You be the judge of your goals. If they are pretty high, go through the carburetor one time with our instructions before dusting off a spot on the mantel to set your trophy. Be sure you have access to drills and general shop tools before starting a serious blueprint job. Once a few initial changes are made, careful tuning at the competitive event will tailor the unit to your application.

Before going into the blueprint recommendations, I will take you through a discussion of the carburetor systems which relate to performance running. It is important that you understand what is taking place in these areas and how it may affect your efforts. Don't get anxious to rip into your carburetor and turn to the blueprint section with tools in hand. Study these next few pages and mentally relate the discussion of each item to your application. There aren't enough pages to discuss every possible happening that may take place in the thousands of varied applications you the reader will delve into.

Ford production-built Quadrajet manifold. The size of the intake ports indicates they had high-RPM HP in mind for this Cobrajet engine.

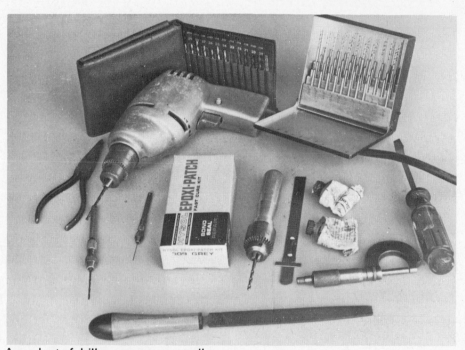

A good set of drills, some epoxy, small hand tools, some measuring devices and instruction from this book are a good start toward the performance you want.

Oldsmobile's 1973 455 cu. in. power plant with a Quadrajet as original equipment.

Take in what we tell you and *think* your way to the winner's circle.

## FUEL SUPPLY

Perhaps the first order of preparation is to be sure adequate fuel gets to the carburetor bowl at all times. The fuel supply system is discussed in detail in chapter sixteen, but some additional tips are provided here relating to performance tuning.

Four to six pounds per square inch (psi) fuel pressure is preferred but the Q-jet can function well under certain conditions with as little as 2 psi.

An ideal situation would be to have a pump and fuel lines of adequate capacity so the fuel pressure never drops below 4 psi or exceeds 6 psi.

Now let's assume during a practice run down the strip or during a fast section of any race course the fuel pressure drops to 2 psi. If this occurs immediately off the line or at the beginning of a straightaway your system is not adequate.

Should this drop come on gradually during a long wide-open throttle (WOT) run and never go below 2 psi, the system is acceptable.

In short—don't give up the race because you have only 2—3 psi toward the end of the full-throttle spurt. Get your trophy, go home and do some more homework on the fuel system for the next tryout.

Before we go on, there is one point you must always consider when evaluating fuel pressure. You are reading the pressure at a point outside (ahead of) the fuel bowl. If fuel demand gets high enough through your jets and the fuel-inlet valve is too small to supply the need, you will be led down the path of poor performance with plenty of pressure showing on the gage. Continue on as we discuss this further. Correctly sizing the fuel inlet is one of the most important phases of hot-rod tuning your carburetor.

I am not an advocate of these so-called super inlet valves on the market that claim to solve all problems. If they are good for one fix they will most likely put you in trouble in some other operating range. Stick with a conventional-type valve and carefully work out the minimum size best suited for your needs.

Frequent removal of the fuel-inlet nut or old-age corrosion sometimes damage the threads in the carburetor body. Tomco Inc. of St. Louis is a supplier of repair units. Other repair nuts are shown in the fuel system chapter.

This cutaway shows the route the fuel travels to get into the fuel bowl. For normal highway use, enough fuel passes through the filter walls (arrows) to supply needs. For high performance don't take the chance! Get rid of the sintered-bronze or replacement filter in the carb inlet.

Perhaps it doesn't look quite as neat to mount the fuel gage outside the driver's compartment, but it's a lot safer. Fuel lines plumbed into the vehicle are one hazard you can do without. This gage is on Jim Mehl's drag-race Camaro.

Once established, stick with it so all your other changes are evaluated with a consistent fuel supply. It would be futile to work diligently perfecting your metering and then, on a whim, change the inlet-valve size or design. Methodical approaches to all changes will get the trophies coming your way.

Think it all through very carefully. What are your exact needs? Determine what they are and work the carburetor out to meet them.

Following is an example of trick tuning occasionally employed by some winning crews.

I have known sharp oval-track mechanics who intentionally design the entire fuel system purposely to lean the air/fuel mixture toward the end of each straightaway. They know a slightly lean power mixture (14 or 15:1) gets maximum performance. For continuous power demands it would not be wise to calibrate lean because it causes high combustion-chamber heat. A mechanic who sets up with a good safe power mixture (rich side) for most of the course with an automatic leaning out at the end of each straightaway gets a double bonus:
1. More power for a few seconds on the straight.
2. Better economy (less fuel consumption). This can mean winning or losing if you are limited on fuel capacity by the race association rules.

Should you choose to lean out the mixture intentionally for short-spurt performance, it is best to do so by altering the fuel-inlet valve size. Gradually decrease the size until you note a slight loss in performance. Monitor this carefully with a stopwatch, on a dynamometer or during practice at the race site.

If you attempt to do this size evaluation on an engine dynamometer, I suggest you rig a sight glass in the bowl so you can see if the smaller valve maintains suitable levels. Remember, dyno steady-state running is different from acceleration conditions which are dynamic (constantly changing). There is no better way to monitor performance than under actual conditions.

Once you get a slight drop in performance come back up in inlet size a few thousandths to reach the ultimate of performance for that particular circumstance.

As mentioned

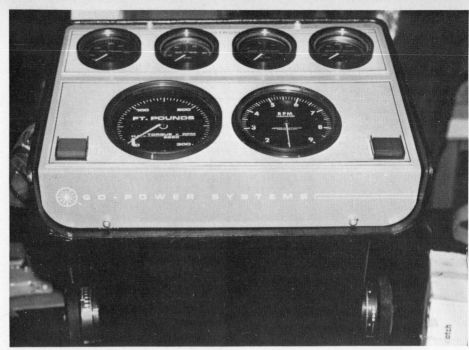

Doug Roe Engineering does a lot of testing with a Go-Power Dyno which attaches to the engine. An engine dyno is a great tool if you remember it doesn't accelerate and brake like a race car. Some things must be checked in the vehicle.

Erick Rice, owner-mechanic of an oval-track-modified racer, says the Quadrajet and wide tires are good partners. His cars run the central New York circuit very successfully. Side forces encountered on these paved circle tracks demand good carburetion or fuel injection.

before, you can go small enough to where metering demand will use fuel from the bowl faster than it's replenished. If there is too much restriction between the gage pickup and the bowl (inlet valve), the gage will be reading good pressure while the nozzles (fuel bowl) starve for fuel.

For competition, always remove the stock sintered or paper fuel-inlet filter. Install a suitable in-line filter ahead of the fuel-pressure gage pickup point. There is more on this in the fuel-system chapter.

## AIR CAPACITY

Because of the air-valve design the Q-jet automatically adjusts air capacity up to its limit by engine demand.

It can be tuned to run extremely well on engines under 200 CID and yet will make one over 400 CID torque with the best of them.

If you want maximum capacity for a large screamer, two sizes have been available since 1971.

The large ones are used on 1971 through 1973 Buick vehicles. Pontiac also uses these units in 1973 which will flow 800 CFM at 1.5-inch Hg vacuum. The other Quadrajets have 50 CFM less flow capacity.

Investigate your needs carefully before buying this larger unit to replace a regular Q-jet which you may already have. Don't be fooled by idle shop talk into believing you need super-big carburetion. In many stock and some modified classes where the engine never reaches super-high RPM, it's likely detrimental to your cause to go too big. It will be a deterrent to low and mid-range torque.

## FUEL BOWL

The Quadrajet fuel bowl has adequate capacity for general driving and most performance uses. It is well shaped and located for good cornering and rough-terrain stability. It does need improvements to prevent fuel spewing from vents under certain conditions.

This spewing generally occurs during WOT accelerations. You will notice a sag (loss of power) which sometimes continues for a number of seconds.

Vent-spewing problems can be eliminated in many cases by extending the vertical vent tubes and/or cutting the top of the tube on a 45° angle

When you plan your performance fuel system be sure an in-line filter is installed just ahead of the carburetor fuel inlet. (Rochester Products does not recommend flex fuel lines under the hood.) There is good quality flex fuel line on the market so be sure you get suitable material and then do a good installation job. Be sure the line doesn't rub on a sharp object, get near the exhaust manifold, etc.

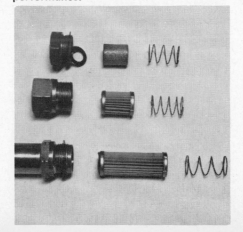

For any serious competition efforts—remove the filter elements from the carburetor inlet. Three Quadrajet types are shown. Install a large-capacity in-line filter between the fuel pump and carburetor. These filters do not have adequate capacity for high performance.

as shown. Try the slash first; it's easiest and may fix your problem. This causes air to pack the vent stack with minor pressure, helping to keep fuel in the bowl. Should this fail to eliminate your problem, increase the length of the tube up to 1/2-inch. You can also add a vent stack to the rearward internal vent if need be. If the air is routed in from the back side of the carburetor, the slash may have to be open to the rear. Some experimentation may be needed to ensure maintaining a positive pressure on each of the vent stacks.

The problem of spewing can be the result of two different situations:

1. Certain air-cleaner or inlet-air plumbing configurations can cause air velocities to act as a depression on a vent area and allow it to feed fuel. This is uncontrolled metering you cannot tolerate.

2. Rapid fuel entry through the inlet valve can cause pressure spewing. This is covered further on in this section under *inlet-needle discussion*.

When a depression (vacuum) exists as in situation 1, the vents are able to grasp fuel that is being sprayed wildly inside the fuel bowl from the inlet needle. This fuel is then expelled upward from the vents, creating a rich-mixture condition. Situations 1 and 2 are often interrelated, but they can occur separately.

A few Quadrajets have a 45° slashed internal bowl vent. In some instances it is beneficial to increase the height of this and add a tube in the existing rear vent hole (arrow). It may also require slashing, depending on the pattern of incoming air. Air entry and internal vents must be compatible with bowl pressure requirements. Experiment until it is right for your installation.

Sometimes special engine adaptations create a need for fabricated air cleaners and unique ducting of air to the carburetor. This can cause detrimental air ram or depression effects on carburetor venting. Some serious thinking and baffle engineering (guess and try) followed by component fabrications can eliminate the problem. This vent spewing can sometimes be alleviated by the installation of 100-mesh screen shaped and installed over the vents.

The screen helps separate the air and fuel, causing the fuel to drop to the bottom of the float chamber more directly than if allowed to wash over the air-horn gasket.

Baffles are sometimes used to serve the same purpose.

If you encounter an interruption of good acceleration performance above 3,500 RPM, it is more likely a rich spewing problem than a lean-out.

Study the accompanying vapor-lock material on page 204, as lean-out can occur when heat and high vapor fuel get together. Be sure you can differentiate between a spewing problem and a lean-out.

The screen mentioned previously can also be effective in reducing or eliminating vapor-lock problems. It tends to condense liquid fuel from highly vaporous gasoline entering the fuel bowl. Study the vapor-lock material again.

Last but not least is the plastic fuel stuffer. If you left it out during initial preparation, keep it handy. Sometimes it is all you need to direct incoming fuel downward to the metering orifices. Personally, I never recommend leaving it out.

## HOT FUEL FLASH

One other situation that arises occasionally is a result of hot soaking during brief stops. It is a fuel-flash condition. If a car is run at expressway speeds in hot weather then stopped for 5 to 20 minutes, restarted and accelerated at WOT back to highway speeds, it may falter and die out momentarily if the fuel is high in vapors.

This occurs when the engine exhaust manifolds, combustion chamber and other large pieces get very hot during continuous running. This condition multiplies rapidly in a race car as continuous high

When additional air ducting is fabricated to the side of an air cleaner as shown here it can cause you bowl-pressure problems. If you run a fresh-air duct like this, leave in the element or plan on some experimenting to combat internal-vent problems. This installation seen at a Baja 1000-mile race got all its air from the added fresh-air duct. The original air-cleaner snorkel was inconspicuously plugged. He did run an air cleaner element in the original location, even though the air was filtered by remote cleaners.

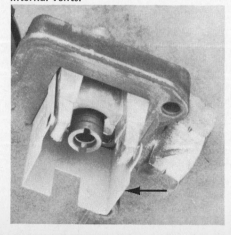

This 2GC illustrates how a baffle around the fuel inlet looks. This tin shield is very effective in forcing the inlet fuel to go downward into the bowl rather than outward and upward where it might get to internal vents.

revs create additional heat. When the vehicle is stopped for this brief period, the 1,000° to 1,800°F. heat from these large iron areas is transmitted to water and engine dress accessories that remain cooler during running because of air flow and radiator cooling.

Gasoline boils at approximately 150°. Consequently, when the fuel pump and carburetor temperatures rise to approximately 190°F., the fuel vaporizes and escapes from the bowl. After you restart and maneuver from the parking lot the bowl fuel is replenished with hot unstable fuel from the pump and lines.

The float, vents and metering normally can handle this situation as long as you do not go immediately to WOT. At best you may have to cope with some nuisance stalling and unstable-idle conditions. This annoying problem is most likely to arise if you do accelerate WOT within a minute or so of restart. Cooler fuel is brought in from the fuel tank and pressurized ahead of the fuel-inlet needle. As you accelerate hard the fuel in the bowl needs to be replenished rapidly. The float drops and a rush of fuel enters. When this hot fuel enters this hot bowl suddenly, it can flash to vapors so violently it will spew from every outlet. The engine cannot cope with this additional unmetered air/fuel mixture and will sag or die. This is the opposite of a vapor-lock sag in that the engine is dying momentarily from excess fuel.

When a vehicle is acting up after a hot soak, it's best to run it a couple of minutes at a fast idle in neutral before putting it in gear. Give the carburetor and other accessories a chance to transmit some of their hot-soak temperature into the cooling water or air stream.

The term *fuel-flash* does not make reference to flame or imply fire hazard. As long as the air cleaner and original equipment pieces are intact, these vaporous fuel flashes are arrested. Should you be foolish enough to be running without an air cleaner, with altered pieces or inadequate bolt-on accessories, a fire could well be the result. Emitting this highly flammable vapor to the under-hood area invites it.

In summarizing fuel-bowl-related problems, it can be seen that a rich fuel flash or a lean vapor lock can cause the engine to lose power or die out following a hot soak. Separating the two can sometimes be frustrating.

The best approach may be to get a friend to follow you through an exercise that produces the problem. If it's fuel-flash richness, a large puff of black smoke will emit from the exhaust during disrupted engine operation. If it's lean vapor lock, smoke is quite unlikely to exist. Heavy blue smoke doesn't count because that is caused by oil burning.

## FUEL-INLET VALVE

For most RPD carburetors there are various sized inlet valves available. In the case of the Quadrajet, United Delco supplies three sizes. Following is a parts listing for these units.

| RPD | DELCO | ORIFICE SIZE |
|-----|-------|--------------|
| 7035130 | 30-130 | 0.110-inch |
| 7035140 | 30-140 | 0.125-inch |
| 7035142 | 30-142 | 0.135-inch |

If your Delco dealer is not able to supply you promptly with suitable parts write to Doug Roe Engineering or visit the Carburetor Shop; these carburetor specialists stock many approved pieces. Doug Roe has inlet valves up to 0.140-inch in approximately 0.005-inch increments. Refrain from using off-brand parts. Be particularly skeptical of

newly designed items for which revolutionary improvement claims are made. Most such items are products of good marketing—not good design.

Before leaving the subject of inlet valves, I want to elaborate further on reasons for not going larger than necessary. Flooding is the most common problem. This causes driving misery in the pits or on the grid at drags or road races and guarantees untold grief for off-road enthusiasts.

The second problem which crops up more often than most people are aware is internal-vent spewing.

Strange things often take place in carburetors when fuel enters the bowl area too rapidly. As the fuel passes the inlet-valve seat it is leaving a pressure area of four to eight pounds and entering a bowl area at near atmospheric pressure (zero). Unless it is cooled adequately with a fuel cooler or is of low volatility, the fuel will likely get very vaporous at this point. It can travel along the top of the bowl-gasket surface and actually feed into the air stream via the internal vent stacks. Remember, we said certain air-entry fabrications can cause depressions at the internal vents. This is quite common during heavy-throttle operation and it is seldom diagnosed correctly.

The flat-dead feel of the engine under these conditions is often called lean-out and search is on for a bigger inlet valve, a drill bit or a magic cure-all aftermarket unit. I have personally made vehicles with this problem go quicker by simply reducing the inlet-valve size. Be aware of this potential fuel feeding (engineering-termed *vent spewing*) from the vents and try an easy fix first—a smaller inlet needle.

The carburetor's capability to control fuel level correctly under most conditions is influenced by four primary factors:
1. Fuel-bowl size and shape
2. Buoyancy and shape of the float
3. Geometry of the float arm to the inlet needle
4. Inlet-needle seat-orifice size (discussed above).

Generally the first three items are pretty well established in original design. Once the bowl is designed major changes in float size are limited. Buoyancy changes can be accomplished to some degree by introducing new materials, new shapes or

adding assist springs. Special requirements on items 1 through 3 will vary depending on intended use. For alteration tips on these components, refer to the section of this chapter which covers your type racing (off-road, drag racing, road racing).

As far as item 4 is concerned keep the orifice size down to handle your needs and no more.

For all-out competition the following inlet-needle/float-level settings are recommended:

| Float | Float Setting (inch)* | Inlet Valve |
|-------|------------------------|-------------|
| 7034454 long fulcrum | 0.325 at 6 psi fuel pressure | 30-140 (0.125-in.) |
| 7037316 short fulcrum | 0.296-in. at 6 psi fuel pressure | 30-134 or 30-142 (0.135-in.) |

*Higher fuel pressure may require increasing these settings to maintain the same fuel level in bowl.

## SECONDARY SYSTEM

The item to be discussed next is very important to the all-out racer who wants to blueprint performance consistency. This concerns secondary system operation of the Quadrajet and applies to any unit that will occasionally see maximum or near-maximum capacity. Let's go on with discussion and recommendations to consider.

First of all the air valve is designed to function only when capacity demands are fairly high (acceleration or high-speed running). Let's assume you are accelerating with a fairly heavy throttle. There is approximately 75° of primary throttle opening from idle to WOT. At approximately 40° of primary opening the secondary throttle blades start to open. Regardless of your speed or RPM a mechanical linkage continues to open the secondaries as the primary is pushed on to WOT. The geometry is such that during the last 35° of primary opening the secondaries open from closed to a WOT position.

The air valve, being independent of the primary or secondary throttle blades, does nothing until engine RPM combined with secondary throttle opening puts a depression under the valve. In short, the air valve doesn't open until the engine needs the capacity. The air valves are offset on the shaft a prescribed amount to cause them to start opening. Depression pulls the wide side (largest area) of the blade downward and open. If you are still hard on the throttle air velocity builds up and continues to push the air valve further open.

That gets the air valve open. Now let's discuss the fuel which must accompany the air through the induction manifolds to the cylinders.

Many Quadrajets have a built-in system which provides a shot of fuel as the air valve starts to open. A secondary pump shot so to speak.

If your carburetor has this system you will note it by observing a hole near the leading edge of the air valve. Many of these feed holes are *above* the blade as shown in the accompanying photo. The large 1971 Buick (800 CFM) Q-jet and a few others have the feed hole just *below* the leading edge of the air valve.

If your carburetor does not have this system and it is to be used on a special performance application (other than street stock), you should change your air horn for one with the feature. Or, you could cannibalize another air horn to get the tubes and then do the necessary drilling to add passages sized and positioned like those in the air horn you are taking the parts from.

Here's how it works:

Extending down from the visible external holes (see photo) are 1/8-inch-diameter fuel-pickup tubes. They extend into small reservoirs of fuel (one on each side of the bowl). As the secondary throttle blades are opened, manifold vacuum is admitted up through the bores to the lower side of the air valve. This pulls an instant shot of fuel if the holes are *below* the blade. If the holes are *above* the blade, the shot comes as the air valve starts to open (as it exposes the holes to vacuum).

Once the fuel reservoirs are depleted, it takes a few seconds for them to refill. Like an accelerator-pump well, there is a contained volume and a refill requirement after use.

If your carburetor has the system described there is one modification I recommend. Drill the lower end of each tube as illustrated. This is further detailed in blueprinting later in this chapter.

Arrow points out the secondary feed hole above the blade. For quicker secondary-fuel response the hole can be located just below the blade. Plugging or redrilling is an acceptable practice provided care is exercised.

The larger tube (arrow) carries the secondary fuel up to the feed hole. Unless you have a Ford Cobrajet carburetor the tubes (one each side) will be solid wall—no holes.

This fuel-feed tube has been modified by drilling four holes through both walls. Details of this are given in the drag race section of this chapter.

## TOOLS REQUIRED

Start with patience! You need more than the feel in the seat of your pants and the speedometer in your car to do serious tuning. Specific tools are required, but you'll also need a patient and methodical approach to the project. This further means you cannot be in a hurry. If you have no intentions of really getting serious about tuning, run your carburetor as it comes out of the box and leave your tool box locked up.

A vacuum gage, fuel-pressure gage and a stopwatch are essential. So is a tachometer. The vacuum and fuel-pressure gages are an extra set of "eyes" to let you look inside of the engine to see what is happening there. For serious competition on a regular basis, an air-density gage can be especially helpful. You'll also want a 1-inch open-end wrench, preferably the MAC S-141 (specifically designed for the fuel-inlet nuts on most carburetors).

## TIMING DEVICES

A stopwatch can be used for some very fine tuning. Pro stock racers often test acceleration over an initial 60 feet to work out starting techniques, tire combinations and carburetion. Most drag races are won in the critical starting period and by the initial acceleration over the first few feet.

## A FEW PARTS WILL BE HELPFUL

It is wise to have spare gaskets, an inlet needle seat assembly in the size you use, an accelerator pump and miscellaneous screws and clips. Purchase these from your United Delco dealer or write Doug Roe Engineering or The Carb Shop.

Now about jets—before you buy main jets, find out what size is already in the carburetor. Once you know what jets you have, buy four sizes lean and four sizes rich for each main jet in your carburetor. A similar varied supply of metering rods should accompany any Quadrajet efforts.

This will handle most engine variations and most atmospheric-condition changes (density).

If your carburetor has been used and perhaps remetered by the previous owner, get it back to the original jetting before you start tuning.

Drag racing often requires richer jets, but decide that at

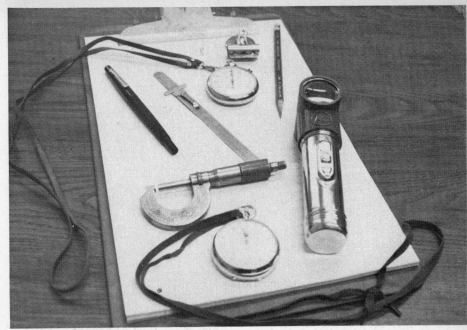

Along with your favorite hand tools you should have a stopwatch, one-inch michrometers, a six-inch scale, an inspection light like the AC unit shown, a clipboard, pencil and paper.

The MAC S-141 wrench shown is a slim tool for getting onto inlet nuts that are difficult to reach.

**Don't drill jets**—Jets have a carefully machined entry angle and a precisely maintained exit angle. Drilling a jet will not provide a jet with the same flow characteristics as a genuine RPD machined jet—even though drilling apparently produces the same size opening.

**Carrying jets**—Store and carry jets in a jet holder such as Holley's punched-out rubber one (Part No. 36-17, holds 42 jets). Or drill and tap a piece of wood, plastic or aluminum with 1/4-28 threads and use that. Some tuners use compartmented plastic boxes or even small envelopes. The practice of carrying jets strung onto a wire is strictly a NO-NO because this will very definitely damage the jets.

**Use correct tool to install jets**—Be sure to use a screwdriver with a wide-enough blade—or use a jet tool—so there will be no danger of burring the slot. Burrs on the edge of the slot can affect main-jet flow.

the strip after making trial runs and looking at the ET slips.

## ONE THING AT A TIME

It never occurs to some would-be tuners—even some old-timers—that changes must be made ONE at a time. It's too easy to be tempted, especially when you are sure that you need a heavier flywheel, different gear ratio, other tires, a different main jet, two degrees more spark advance and a different plug heat-range. Changing one of these at a time would be just TOO SLOW.

Don't you believe it. Stay on the track. Remember, patience is a necessary ingredient. When you are trying to tune your carburetor—a complex piece of equipment in itself—there's all the more reason to *make one change at a time* and then check it against known performance by the clocks, or by your stopwatch and tachometer.

Anytime that you change more than one thing, you no longer have any idea of which change helped—or hindered—the vehicle's performance. It's even possible that one change provided a positive improvement—cancelled by the negative effects of the other change. The net result *seemed to be* no change. And, you've thrown out the good idea with the bad—which is unfortunate!

When tuning, follow a procedure and stick to it. The less help (and therefore, advice) that you have, the better. Concentrate. Be deliberate. And don't be surprised when the process eats up more time than you ever thought it could. If you are planning other changes soon, such as a different manifold, camshaft, air cleaner, distributor-advance curve, cylinder heads or exhaust system—put off tuning until the car and engine are set up as you expect to run it. Otherwise, your tuning efforts will be wasted and you can look forward to a repeat performance of the entire tuning process.

Whatever you do, don't fall into the common trap of rejetting the carburetor to some specialized calibration that you read about in a magazine article or heard about at the drag strip last week. RPD spent thousands of dollars getting the calibration correct and there is good reason to believe that they used more engineers, technicians, test vehicles, dynamometers, flow benches, emission instrumentation and other equip-ment and expertise than you may have available as you start your tuning efforts. Remembering these hard facts will save your time, effort, money and temper.

When you have made jet changes from standard, use a grease pencil or a marking pen to mark the carb with the main-jet sizes which you've installed. Some tuners use pressure-sensitive labels on the carb for noting which jets are installed. Others write the information on a light-colored portion of the firewall or fenderwell where it won't be wiped off during normal tuning activities such as plug changes, etc.

Noting what jets have been used saves a lot of time when tuning because no time is wasted in disassembly and reassembly to see what jets are in the carburetor/s. Even the best memories are guaranteed to fail the jet-size memory test. Stagger metering multiplies the problem.

## BEFORE YOU START TUNING

Now that you've gathered the tools, spare jets and other paraphernalia, there are a few more details to take care of. *Be sure that the fuel level is correct before proceeding.* Remove the air cleaner (temporarily!) so that you can peer into the air horn. Have someone else mash the pedal to the floor as you check with a flashlight to make sure that the throttles fully open (not slightly an-

**Keeping records**—Lots of active people want to tune and drive. Paperwork seems like a bother and a waste of time. It's the other way around. Keeping records will positively save you time in the long run and will allow you to tune better. If you have not been doing it, don't judge the method until you have tried it. Personal experience of the authors has shown that it is impossible to remember the changes made earlier on the same day unless they are written down.

*Change your test procedures as often as you like—provided progress and accomplishment are unimportant.*

*—Doug Roe*

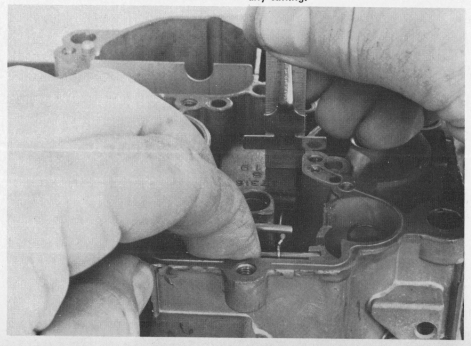

**Check float setting with a scale or other appropriate gage before attempting to do any tuning.**

gled). If they're not opening fully, figure out why and fix the problem. Anytime you remove and replace the carburetor—check again to ensure that you have a fully opening throttle. It is the easiest thing to overlook and the cause of a lot of lost races or poor times. Any honest racing mechanic will admit that he's been tripped up by a part-opening throttle *at least once.*

For drag racing, air cleaners are generally not used. For any other competition where dust is involved, be sure to use a low-restriction paper-element cleaner, such as a tall open-element AC type and replace it often. Combination paper-element and oiled-foam cleaners should be used for very dusty conditions. If the cleaner is removed, the engine can suck in a lot of abrasive dirt by merely running back down the return road. If the engine has to last, then stop at the turn off and put the air cleaner on—or get the car pushed back to the pits. Dust is a fast-acting abrasive—don't gamble with it.

The air-cleaner directs the air into the carburetor in such a way that the vents work correctly and so that the air gets into the air-bleed jets in the correct fashion. You have probably read a lot of articles which said to be sure to leave the air-cleaner base in place, even if you had to remove the air cleaner for some obscure reason. Lest you think that the writers have been kidding you, one

dyno test series showed a 3 HP loss by removing the air-cleaner base. This little item causes air to flow into the carburetor with less turbulence. It can be essential! Check it out.

Air cleaners also protect against fires caused by starting "belch-backs"—reduce intake noise—and reduce engine wear. Intake noise can be horrendous—even worse than exhaust noise—on an engine turning up a lot of RPM.

Before leaving the subject of air cleaners, let us remind you to look at the way the stock cleaner is designed before you invest in some flat-topped short cleaner because it looks good. Note that the high-performance cleaner stands high above the carburetor air inlet. Adequate space allows the incoming air to enter correctly and with minimum turbulence. With a flat filter sitting right on top of the air inlet you may get HP loss. For racing take the sintered-bronze Morraine filter out of the inlet to the fuel bowl. Make sure there is an in-line filter in the line between the pump and the carburetor/s.

## ACCELERATOR-PUMP TUNING

The ability of the engine to come off the line "clean" indicates a pump "shot" adequate for the application. A common complaint—heard again and again—"It won't take the gas." Actually it's the other way around because the engine is not getting enough gas and a "bog" is occurring. Two symptoms often appear. The first of these is that the car bogs—then goes. This can be caused by pump-discharge nozzles which are too small so that not enough fuel is supplied fast enough. The second symptom is one of the car starting off in seemingly good fashion—then bogging—then going once more. We are talking about drag-race starting here, of course. In this second situation, the pump-discharge nozzles may be correctly sized but the pump is not big enough to supply sufficient capacity to carry the engine through.

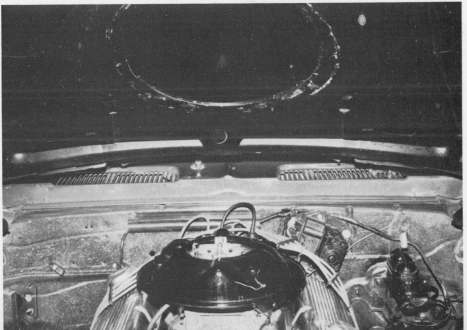

**Jim Mehl often runs down the strip with the top and filter element removed. He secures the bottom with two pull springs 180° from each other.**

**3 X 2 installation on Olds had this gigantic cleaner assembly with two oil-bath cleaners—one at each side of the engine.**

Solving the first problem may mean enlarging the pump-discharge nozzles—which may subsequently lead to the need for increasing the pump stroke. More details on altering the accelerator pump are given under the section on drag racing.

Other areas of the carburetor metering can have a great effect on pump requirement. Study all of this chapter if you are prepping for performance. Metering is detailed in the economy, drivability and general performance section.

As a general rule, the more load that the engine sees, the more fuel that is needed in rate and volume. For instance, if an 1800 RPM stall speed converter is replaced with a 3000 RPM converter in an automatic transmission, the shooter (discharge-nozzle) size can be reduced. The same would be true when replacing a light flywheel with a heavier one. This is because the engine sees less load as the vehicle leaves the line in each instance.

Valve timing affects the pump-shot requirement. Long valve timing (duration) and wide overlap create a need for more pump shot than required with a stock camshaft.

Carburetor size and position also affect pump-shot requirements. More pump shot is needed when the carburetor is mounted a long way from the intake ports, as on a plenum-ram manifold or a center-mounted carburetor on a Corvair or VW. The larger the carburetor flow capacity in relation to engine displacement and RPM—the more need for a sizeable pump shot to cover up the "hole" caused by slamming the throttles wide open. This is especially true with mechanically operated secondaries. The primary side of the Quadrajet is small and efficient so the pump is generally quite adequate with minimum alterations.

When a carburetor is changed from one engine to another, the engine is changed into a different vehicle, or drastic changes are made in the vehicle itself—work may be needed to get the pump system to perform as you'd like. The main problem will always be tip-in performance (opening throttle from idle).

Shooter size tuning is best done by increasing the nozzle diameter (or decreasing it) until a crisp response is obtained when the throttle is "winged" (snapped open) on a free engine (no-load).

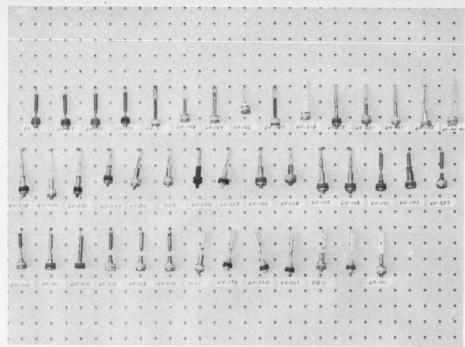

There are a few pumps!

Vent clearance—Allow at least 3/8-inch between any vent and the underside of the air cleaner. If you get the vents much closer than this it can create a high-speed miss you'll swear is an ignition problem. Also, air cleaner element heights vary, so measure the vent clearance carefully—or else check any new cleaner against your old one to make sure the new one is not too short.

A plenum-ram manifold typically requires greater pump capacity.

A cross-H manifold requires less pump capacity because stream velocity in each side of the manifold tends to be higher than in an open-plenum type.

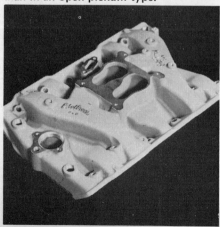

When crisp response is obtained, increase the nozzle size another 0.002-inch and the combination will probably be drivable for a drag application. Don't overlook the pump duration springs. A slightly stiffer duration spring will also give you a quicker responding shot.

Getting a boat engine to take the throttle can require a greater capacity than any other application. Some alterations may be needed to overcome the constant load imposed by the propeller.

Pump-shot duration can be timed by shooter size, pump duration and return springs, pump capacity and pump rod to lever setting. The duration spring on the shaft is a safety valve which "gives" (compresses) when the throttle is slammed open. The compressed spring force against the pump cup establishes delivery pressure in the pump system. Thusly, delivery rate depends on both system pressure and shooter size.

The duration spring must never be adjusted or of a wire size to get it into a coil-bound (solid) condition where it cannot be compressed. And, do not attempt to replace the spring with a solid bushing in hopes of improving pump action. Either course of action will net you a badly bent pump linkage because gasoline is incompressible. Something has to give or break if the throttle is slammed open and there's no way for the system to absorb the shock.

Carburetors can sometimes be modified for added capacity by lever/linkage changes to get maximum pump stroke. The pump piston should bottom in its well. Stroke increases may be gained by lifting the piston to a higher starting position but not past the fill slot or pump-well entry. Each

**Accelerator-pump shooters**—These "nozzles" are drilled into RPD carburetor air horns. The air horn must be removed to allow drilling these passages. Take it easy! Make your changes in small increments because if you go too large, the holes will have to be bushed or plugged with epoxy and redrilled. Do not change the angle of the hole as this targets the pump shot—usually against the booster. In the case of a high-ram manifold or a large open plenum you might want to consider retargeting the shooters to point down toward the throttle blades. This could provide a helpful improvement in throttle response as the targeting against the boosters actually delays the pump-shot effect, while helping to ensure that the fuel is vaporized on its way through the carburetor.

carburetor must be examined to see whether *any* changes can be made.

The limit of pump travel in a Q-jet pump bore is 21/32-inch from the bottom of the cup to the lower edge of the fill slot. Anytime the pump cup rises above the lower edge of the fill slot, the pump action will be inefficient because a portion of the pump travel will be wasted while it expels fuel through the fill slot into the bowl, resulting in an intolerable pump lag.

Q-jet air horns vary as you can see from the accompanying photos. Some of them (mostly early models) are flat and level with the gasket surface around the pump-rod bore. Others (mostly '67 and later) have a recess in the air horn, permitting the pump-duration spring retainer to rise *above* the gasket level.

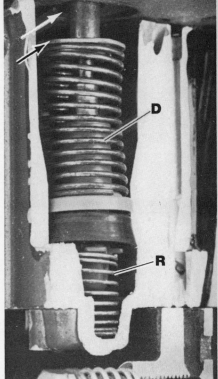

Spring retainer (arrow) fits air-horn recess (white arrow) to limit pump's upward travel. Some careful tuners have moved the retainer down to gain added stroke. Care is needed to avoid locating the pump-cup lip above the fill slot. D is duration spring. R is pump-return spring.

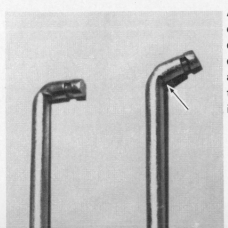

Arrow indicates broken pump-rod end caused by attempting to adjust the rod end for a better fit in the pump arm. The ends are hardened against wear and any attempt to straighten them is sure to break the rod. Always adjust these rods by bending in the long main areas.

Fill-slot bottom can sometimes be raised about 1/16-inch to allow more pump stroke. This is easiest to do when the pump is bored for the large 0.730-inch cup.

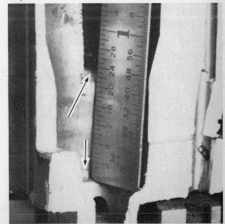

Numerous articles have been published indicating a certain amount to be ground off the end of the pump rod to get more travel. After checking numerous Q-jets it became obvious that no magic single dimension could be given. At first it was thought 3/32-inch could be ground off of every Q-jet pump rod. Further checking showed some early models could be adjusted for the full 21/32-inch travel by grinding (or filing) a mere 1/32-inch from the rod top. I did not find any models which allowed more than 3/32-inch to be ground off the rod end without raising the cup above the fill slot.

The easiest way to check pump travel is on an assembled carburetor. Use a small screwdriver to push the pump to the bottom of its bore. When it is at the bottom, scribe a mark on the pump rod even with the air horn. The small amount of the rod protruding above the air horn may be as little as 1/32-inch in some cases. If so, the maximum to be removed from the end of the rod cannot exceed 1/32-inch because the pump linkage cannot push the pump rod below the level of the air horn casting. In fact, grinding off more than this could *reduce* the pump travel if the rod top ends up below the top edge of the bore in the air-horn casting. But, before getting a file out of your tool chest, check one more thing, please! With the pump actuating rod removed from the pump lever, release the pump lever and measure the height of the pump rod above the casting. If it measured 1/32-inch above with the pump at the bottom and the rod now stands 22/32 (11/16) above the air-horn casting, the difference between the two measurements is:

$$22/32 - 1/32 = 21/32\text{-inch}$$

*the maximum allowable travel for the pump!!* If the top of the rod stands 21/32 above the air horn, then grinding 1/32 off the rod end will allow the full 21/32 travel *if* the rest of the pump linkage is adjusted to get this *full* travel.

Assume you push the pump to the bottom of its bore and 3/32 protrudes from the air horn. Then you release the rod and it protrudes 21/32. $21/32 - 3/32 = 18/32$ pump travel so 3/32-inch cut from the top of the rod would give full travel *if* the linkage is adjusted to take full advantage of the travel.

Underside of air horns reveals one with counterbored area (arrow) for pump-rod retainer. One which is flat in this area may not allow same travel. Keep this in mind when working to get maximum pump travel.

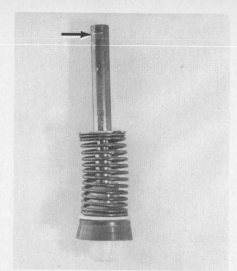

When checking for travel, coat the tip of the pump rod with machinist's dye. Then you can scribe a mark to indicate the lowest point of travel.

With the pump at the bottom of its bore, measure the exposed portion of the pump rod. In this case it was 1/16-inch. Check that the pump is really at the bottom of its bore by using a screwdriver to see if it will go any further down (don't force it).

Then measure exposed portion of the pump rod at the top of its travel. Here it was 9/16-inch. This says that the pump in this carburetor was travelling 9/16 - 1/16 = 1/2-inch instead of the desired 21/32-inch. 21/32 - 16/32 = 5/32-inch travel not available with this arrangement of linkage, etc. 1/16-inch can be gained by cutting off the top of the pump rod so it is even with the top of its bore in the air-horn casting. Getting the remaining 3/32-inch travel may require other modifications.

230

One important thing about setting the pump linkage: It must be preloaded at idle so there is no slack. The pump lever must push down on the rod slightly. Any movement of the throttle must continue the pump's downward movement in its bore. I prefer to make this setting so the linkage pushes the pump rod about 1/64 down in its bore, thereby ensuring the pump cup lip is just slightly *below* the fill slot. The most you will have to do here, once you have the 21/32-inch stroke arranged, is to bend the actuating rod to get this preload.

If the accelerator-pump linkage on your carburetor is the B type requiring a pin to be driven out of the lever to get the air horn off, consider getting the A style pump link from an old Q-jet. Replace the bent-end link with the one requiring a C clip. This will make all your work on the carburetor much faster, especially where the air horn has to be taken off or where you are modifying the pump linkage. Refer back to page 136 for photos of the A and B style pump linkages.

Capacity can also be increased by raising the bottom of the fill slot by up to 1/16-inch. Care must be exercised so the cup lip does not end up above the slot with the pump assembly in the up position. With this modification the spring can be shimmed to a shorter length (compressed). This allows the assembly to lift higher before it is stopped by the air-horn casting. And, it changes the spring characteristic so that the pump shot is discharged faster.

**Save your old carburetors**—Spare links and levers can be invaluable to the man tuning a Q-jet. When you need an accelerator-pump rod or lever or some other link in a hurry, the best source for these parts is a junked carburetor. Fortunately, the enormous quantities of Q-jets produced over the years make them almost giveaway items at the junkyard. These are tough to obtain on short notice from a Delco dealer or your local auto dealer's parts counter.

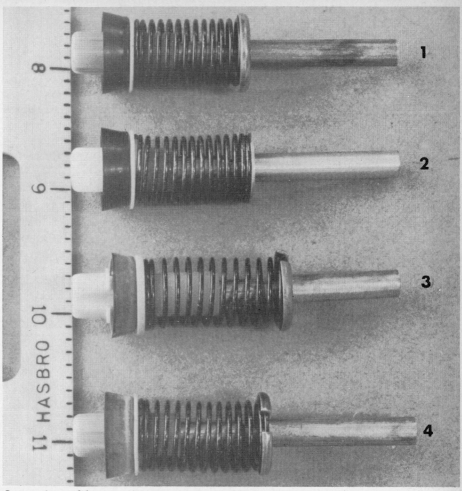

Comparison of four accelerator pump assemblies:
1—1967 Cadillac, 7037504, measures 1-1/4 inch from top of spring retainer to top of rod.
2—1966 Chevrolet, 7034111, measures 1-7/32 inch from top of spring retainer to top of rod.
3—1970 Chevrolet, 7037327, measures 7/8-inch from top of spring retainer to top of rod.
4—1972 Oldsmobile, 7035031, measures 1-5/32 inch from top of spring retainer to top of rod.
All of the assemblies measure about 2-11/16 inches overall. Units shown here do not represent all of the available variations. You may find others more suited to your particular application.

Another alternative is to increase the capacity of the pump-well cavity and use a larger pump cup. This will mean a trip to your local competent machinist and probably a fee of $10 to $15 to bore the pump well to the proper diameter. Your present Q-jet pump well should measure approximately 0.640-inch diameter at the lower part of the well and the pump rubber will measure approximately 0.645-inch diameter.

By using the 0.735-inch diameter pump cup from a late-model big-bore (1-1/2-inch) Rochester two-barrel pump assembly, you can increase the capacity by more than 30 percent. To do this you must bore and polish the pump well to 0.730-inch diameter. You can get this 3/4-inch pump cup from the two-barrel (for enlarging the Q-jet pump) from your Delco dealer, or through a new car dealer's parts counter as:

| | |
|---|---|
| 7036282 | Chevrolet, Pontiac, Buick |
| 7046028 | Oldsmobile |
| 7037562 | Oldsmobile |
| 7037561 | Pontiac |

There are many more.

Other late-model 1-1/2-inch 2G carburetor pump assemblies are OK. In a hurry and the dealer doesn't have a pump assembly? Spend $7.50 for a Delco Power Carburetor Kit such as the 9251 for 1973 Chevy 350 and 400 CID. Cost is a bit high because—in addition to a pump cup—this gets you gaskets, inlet needle and seat, float gage, etc. (all for a two-barrel!).

If you want to buy the cup separately, contact Doug Roe Engineering or the Carb Shop (Part 64-173). When the pumps are purchased separately, the cost is nominal, so the idea of having some spares on hand is realistic. These firms also offer the small-size pump cups.

There are many applications even in competition where the production pump capacity is adequate. Their very efficient primary venturis do not require much pump.

Drilling out the accelerator-pump shooters on RPD carburetors requires disassembling the carburetor. Before disassembly, observe a few pump shots to determine best access to the shooters in your carburetor.

## POWER-SYSTEM TUNING

You may immediately say, "Ahah—I know just what to

Two Rochester pump cups. Stock Q-jet at left. 0.735-inch-diameter cup at right is from a large 2G. Big one fits Q-jet when simple machine work is accomplished as shown in the accompanying photo and text.

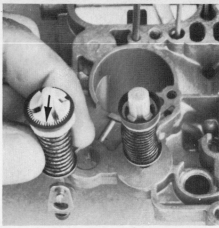

Arrow indicates spring which must be removed from 2G pump assembly to get the pump cup off. Spring is not used when cup is installed onto Q-jet pump assembly. Note size difference between these two pump cups.

Q-jet clamped rigidly to mill table after centering accelerator-pump bore under milling-machine collet. Boring bar is shown in collet. Finished bore should be polished with crocus cloth to ensure seal between pump cup and bore.

do—take it out!" No, regardless of all the magazine articles written to the contrary, *there is seldom ever any real reason to deactivate the power system.*

Power-system tuning requires using a vacuum gage. Let's take the example of a car equipped with a camshaft which provides such low manifold vacuum at idle or part-throttle that the power system operates—or perhaps turns on and off due to vacuum fluctuations. In this instance, a power system which opens at a still lower vacuum is needed. If a 7-inch Hg manifold vacuum power system is in the carburetor and the vacuum occasionally drops to 5 inches Hg at idle or part-throttle, the power system actuating spring will need to be weakened or shortened so it actuates the system at 4 to 5-inches Hg vacuum.

You can see that it is essential to know what the manifold vacuum is at idle. It is also one situation where it is necessary to have a gage which is not highly damped, that is, the needle will have to "jump" to follow the vacuum fluctuations or you won't know how low the vacuum is getting.

Another application which demands using a vacuum gage is racing in a class where carburetion is limited to a certain carburetor type or size which is really too small for the engine which it is feeding. A class demanding the use of a single two-barrel would be typical. Here, the vacuum in the manifold may remain fairly high as the car is driven through the traps at the end of the quarter (that's just one example, of course). The power system should always have a higher opening point than the highest manifold vacuum attained during the run, especially at the end. If this is not the case, the power system will close and the engine will run lean. Disaster may result—usually in the form of a holed piston. Sometimes it will get lean enough to cause "popping" sounds from the exhaust. For example, a carburetor altered to have a 3.0-inch Hg vacuum power system will go lean if the manifold vacuum is 4.0-inches Hg through the high-speed portion of the course.

**How to change the cut-in point**—Changing the power-system (power-piston) cut-in or operating point requires changing the power-piston spring (Q-jet, Monojet, B) or the power-piston assembly (2G, 4GC). The accompanying tables show springs/assemblies which can be used for making changes.

When different power-piston springs/assemblies are not easily obtainable, changes can be accomplished by removing two to three full coils of the power-piston spring. NOTE: Vega-style Monojet and 2GC and the older Corvair H carburetors have weighted power valves as described in specific chapters for those carburetors. These velocity-actuated power valves should be tuned by observing engine performance at 4-inch Hg crowds. The use of "crowds" is discussed in the following section, "Tuning in the Vehicle." Increasing the weight of the velocity-actuated valve causes the power system to come in at a higher RPM. Note the caption on the accompanying photo of Vega power valves (page 234).

Observe and check cut-in point of conventional power systems with an adjustable regulated vacuum source. This is usually a feature of a distributor testing machine. Or, vacuum can be tapped off the manifold of an idling engine. In either case, a vacuum gage must be included in the plumbing so you can see when the power system starts to operate and when it is fully actuated. The vacuum source must be applied to the base of the carburetor, which may require making an adapter plate to apply the vacuum without leaks.

Serious tuners will cut away the bowl of an old carburetor, taking care not to disturb any vacuum passages. This allows observation of power-piston operation. In the case of the 2G and 4GC it also allows observing power-piston operation.

If you are working on a 2G or 4GC, remember that the power piston is held up by manifold vacuum and the spring tries to overcome this vacuum. As vacuum is decreased, the spring overcomes the vacuum force so the actuating rod is driven down against the power-valve pin to turn power-system fuel on. Start with about 10 to 12-inches Hg vacuum and reduce this so the power-piston actuating rod descends to move the center pin of the power valve ever so slightly. Record the vacuum which causes this to occur, as this is where the power system starts to operate. Then decrease the vacuum still further until the rod end bottoms against the power valve, indicating that the valve is fully open. Record this figure to indicate the vacuum at which the power system is fully on.

On the Q-jet and Monojet the power piston is held down by manifold

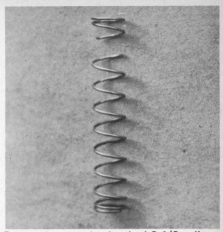
Power-piston spring has had 2-1/2 coils cut off (count the 1st or end coil as 1).

*In our daily tasks it's easier to get the answer wanted than the answer needed.*
*—Doug Roe*

Power spring drops in before installing Q-jet or Monojet power piston. Make sure cavity bore is smooth and free from scratches. Also polish the vacuum piston carefully to ensure that it will operate smoothly without sticking.

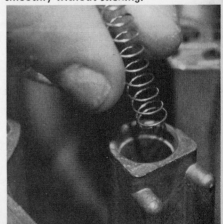

Comparison of Vega Monojet velocity-actuated power valves. 1973 type on left weighs 1 gram; 1971-72 type on right weighs 1.25 grams. Use of the heavier valve causes power-system operation about 10 MPH later (65 MPH instead of 55 MPH in high gear when holding a constant 4 in. Hg manifold vacuum while accelerating) than with lighter '73 valve, and thereby improves economy. A tiny lead ball (0.345 grams, 0.158-inch diameter) can be dropped onto the top of the heavier valve to bring the system in at a higher speed than 65 MPH. Never add more than 0.34 gram weight to the 1.25-gram valve or the power system will never come in, even at WOT. 1974 Vega carburetors have lighter aluminum power valves (instead of brass as used 1971-73) to bring the power system into operation at lower engine speeds. Changing the valve on the 1971-72 Vega 2G shown below is tougher because the assembly is closed with a plug which must be pried out with an awl or ice pick. Weight can then be added to the 0.75-gram valve and the cup epoxied back into position (lip even with the bottom of screwdriver slot). Never exceed 1 gram total weight of valve in the 1971-72 2GC.

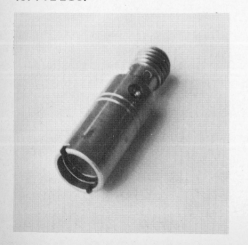

vacuum and the spring tries to overcome this vacuum. As manifold vacuum decreases, the spring overcomes the vacuum force, raising the piston and pulling the metering rods up so their minimum diameter is in the primary main-metering jets, thereby providing power-system fuel.

The Q-jet and Monojet power systems should always be fully up by 3.0-inches Hg vacuum. If not, then the spring is too weak, or you have cut too many coils off the power-piston spring.

Although power-piston and metering-rod operation is most easily observed with a cutaway carburetor, a feeler can be inserted through the forward vent of a Q-jet so that it lightly rests against the top of the power piston. The feeler (soda straw or match stick) should be stiff enough so you can use it to push the power piston down (representing the situation when vacuum is applied) so the largest portion of the primary metering rods are inserted into the primary jets. Mark this position on the feeler. Also mark the full-up position (spring holding piston in full-up or power-system-on position). Apply vacuum at about 10 to 12-inches Hg and reduce this so the power piston raises. Record the vacuum at which the piston starts to raise and the vacuum at which the piston reaches its full-up position. The first value will approximate the start of power-system operation but it is difficult to tell whether this is really the start unless you can observe the position of the tapered portion of the metering rods in the main jets through the cutaway section of an old carburetor.

On the B models, the power piston was held up by manifold vacuum and pushed down by the spring to open a ball-type valve when manifold vacuum decreased.

## TUNING IN THE VEHICLE

A lot of engine-development work related to getting carburetion correct can be done on an engine dynamometer. However, some of the tuning has to be done with the engine installed in the car or boat. In general, it is safe to jet up (richer) one or two jet sizes when moving the engine from the dyno to the chassis or boat.

For tuning—you need a place that is always available whenever you want to use it for tuning purposes. You also need a vacuum gage and a stopwatch. The reason for

### Altitude Kits for 1969 Oldsmobile (5,000 ft.)

Two-barrel carburetors 7029158 and 7029159 use kit 7036528 with 61 main jets and 7009718 6-ounce spring assembly. Carburetor was originally released with 62 main jets and a 7.5-ounce spring.

Quadrajet carburetor 7029250 uses kit number 7036731 with 69 primary jets and 7029922 6-ounce spring. Carburetor was originally released with 70 primary jets and an 8.3-ounce spring.

**NOTE:** Altitude kits are not available for all models because altitude compensation has become less of a problem with the leaner calibrations used in emission-type carbs. Altitude kits may become available again as emission requirements for specific high-altitude areas become more restrictive.

### Power Piston Assemblies for 2G 1-1/2-inch carbs

| Assembly Number | Cut-in point (inches Hg) |
|---|---|
| 7035603 | 11-8-6 |
| 7006323 | 9-7-5 |
| 7009718 | 8-6-4 |

### Power Piston Springs for Quadrajet carbs

| Assembly Number | Cut-in point (inches Hg) |
|---|---|
| 7037734 | 14-4 |
| 7032758 | 11-5 |
| 7037305 | 10-6 |
| 7036019 | 8-4 |
| 7029922 | 7-3 |
| 7037851 | 6-3 |

using the same place is that subtle changes in roads can really throw off your best tuning efforts—unless you always use the same strip. A road can look perfectly level and yet include a substantial grade of several percent. You won't be able to tell this with your naked eye. Surveyor's apparatus would be needed to make this kind of judgment.

If you are tuning for top-end performance, and you know the engine RPM at the end of the quarter mile, a lot of tuning can be accomplished with a stopwatch. Start the watch as you accelerate past an RPM point which is two or three thousand below where you want to be at the top end, then stop the watch when you reach the RPM which marks the end of the range in which you are interested. Use the highest gear so that you eliminate gear changes. By eliminating these, you get rid of one more variable in the tuning procedure. Stopwatching runs from 3500 RPM to the peak RPM which will be used gives a very accurate indication of whether a change is helping or hurting performance. High gear stretches the time required to pass through the RPM range of interest and keeps away from wheelspin (which often occurs at gear changes, even though it may not be obvious) so that you can see the effects of changes in metering.

The race course is a poor place to do tuning when competition is going on. There is never enough time to get the combination running exactly right and there is always a lot of confusion in the busy, exciting and emotion-charged atmosphere of the racing situation. The competitor who has to do any more than "fine" tuning or adjustments is literally not ready to race. Further, at a drag race, conditions constantly change so that the times change. For instance, as more rubber is laid down at the starting line, traction "out of the hole" improves and the times get faster . . . without changing the car.

If you can use the drag strip where you ordinarily race, this is an excellent place to tune, especially if the timing devices can be installed and operating to give you instant feedback on how your tuning efforts are succeeding. The idea here is to use the timing devices and the convenience of an unchanging stretch of road to get things right without any time pressures from competitive action.

Standardize your starting procedure, preferably eliminating any standing starts as wheelspin at the line can make a lot of difference in your times—and will confuse your best tuning efforts. If you want to work on your starting techniques, do that separately—when you have the car tuned to your complete satisfaction.

The times or speeds that a car is turning (assuming good, consistent starting techniques) are a good indication of whether the mixture ratio being supplied by the carburetor/s is correct. If a jet change makes the vehicle go faster, the change is probably being made in the correct direction, regardless of what the spark-plug color tells you. As long as a change produces improvement keep making changes in that direction until the speed falls off, then go back to the combination that gave the best time.

Unless the plugs indicate that the mixture is rich, keep making your moves in the richening direction until times start to fall off. Plug color can indicate perfect mixture, even though the engine is getting into a detonation condition at the upper end.

Because the drag-race engine spends such a small portion of the run at a peak-power condition, spark plugs—especially in a quench-type (wedge) combustion chamber—may need to be bone-white for the best times. Don't be concerned about plugs not coloring in drag events so long as the times keep improving as you make one change at a time—and *only one!*

When reading spark plugs, remember that plugs have tolerances, too. One set of plugs may read one way—and another set with the "same heat range" may read differently. If you can't get the plugs to read correctly, try one step colder plugs and then one range hotter. Plugs can often give you a clue as to how the mixture is being distributed to the various cylinders—provided that the engine is in good tune and the compression is the same in all cylinders—and all valves are seating.

## STAGGER JETTING

If you are working with an engine which has stagger jetting, then make jet changes up or down in equal increments. Keep everything moving equally.

If the engine has "square" jetting (same size in all four holes or same size in

primary barrels with a different "same size" in the secondaries), make the changes equally for each jet position.

A change in manifolding will likely dictate significant metering changes.

*When you are tuning, watch the results. If the results disagree with your theory—believe the results and invent a new theory!*
*—Mike Urich*

## Throttle/transfer-slot relationship

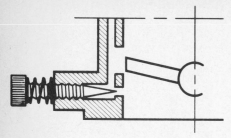

**A** — Transfer slot in correct relation to the throttle plate. Only a small portion of the slot opens below the plate.

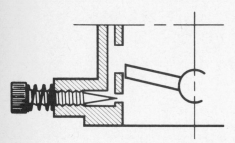

**B** — Throttle plate closing off slot gives smooth idle with an off-idle flat spot. Manifold vacuum starts transfer-fuel flow too late. Cure requires chamfering throttle-plate underside to expose slot as in **A**.

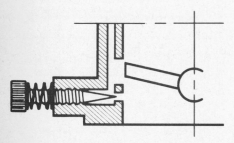

**C** — If slot opens 0.040 inch or more below throttle, rough idle occurs due to excessive richness and little transfer flow occurs when throttle is opened. A long flat spot results.

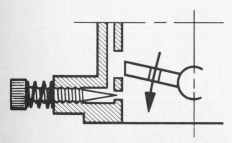

**D** — Correcting condition in **C** requires re-setting throttle to correct position **A** and adding hole in throttle as described in text.

## SPECIAL PROCEDURES FOR "WILD" CAMSHAFTS

The next few paragraphs are very specialized and really apply only to the pro racer. Normally such modifications are not required, even with a wild camshaft. So be sure that you try the carburetor in its "box-stock" condition before you proceed with such changes.

A wild racing camshaft with lots of valve-timing overlap can cause seemingly insurmountable tuning problems. Fortunately, solutions are available, although they are not widely known. If you have gone so far down the racing road that you are using a wild "bumpstick," you should not mind the extra effort required to ensure that the engine will idle at a reasonable speed and will not load up the plugs when the car is run slowly—as on warm-up laps, caution laps or running back to the pits after a drag run. This work also makes the engine more controllable in approaching the staging lights at the drags, which in itself could make the effort worthwhile.

The first thing to do is to follow the normal pre-installation procedures of checking the accelerator-pump setting and making sure that the bowl screws are tight. Look at the underside of the carburetor with the throttle lever held against the curb idle stop (not against a fast-idle cam). Note the position of the primary throttle plates in relation to the transfer slots or holes. This relationship has been established by the factory engineers to give the best off-idle performance. The reasons for these slots/holes are described in the idle system explanation earlier in the book.

Check the throttle-plate-to-throttle-bore clearance with a feeler gage or pieces of paper as you hold the throttle lever against the curb idle stop. Note this clearance in your tuning notebook. Record everything as you proceed, regardless of how good your memory may be.

Install the carburetor on the engine and start the engine. If you have to increase the idle-speed setting to keep the engine running, note how many turns or fractions of a turn open the throttle to this point. Adjust the idle-mixture screws for the best idle. If the mixture screws do not seem to have any effect on the idle quality, note that fact.

Use a respon-sive (not highly damped) vacuum gage to measure the manifold vacuum at idle. If the engine idles with the manifold vacuum occasionally dropping to a value lower than that required to open the power valve, a power valve which will remain closed at idle must be installed before proceeding with the other changes described in the next paragraphs.

Take the carburetor off of the engine. Turn it over and note where the throttle plates are in relation to the transfer slot. If you can see more than 0.040-inch of the off-idle (transfer) slot between the throttle plate and the base of the carburetor, drill a hole in each of the primary throttle plates on the same side as the transfer slot. If holes already exist in the throttle plates, enlarge these holes. Smaller engines require smaller holes than big displacement engines. Start with a 1/16-inch drill on your first attempt and then work up in 1/32-inch steps. Before reinstalling the carburetor, reset the idle to provide the same throttle-plate-to-bore clearance which you measured at the beginning of this procedure.

Start the engine. If the engine idles at the desired speed, the holes are the correct size. Too slow an idle indicates that the holes need to be larger; too fast an idle indicates the need for smaller holes. If holes have to be plugged and redrilled, either solder the holes closed or close them with Devcon "F" Aluminum. When you have the holes at the correct size (which may require the use of number or letter drills to get the idle where you want it), note the size. Where there are two throttle plates on the primary side, both plates should have the same size hole.

Do the idle mixture screws provide some control of the idle quality, that is, do they cause the engine to run rough as the idle needles are opened? If not, check the size of the curb-idle discharge ports into the throttle bore. You can check this simply by unscrewing a mixture needle and using your number-drill set. If the port is not within 0.030-inch of the largest part of the needle—say 0.095-inch port with a 0.125-inch needle—first drill out the discharge ports for each barrel equipped with a mixture screw. If this does not give you idle control so you can get a "rich roll" when backing out the mixture screws, then open up the idle-feed restrictions approximately 0.002-inch at a time until some control is achieved. Correct control is indicated when the engine runs smoothly as possible and turning the idle-mixture screw either way causes the engine RPM to drop and roughens the idle as the mixture is leaned or richened.

Because you are working here with very small holes (requiring a "wire-drill" set), you must proceed in very small increments. Even a 0.002-inch increase in an idle-feed restriction of 0.028 is an area—and hence, flow—increase of 15%. Wire drills, incidentally, are not used in a power drill, but must be held in a pin vise. Pin vises are available where you buy wire drills, namely at precision tool supply houses, model shops, or through the Sears Precision Tool catalog.

An adequate accelerator pump set-up usually eliminates any need to work on the off-idle transfer fuel mixture (controlled by the idle fuel feed restriction). The mixture can be checked by opening the throttle with the idle screw until the main system just begins to start, then backing off the screw until it just stops. If idle and off-idle performance turn out to be unacceptable, increase the idle-feed restriction slightly.

Make changes in very small increments, remembering that it is easy to drill out holes, but there is no easy way to get back to the starting point. Sometimes it may require a new throttle body.

And, it's easier to take the carburetor off to increase a hole size in the throttle plate/s than it is to epoxy the hole closed and drill another one. This procedure is effective if done with care.

Olds Toronado on the way up Pike's Peak to set a 14:09.9-minute record on October 7, 1965. The car, driven by Bobby Unser, recorded the best time to that date—and did it with an automatic transmission. Carburetion was with a single Rochester Quadrajet.

NOTE: Never back off the idle-setting screw or change the fixed or spring stop to allow the primary throttle/s to seat against the throttle bores. Always maintain the original clearance between the plates and bores. Secondaries will be seated against the bores and this is correct and normal practice in RPD carburetors.

**Emission-control carburetors on engines with wild camshafts**—As discussed on page 32, emission carburetors may have reduced-size curb-idle ports. Before increasing the idle-fuel orifice to make a carburetor work with a wild cam, always open up the curb-idle discharge ports to a pre-emission-control size. On the Q-jet this is about 0.095-inch.

# Economy—Driveability & General Performance

This book has been written to fulfill many objectives: Save you money, destroy myths concerning many phases of carburetors, simplify carburetion endeavors for you by illustrating and explaining things in a new manner, etc. This section is going to delve into areas of fine-tune metering tips to provide you capability to calibrate your Quadrajet for any purpose or use. Many things we say here relate to driving situations you will encounter in family- or business-vehicle use. We must reiterate previous statements concerning emission regulations.

Since 1970 it has been taboo to alter or delete emission-related components of the vehicle. This applies to most engine-dress items, ignition-timing settings, carburetor idle or specification changes, etc. We do not advocate the attitude many people have taken in just removing and altering things because someone said so. I do however have a lot of sympathy with grievances from motorists who have vehicles that spit, cough, die, idle poorly, accelerate like a snail, and/or eat gas like a broken trunk line. Sympathy, but no right to recommend a course of action other than to suggest you give your dealer every opportunity to cure your problems. In many cases he can help. If he does not tune your vehicle to reasonable drivability and performance, you will have to be the judge in the application of metering techniques discussed here.

By applying some thought to your own needs—whether it be slow driving in the city, flat-out hauling, off-road running, or whatever—these procedures will get you results.

Suppose you have only one vehicle and a strong desire to do some "weekend warrior-ing" in motorsports events: slaloms, gymkhanas, rallies, road races, drag races, hill climbs, etc. You know a bone-stock engine hasn't a prayer of being competitive in any type of off-road event where modifications are permitted. You also know that you must sacrifice some drivability and loss of operating economy on the street to be reasonably competitive in the off-road event of your choice. And you are willing to accept these sacrifices—up to a point. But where is that point?

A good question and a difficult one compounded by increasingly stringent exhaust-emission limits. You may be called upon at any time for a roadside emissions check by state or city authorities, a practice that is not only entirely legal but one which is increasing in frequency all over the U.S. It's a very safe bet that if your engine is modified, by 1975-76 any exhaust "sniffer" will turn thumbs-down on the exhaust-emission levels. Then you'd best be prepared to convert the engine to stock condition and submit your vehicle for a recheck, or face the consequences and penalties of the law.

The best advice is to keep all emission-control devices hooked up and operative when the vehicle is driven on any public road. These devices *do* help reduce exhaust emissions and if nothing else, they give tangible evidence to any vehicle-inspection officer that your implied intent was conformity with the law, even if the exhaust-emission levels are unacceptable. Let's listen in on a roadside inspection: Guy, "I've been thrashin' it pretty hard lately. Guess it needs a sharp tune-up." Inspection official, "Yeah. Sign your citation here, and *do* visit our humble inspection station again. Within the mandatory 30-day period, of course. And with the emission levels *right*." Sound ridiculous? Perhaps. But it's happening every day!

Because most of our readers will be interested in getting more economy from a Q-jet, let's start off with a discussion of what might be done to improve the economy, drivability and general all-around performance of that carburetor. Later on we'll get to the two-barrel and the Monojet. Much of what we discuss for the Q-jet will apply to those, so pay attention as you read through this section as we won't repeat important information in the 2G and Monojet sections.

**Idle Systems**—Do not alter the idle systems unless off-idle or light-throttle tip-in acceleration sags exist. Nearly any modification you make in this area will have a significant effect on the ability of your engine to pass any of the current emissions inspections. Should you decide your situation demands

Idle tubes 1 have restrictions 1-21/32 inch below the upper end. A very special long drill or special reamer is required to enlarge the metering restrictions without removing the tubes. If the tubes are to be removed, new ones must be drilled to the desired opening and installed. Sometimes the old tubes can be used, as described in the text. Channel restrictions 2 can be tailored to provide desired mid-range driving characteristics.

Here a special drill is being inserted to enlarge the idle-tube metering restriction. Such modifications destroy the engine's ability to pass emission tests at speeds above curb idle.

a change, the accompanying photos point out the location of the channel restrictions and idle tubes which can be modified to increase idle and off-idle fuel. *Before* making any changes to the idle feed orifices or to the channel restrictions, re-read the information on curb-idle discharge port size in the idle system discussion of chapter two. Increasing the size of the curb-idle discharge ports to pre-emission size should always be a first step prior to enlarging the idle-feed restrictions (orifices).

       The channel-restriction orifices can be easily altered if you have a suitable set of small drills. The only reason to do this is if light-throttle weakness and surge exist. Increasing the diameter a few thousandths of an inch will richen the off-idle air/fuel mixture. Never exceed a 10 percent increase in orifice area. Use a tacky grease on the bit to collect the tiny chips as you drill the thin orifice. A 10 percent increase in channel-restriction orifice area is equivalent to approximately a 0.003-inch increase in orifice diameter.

       Q-jet idle tubes are very difficult to alter. They are generally in the 0.030-inch to 0.035-inch-diameter range and regular drill bits that size are approximately 1-1/2-inch long. The orificed part of the tube is 1-21/32 inch below the top of the tube. It is obvious the drill will not reach the calibrated orifice and enlarging the upper portion of the tube does not help. Extra-long drills in the required size range are very difficult to find. Idle tubes are literally impossible to remove without a special tool. They are not a dealer-service item so expect zero help from your regular parts man or service manager. They do not have the parts or the required tools.

**Strictly for the professionals**—If you are intrigued by the idea of using the tapered jet reamers discussed here, make sure you understand what you are doing and have access to micrometer and a good machinist's scale or calipers. You may want to saw up a carburetor to make precise measurements of where the orifice area is located.

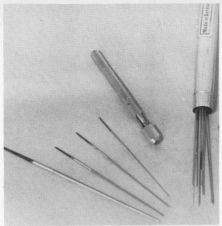

Jet reamers (available from foreign car automotive tool supply dealers) can be used to open up the idle tubes on a Q-jet IF you first open the top of the idle tube (where it is held in the collar) with a 0.060-inch drill to a depth of perhaps 3/8-inch. Wiggle the drill a bit and you should end up with sufficient clearance for the larger upper portion of the tapered reamer. The reamers also allow you to measure an in-between size orifice which doesn't happen to be the same size as one of your number drills. Just insert reamer until it stops, mark that point on the reamer and mike it to get the orifice size.

Here a tapered jet reamer is being used to open the channel restriction ever so slightly to cure mid-range problems as discussed in the text.

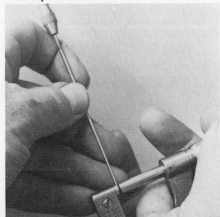

Miking the tip of a tapered reamer to find the portion providing the desired dimension. If this point is more than 3/32-inch from the tip, the tip of the reamer will have to be ground off as there is less than 1/8-inch clearance under the idle orifice in a Q-jet. Be sure to mike across the high points of the reamer. Reamer vise comes with tapered reamer set.

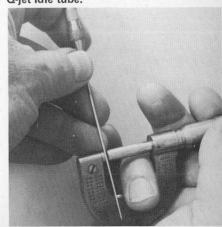

Once you have found out where the desired dimension is on the tip of the reamer, mark that with a pen and measure 1-21/32 inch up the reamer to see whether the larger portion of the reamer will clear the 0.060-inch area you created in the top of the Q-jet idle tube.

I am going to suggest a few steps that should let you remove Q-jet idle tubes without damage. It will take some time for fabrication, but this time will be regained many times over if you are seriously engaged in a tuning program—or helping your buddies with their carburetors.

Get a 5/64-inch-diameter drift punch or similar round stock and carefully file or grind the end to create a center pilot about 1/16-inch long. The pilot or tip should fit nicely into the center of the top of the idle tube. This will allow the shoulder of the drift punch to sit squarely on the idle tube. With short quick taps, drive the idle tube downward until it hits bottom. It will hit bottom before freeing itself from its mounting collar.

Now select a sheet-metal screw. A no. 8 about 1/2-inch long should be about right. Now get a 5- or 6-inch long 1/4-inch capscrew and place a heavy nut on it before you weld the end of the bolt to the sheet-metal screw (or braze or silver solder). The nut serves as a knocker which can be banged upward against the bolt head.

Firmly screw the tool into the idle-tube collar and begin trying to pull it out of the carburetor by jarring the tool upward with the knocker nut.

Usually these tools and this approach to the problem will allow you to remove the two-piece idle-tube assembly. No guarantees, my friend, you are on your own!

The tube and collar generally come out as an assembly. In this case you drill out the lower orifice end to suit your purpose. Before installing the idle tube back into the carb, tap the tube back into its collar until the top edge of the tube is flush with the collar. Do this with a piece of hard wood to protect the orificed end. Recheck your orifice and replace the tube in the carburetor.

It is smart to have some extra idle tubes on hand just in case you damage the assembly in getting it out. These are available from Doug Roe Engineering.

If you don't want to take the time to build the tool, then having the new idle tubes on hand is absolutely essential. In this case, you can drill out the idle tube—which destroys it—and then remove the remaining collar with an Ezy-Out or an awl, taking care not to damage the bore into which the collar fits.

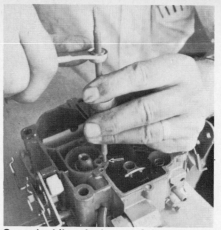
Once the idle tube is out of the collar, the threaded end of the special tool you have made can be screwed into the idle-tube collar.

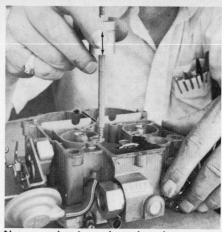
Now a socket is used as a knocker to cause the tool to move upward, pulling the idle-tube collar out.

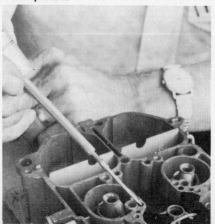
Here the idle tube and collar are out of the carburetor. The hardest part of the project is completed.

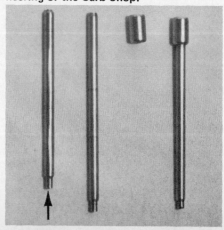
Q-jet idle tubes left to right: replacement idle tube (shorter than stock tube, but does not affect performance), idle tube, idle-tube collar, and complete two-piece idle tube (stock parts). Replacement tube (arrow) is available from Doug Roe Engineering or the Carb Shop.

## CHART A
## METERING-AREA QUICK REFERENCE CHART

$$\pi \left(\frac{d}{2}\right)^2 = A$$

$$d^2 \times \frac{\pi}{4} = A$$

$$d^2 \times 0.7854 = A$$

With this chart you can read the calculated area determined for each of the listed diameters.

| INCH | MM | SQ. IN. | DRILL SIZE | INCH | MM | SQ. IN. | DRILL SIZE |
|------|------|-------------|----------------------------|------|-------|-------------|------------------------------|
| .001 | .025 | .000000785 | | .071 | 1.803 | .003959189 | |
| .002 | .051 | .000003142 | | .072 | 1.829 | .004071501 | |
| .003 | .076 | .000007069 | | .073 | 1.854 | .004185383 | 49 |
| .004 | .102 | .000012566 | | .074 | 1.880 | .004300837 | |
| .005 | .127 | .000019635 | | .075 | 1.905 | .004417861 | |
| .006 | .152 | .000028274 | | .076 | 1.930 | .004536456 | 48 |
| .007 | .178 | .000038484 | | .077 | 1.956 | .004656622 | |
| .008 | .203 | .000050265 | | .078 | 1.981 | .004778358 | 5/64 (.0781)—47 (.0785) |
| .009 | .229 | .000063617 | | .079 | 2.007 | .004901666 | |
| .010 | .254 | .000078540 | | .080 | 2.032 | .005026544 | |
| .011 | .279 | .000095033 | | .081 | 2.057 | .005152993 | 46 |
| .012 | .305 | .000113097 | | .082 | 2.083 | .005281013 | 45 |
| .013 | .330 | .000132732 | 80 (.0135) | .083 | 2.108 | .005410603 | |
| .014 | .356 | .000153938 | 79 (.0145) | .084 | 2.134 | .005541765 | |
| .015 | .381 | .000176714 | 78 — 1/64 (.0156) | .085 | 2.159 | .005674497 | |
| .016 | .406 | .000201062 | | .086 | 2.184 | .005808800 | 44 |
| .017 | .432 | .000226980 | | .087 | 2.210 | .005944673 | |
| .018 | .457 | .000254469 | 77 | .088 | 2.235 | .006082118 | |
| .019 | .483 | .000283528 | | .089 | 2.261 | .006221133 | 43 |
| .020 | .508 | .000314159 | 76 | .090 | 2.286 | .006361720 | |
| .021 | .533 | .000346360 | 75 | .091 | 2.311 | .006503876 | |
| .022 | .559 | .000380132 | 74 (.0225) | .092 | 2.337 | .006647604 | |
| .023 | .584 | .000415475 | | .093 | 2.362 | .006792903 | |
| .024 | .610 | .000452389 | 73 | .094 | 2.388 | .006939772 | 42 (.0935)—3/32 (.0937) |
| .025 | .635 | .000490873 | 72 | .095 | 2.413 | .007088212 | |
| .026 | .660 | .000530929 | 71 | .096 | 2.438 | .007238223 | 41 |
| .027 | .686 | .000572555 | | .097 | 2.464 | .007389805 | |
| .028 | .711 | .000615752 | 70 | .098 | 2.489 | .007542957 | 40 |
| .029 | .737 | .000660519 | 69 (.02925) | .099 | 2.515 | .007697681 | 39 (.0995) |
| .030 | .762 | .000706858 | | .100 | 2.540 | .007853975 | |
| .031 | .787 | .000754767 | 68—1/32 (.0312) | .101 | 2.565 | .008011840 | 38 (.1015) |
| .032 | .813 | .000804247 | 67 | .102 | 2.591 | .008171275 | |
| .033 | .838 | .000855298 | 66 | .103 | 2.616 | .008332282 | |
| .034 | .864 | .000907919 | | .104 | 2.642 | .008494859 | 37 |
| .035 | .889 | .000962112 | 65 | .105 | 2.667 | .008659007 | |
| .036 | .914 | .001017875 | 64 | .106 | 2.692 | .008824726 | 36 (.1065) |
| .037 | .940 | .001075209 | 63 | .107 | 2.718 | .008992015 | |
| .038 | .965 | .001134114 | 62 | .108 | 2.743 | .009160876 | |
| .039 | .991 | .001194590 | 61 | .109 | 2.769 | .009331307 | 7/64 (.1094) |
| .040 | 1.016 | .001256636 | 60 | .110 | 2.794 | .009503309 | 35 |
| .041 | 1.041 | .001320253 | 59 | .111 | 2.819 | .009676882 | 34 |
| .042 | 1.067 | .001385441 | 58 | .112 | 2.845 | .009852026 | |
| .043 | 1.092 | .001452200 | 57 | .113 | 2.870 | .010028740 | 33 |
| .044 | 1.118 | .001520530 | | .114 | 2.896 | .010207025 | |
| .045 | 1.143 | .001590430 | | .115 | 2.921 | .010386881 | |
| .046 | 1.168 | .001661901 | | .116 | 2.946 | .010568308 | 32 |
| .047 | 1.194 | .001734943 | 56(.0465)—3/64 (.0468) | .117 | 2.972 | .010751306 | |
| .048 | 1.219 | .001809556 | | .118 | 2.997 | .010935874 | |
| .049 | 1.245 | .001885739 | | .119 | 3.023 | .011122013 | |
| .050 | 1.270 | .001963494 | | .120 | 3.048 | .011309723 | 31 |
| .051 | 1.295 | .002042819 | | .121 | 3.073 | .011499004 | |
| .052 | 1.321 | .002123715 | 55 | .122 | 3.099 | .011689856 | |
| .053 | 1.346 | .002206182 | | .123 | 3.124 | .011882278 | |
| .054 | 1.372 | .002290219 | | .124 | 3.150 | .012076271 | |
| .055 | 1.397 | .002375827 | 54 | .125 | 3.175 | .012271836 | 1/8 |
| .056 | 1.422 | .002463007 | | .126 | 3.200 | .012468971 | |
| .057 | 1.448 | .002551756 | | .127 | 3.226 | .012667676 | |
| .058 | 1.473 | .002642077 | | .128 | 3.251 | .012867953 | |
| .059 | 1.499 | .002733969 | 53 (.0595) | .129 | 3.277 | .013069800 | 30 (.1285) |
| .060 | 1.524 | .002827431 | | .130 | 3.302 | .013273218 | |
| .061 | 1.549 | .002922464 | | .131 | 3.327 | .013478206 | |
| .062 | 1.575 | .003019068 | 1/16 (.0625) | .132 | 3.353 | .013684766 | |
| .063 | 1.600 | .003117243 | 52 (.0635) | .133 | 3.378 | .013892896 | |
| .064 | 1.626 | .003216988 | | .134 | 3.404 | .014102597 | |
| .065 | 1.651 | .003318304 | | .135 | 3.429 | .014313869 | |
| .066 | 1.676 | .003421192 | | .136 | 3.454 | .014526712 | 29 |
| .067 | 1.702 | .003525649 | 51 | .137 | 3.480 | .014741125 | |
| .068 | 1.727 | .003631678 | | .138 | 3.505 | .014957110 | |
| .069 | 1.753 | .003739277 | | .139 | 3.531 | .015174665 | |
| .070 | 1.778 | .003848448 | 50 | .140 | 3.556 | .015393791 | 28 (.1405)—9/64 (.1406) |

This chart prepared by Doug Roe Engineering, supplier of metering components for Rochester carburetors.
P. O. Box 26848, Tempe, AZ 85282

# Quadrajet Main Metering Jets, Primary Metering Rods & Secondary Metering Rods

**Jets and rods for other Rochesters**—Monojet main metering jets and rods are detailed in a table on page 52. Main metering jets for Models B, 2G, 4G and H are shown in a table on page 214. Jets should not be mixed, that is, jets for a Quadrajet should only be used in that type carburetor, etc.

## JET & ROD SUPPLIERS

Jets and rods for Rochester carburetors are available through your Delco (United Delco) dealer or GM automotive and truck dealers. If these sources are out of stock on the specific items you need, write or call:

Doug Roe Engineering, Inc.
P. O. Box 26848
Tempe, AZ 85282

Enclose self-addressed stamped business-size envelope (at least 4 x 9-3/8 inches) or 25¢ for price sheet.

The Carburetor Shop
1942 Harbor Blvd.
Costa Mesa, CA 92626
(714) 642-8286

No correspondence please. Visit or call this firm if you live in Southern California.

## QUADRAJET MAIN METERING JETS (MODELS 4MC AND 4MV)

Primary jets used in the Q-jet differ from and should not be interchanged with those in other models. They are identifiable by two curved lines stamped opposite the identification number on the jet face. Number indicates the orifice size in thousandths of an inch.

Example:

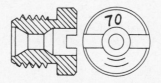

| Part No. | Stamped On Jet | Diameter Of Orifice |
|---|---|---|
| 7031970 | 70 | .070'' |

### MAIN METERING JETS—PRIMARY

| Part No. | Stamped On Jet |
|---|---|
| 7031966 | 66 |
| 7031967 | 67 |
| 7031968 | 68 |
| 7031969 | 69 |
| 7031970 | 70 |
| 7031971 | 71 |
| 7031972 | 72 |
| 7031973 | 73 |
| 7031974 | 74 |
| 7031975 | 75 |
| 7031976 | 76 |
| 7031977 | 77 |
| 7031978 | 78 |

## QUADRAJET PRIMARY MAIN METERING RODS

Two primary metering rod types are used. Some have a single taper at the metering tip; others have a double taper at the metering tip. Both rod types use a similar two-digit numbering system. The number indicates the metering rod diameter at point "A" and is the last two digits of the part number. Models with a double taper have "B" stamped after the two-digit number on the rod. All have a 0.026-inch-diameter tip at the small end of the rod, except rod 50D which has a 0.036-inch tip.

Example:

### 1965-67 RODS
Diameter "A"
Single taper

| Part No. | Stamped No. | Diameter at "A" |
|---|---|---|
| 7031844 | 44 | .044'' |

### MAIN METERING RODS—PRIMARY SINGLE TAPER

| Part No. | Stamped |
|---|---|
| 7031833 | 33 |
| 7031835 | 35 |
| 7031836 | 36 |
| 7031837 | 37 |
| 7031838 | 38 |
| 7031839 | 39 |
| 7031840 | 40 |
| 7031841 | 41 |
| 7031842 | 42 |
| 7031843 | 43 |
| 7031844 | 44 |
| 7031845 | 45 |
| 7031846 | 46 |
| 7031847 | 47 |
| 7031849 | 49 |

Example:

### 1968 RODS
Diameter "A"
Double taper

| Part No. | Stamped No. | Diameter at "A" |
|---|---|---|
| 7034844 | 44B | .044'' |

### MAIN METERING RODS—PRIMARY DOUBLE TAPER

| Part No. | Stamped |
|---|---|
| 7034836 | 36B |
| 7034838 | 38B |
| 7034839 | 39B |
| 7034840 | 40B |
| 7034841 | 41B |
| 7034842 | 42B |
| 7034843 | 43B |
| 7034844 | 44B |
| 7034845 | 45B |
| 7034846 | 46B |
| 7034847 | 47B |
| 7034848 | 48B |
| 7034849 | 49B |
| 7034850 | 50B |
| 7034851 | 51B |
| 7040699 | 48C* |
| 7040701 | 52C* |
| 7046338 | 50D** |

*Special triple-taper rods—used 1970 Oldsmobile only.

**Special rod used in GMC Motorhome—has 0.036-inch power tip.

## QUADRAJET SECONDARY METERING RODS
**The secondary rods are identified with a two-letter code.**

| Ident. | Part No. | A | B | C | Power Tip Length |
|---|---|---|---|---|---|
| AD* | 7033772 | .0950 | .0225 | none | — |
| BV* | 7040724 | .1350 | .1130 | .0300 | S |
| CB* | 7042335 | .1350 | .1130 | .0300 | S |
| DC | 7047816 | .1350 | .1095 | .0303 | M |
| **CC** | **7042356** | **.1350** | **.1095** | **.0303** | **M** |
| BY* | 7040856 | .1290 | .0950 | .0320 | M |
| CF | 7044775 | .1350 | .0964 | .0340 | M |
| CS | 7045924 | .1333 | .1130 | .0400 | S |
| CM | 7045840 | .1345 | .1060 | .0400 | M |
| BF | 7034400 | .1328 | .1118 | .0400 | S |
| BG | 7034822 | .1335 | .1060 | .0400 | M |
| BH | 7035916 | .1345 | .1060 | .0400 | M |
| BP | 7038034 | .1340 | .1130 | .0400 | S |
| CA | 7042304 | .1325 | .1070 | .0400 | M |
| AX | 7033549 | .1342 | .0964 | .0400 | S |
| BB* | 7034335 | .1342 | .0964 | .0400 | S |
| BJ | 7036077 | .1336 | .1118 | .0400 | S |
| BK | 7037295 | .1328 | .1118 | .0400 | S |
| BM* | 7037744 | .1110 | .1060 | .0400 | M |
| BW | 7040767 | .1320 | .0980 | .0400 | M |
| **CJ** | **7045780** | **.1342** | **.0964** | **.0400** | **S** |
| CE | 7043771 | .1350 | .0964 | .0413 | L |
| BN* | 7036671 | .1332 | .1064 | .0413 | S |
| BL | 7037733 | .1332 | .1114 | .0413 | S |
| **BE** | **7034377** | **.1332** | **.1114** | **.0413** | **S** |
| **CY & DA** | **7046004** | **.1334** | **.0964** | **.0443** | **M** |
| AD | 7033772 | .1349 | .0950 | .0450 | S |
| CK* | 7045781 | .1348 | .1051 | .0530 | L |
| CV | 7045984 | .1335 | .1051 | .0530 | L |
| **AU** | **7033655** | **.1348** | **.1051** | **.0530** | **L** |
| AH | 7033812 | .1348 | .1051 | .0530 | M |
| CR | 7045923 | .1342 | .1130 | .0550 | S |
| BU | 7040725 | .1350 | .1130 | .0550 | S |
| AR | 7033171 | .1355S | .1060 | .0570 | S |
| CD | 7042719 | .1354S | .0990 | .0570 | L |
| BA | 7034337 | .1335 | .0987 | .0570 | S |
| AZ | 7033889 | .1353 | .1060 | .0570 | L |
| AY | 7033830 | .1355S | .1060 | .0570 | L |
| CH | 7045779 | .1353 | .1060 | .0570 | S |
| CX | 7045985 | .1349 | .1060 | .0570 | L |
| CP | 7045842 | .1323 | .0987 | .0570 | S |
| CN* | 7045841 | .1343 | .1060 | .0570 | S |
| AV | 7033182 | .1341 | .1046 | .0570 | M |
| AL* | 7033680 | .1352 | .0987 | .0570 | S |
| AK | 7033104 | .1353 | .1060 | .0570 | S |
| AJ* | 7033628 | .1348 | .1046 | .0570 | M |
| AP* | 7033981 | .1348S | .1046 | .0570 | M |
| BZ* | 7042300 | .1355S | .1060 | .0570 | L |
| **BD** | **7034365** | **.1335** | **.1068** | **.0580** | **M** |
| BC | 7034300 | .1355S | .1089 | .0584 | S |
| BT | 7040601 | .1341 | .1046 | .0600 | M |
| **AT** | **7033658** | **.1353** | **.1082** | **.0670** | **L** |
| CL* | 7045782 | .1353 | .1082 | .0670 | L |
| AN | 7034320 | .1355S | .1103 | .0700 | S |
| BX* | 7040797 | .1355S | .1103 | .0700 | S |
| DB | 7047806 | .1337 | .1103 | .0700 | S |
| CT | 7045983 | .1340 | .1082 | .0777 | M |
| **CG-AS** | **7045778** | **.1353** | **.1082** | **.0777** | **M** |
| DE | 7048092 | .1340 | .1108 | .0877 | M |
| BR* | 7038910 | .1353 | .1082 | .0900 | L |
| AW* | 7033194 | .1353S | .1121 | .0908 | M |
| BS | 7038911 | .1353 | .1060 | .0950 | L |
| **CZ** | **7045986** | **.1348** | **.1060** | **.0950** | **L** |
| DD | 7048091 | .1348 | .1078 | .1050 | L |

Rods shown in bold-face type will handle 90% of all metering requirements for the performance tuner. The chart starts at the top with rich WOT metering C and goes toward leaner WOT metering at the bottom C. 50° of air-valve opening moves the rod approximately one-half of its travel. **S** after dimension **A** indicates that there is a slot in that portion of the metering rod. **S** is a short power tip (indicates that taper from **A** to **C** is longer). Power tip is reached at 90° to 100° of air-valve opening. **M** is a medium-length power tip. Power tip is reached at 80° to 90° of air-valve opening. **L** is a long power tip (indicates that taper from **A** to **C** is short). Power tip is reached at 70° to 80° of air-valve opening.

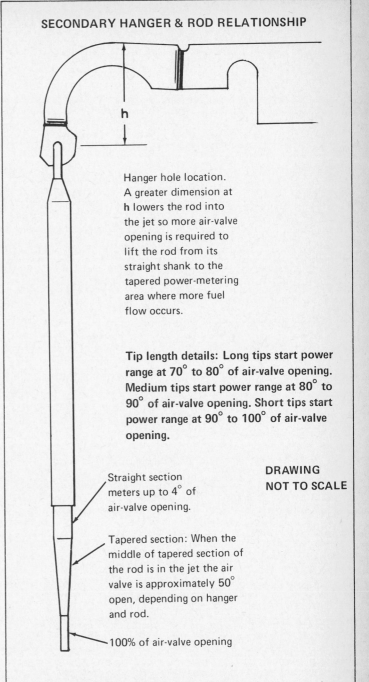

## SECONDARY HANGER & ROD RELATIONSHIP

Hanger hole location. A greater dimension at **h** lowers the rod into the jet so more air-valve opening is required to lift the rod from its straight shank to the tapered power-metering area where more fuel flow occurs.

Tip length details: Long tips start power range at 70° to 80° of air-valve opening. Medium tips start power range at 80° to 90° of air-valve opening. Short tips start power range at 90° to 100° of air-valve opening.

Straight section meters up to 4° of air-valve opening.

Tapered section: When the middle of tapered section of the rod is in the jet the air valve is approximately 50° open, depending on hanger and rod.

100% of air-valve opening

**DRAWING NOT TO SCALE**

**WHICH SECONDARY RODS?** For drag racing, the longer power tips would probably be helpful because this gets the power mixture in earlier (less opening of the air valve).

For off-road, the shorter power tips would help to provide better economy and short-spurt performance.

For street use or general driving, the short tips will help economy, providing that there are no sags or tip-in hesitations.

When wide-open mixture is being tailored to the engine, use of the longer power tips makes it less important to have a full range of hangers available for your tuning efforts. You have more assurance that you will reach the wide-open tip when the air valve opens.

Wherever there is to be long-term full-power operation of the engine, use the long power tips to ensure that the air valve lifts the rod to the point where the power tip will provide a power mixture.

There are two reasons why you might want to increase the idle-tube orifice size. Firstly, when curb idle is not satisfactory with the idle-mixture needles turned well out (counterclockwise), larger idle tubes will give you more mixture control. Secondly, Q-jets are sometimes plagued with nozzle drip at idle which causes poor idle stability. This can be observed by taking off the air cleaner and looking into the carb while the engine is idling. Increasing idle-mixture richness with the adjusting screws will cure many cases of nozzle drip, but it takes larger idle orifices in occasional instances. Larger orifices also help light-throttle driving feel during maneuvers from idle up to 30 MPH. Should you find a means of enlarging the orifices, be sure to get all the chips out. Remove the primary jets and force fuel and air back and forth through the idle tube via the jet wells. When enlarging the orifices do not exceed the stock diameter by more than 0.005-inch. Remember 0.001" to 0.002" is usually enough of a change to raise CO emission beyond legal limits at idle and just off idle.

**Primary Metering**—Quadrajet carburetors have small primary bores with a triple venturi in each one. These provide for fine calibrations of the carburetor. The Q-jet is the most versatile and adaptable four-barrel carburetor ever built. For this reason, the amateur carburetor mechanic can often diagnose problems and make changes in the Q-jet that would be difficult or impossible on any other four-barrel.

The following metering exercises demonstrate the latitudes provided by these units.

This example illustrates a means of improving 30 to 60 MPH drivability by increasing part-throttle metering and improving gas mileage by reducing heavier throttle fuel. The accelerating performance in this instance did not change after alterations. The owner wanted good drivability and improved economy. To accomplish this, the final calibration for WOT fuel was on the lean side. Although these changes are not recommended for continuous high-speed driving or pulling a trailer, they provide an excellent combination for general transportation use.

Our test car was a completely stock 1972 Blazer 350 CID with automatic transmission. The first step was to check and tune the electrical system including setting dwell and timing. Too

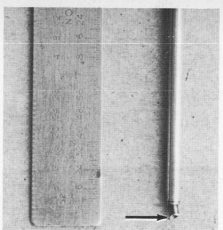

**Proof that the knock 'em down then pull 'em out technique is not fool-proof. There is very little room under the idle orifice and driving the tube part way through the collar can bend the tip as I did here.**

many so-called "carburetor problems" are corrected in the ignition system after a very expensive carburetor rebuild has not done the job.

Metering area (orifice area minus rod area) is the only way to make effective comparisons of various combinations of metering rods and jets without elaborate flow equipment.

**Original Metering**—After the ignition had been completely checked, the carburetor was removed and disassembled to find out what the stock metering areas were. Identification numbers are marked on the jets and rods. Chart A has been included as a quick reference for finding areas once the diameter is known (based on the formula: $\pi \times (\frac{d}{2})^2$ = area of circle).

### Part-Throttle Metering

| Part-Throttle Metering | |
|---|---|
| **Production Released** | **Area (in.$^2$)** |
| Primary metering jet— 0.073-inch diameter | 0.0041854 |
| Part-throttle metering rod dimension— 0.042-inch diameter | - 0.0013854 |
| Part-throttle metering area— (one jet only) | 0.0028000 |

All Quadrajet primary metering rods have a wide-open (small-end) diameter of 0.026-inch. A vacuum-operated power system automatically moves this small part of the metering rod to the jet orifice when manifold vacuum is below approximately seven inches Hg (WOT). To reduce wide-open throttle fuel and increase economy we must reduce the jet size. A couple of changes, followed by stopwatch-monitored accelerations determined this vehicle would tolerate a primary jet decrease to 0.070-inch without hurting performance. This was a stock vehicle and quick times were not the goal.

WOT Accelerations/0—1/4-mile—2-way average— 18.5 seconds
WOT Accelerations/30—70 MPH—2-way average— 11.1 seconds

Surge was noticed near the end of the 1/4-mile with the 0.070-inch jet but the time compared to base runs within a tenth of a second. The "lean limit" for WOT had been reached. Steady-state full-power runs on a dynamometer would show that the 0.070-inch jets are too lean for continuous high speed or trailer pulling. Running at WOT with this kind of metering could be expected to damage pistons and valves, and the spark plugs would not last very long at all.

Part-throttle driving at this point was intolerably lean because the 0.003-inch leaner jets, along with the original 0.042-inch-diameter part-throttle portion of the primary metering rod, did not leave enough flow area. Using the area chart we determined that a 0.036-inch diameter for the part-throttle portion of the rod in a 0.070-inch main jet would provide an increase in part-throttle metering area of 1.09 percent, compared to original specifications. You might wonder why we settled for this 1.09 percent increase. We carefully evaluated the vehicle for drivability before starting any work. It was apparent this area needed a bit more fuel because the stock metering had caused somewhat objectionable part-throttle surge. It is also known that big changes in this area are not tolerable (3% would be considered a big change!), so when the area calculations showed a minor increase (namely 1.09 percent) we bought it. Had it driven poorly on the reevaluation-ride check, a 0.035-inch rod would have been the next step. You can't always hit "the specification" the first time.

These changes affect all light- to medium-throttle driving from 30 to 60 MPH. Once you throttle hard enough to lower manifold vacuum below about 7 inches Hg, you are metered by the 0.070-inch jet and the 0.026 portion of the rod as mentioned above.

The installation was made and a modest improvement in part-throttle driving was noted. Once again this confirmed the validity of using metering area as a calibration tool and thereby finalized the primary-metering specifications. In cold climates another two percent throttle-metering area would be beneficial to cold start and warm-up operation. In such a case a 0.035-inch part-throttle rod diameter would have been the choice.

| Final Primary Metering | Area (in.$^2$) |
|---|---|
| Primary metering jet—<br>0.070-inch diameter | 0.0038485 |
| Part-throttle primary metering rod—<br>0.036-inch diameter | - 0.0010179 |
| Final part-throttle metering area— | 0.0028306 |
| Original part-throttle metering area— | - 0.0028000 |
| Area increase— | 0.0000306<br>(+1.09%) |

$$\frac{\text{New area}}{\text{Original area}} \times 100\% = \% \text{ change (+)}$$

| Final WOT Metering | |
|---|---|
| Original metering jet—<br>0.073-inch diameter | 0.0041854 |
| Metering rod power tip—0.026'' | - 0.0005309 |
| Original WOT metering area | - 0.0036545 |
| Final metering jet—<br>0.070-inch diameter | 0.0038485 |
| Metering rod tip—<br>0.026-inch diameter | - 0.0005309 |
| Final WOT metering— | 0.0033176 |
| WOT area decrease— | 0.0003369<br>(- 9.22%) |

$$\frac{\text{Original area}}{\text{New area}} \times 100\% = \% \text{ change ( - )}$$

Normally 8 to 10 percent is the maximum tolerable reduced primary-metering area for near-sea-level-stationed vehicles. This Blazer ended up driving better and the economy was improved 20 percent (approximately three mpg in this instance). High-altitude power and economy would also benefit from this calibration. As mentioned previously, the only shortcoming with this calibration is WOT leanness. If WOT running is limited

to passing maneuvers, the car will run better and more economically with no ill effects.

As a matter of interest, these particular changes will raise CO emissions slightly at modest speeds and greatly reduce it at heavier throttle maneuvers. HC could go a little either way and $NO_x$ will be reduced at low speeds and increased during heavier throttle. In most cases these metering changes will not influence current highway surveillance checks and in my humble opinion, this type metering fix will put out less pollutants overall than original metering. Regardless—if your state has highway checks, it would be wise for you to use an exhaust analyzer while calibrating your carburetor.

At any rate try it for yourself and if you don't get the results you are looking for, you can always go back to stock.

1. Diagnose the problem; i.e., idle, off-idle, part-throttle, rich/lean, etc.

2. Use metering areas to determine changes and keep track of base.

3. Use correct tool for the job and always check drill sizes with a micrometer *before drilling*.

4. Do before-and-after performance and economy tests to confirm the validity of the changes.

5. Use Delco parts whenever possible for high quality at reasonable cost.

6. If seemingly unsurmountable problems arise contact the carburetor experts at Doug Roe Engineering.

**Secondary Metering**—Once you try metering the Q-jet to meet specific standards it becomes an enjoyable challenge. This carburetor has many primary rods, secondary rods and hangers available which provide the means for making infinite variations in the air/fuel ratio to meet engine demands. Suppose you wanted to improve low-speed drivability and economy as we did with the Blazer, but you also want to be able to run the engine at near WOT for extended periods of time. Because the primary metering cannot be changed without losing economy and drivability during normal conditions, the extra fuel must come from the secondary metering system. The secondary metering jets are a fixed size and not alterable, but there are 65 different sizes/shaped rods available with various tapers and tip diameters. For extreme fine tuning at the factory there are also 20 different height rod hangers. (Only one is available from

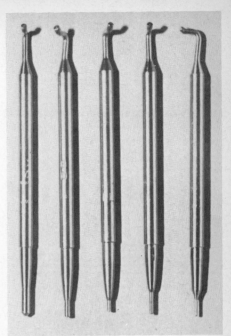

Secondary metering rods vary greatly at the small power end as shown in this slightly-larger-than-actual-size photo. Super-lean rod at left is CZ, followed by BD, BL, AX and CC (super-rich).

Delco as a replacement part). Doug Roe offers three: a high, medium and low. Secondary metering is a matter of selecting the correct pieces to do the job.

On the Blazer we left off with a lean-limit power calibration amounting to a 9.22 percent area reduction from stock. Now, so you will be able to run at full throttle for extended periods, we will need to get some additional fuel from the secondaries. By doing it this way you can retain the desirable economy potential and eliminate the chance of encountering a medium- to high-speed lean-heat problem. All secondary metering orifices are fixed at 0.135-inch diameter so the area will be constant at 0.0143139 square inches. Let's assume your secondary rods are stamped BA. By referring to a factory chart we find this is part number 7034337 with a 0.0570-inch-diameter power (WOT) tip and an intermediate (middle-of-taper) size of 0.0987-inch.

At wide-open throttle, approximately 60 percent of the total mixture flow through the carburetor is supplied by the secondaries. For that reason, a given percentage change in secondary metering would have half again as much effect on the overall air/fuel ratio as an identical-percentage change in primary metering. (The secondary is approximately 1.5 x the primary area). This factor must be considered when changing secondary metering. Determining the ideal dimensions for a secondary metering rod to compensate for the 9.22 percent decrease in primary metering is done as follows:

| Original Secondary Metering | Area (in.$^2$) |
|---|---|
| Jet—0.135-inch— | 0.0143139 |
| Rod (power tip)—0.0570-inch— | - 0.0025518 |
| Original metering— | 0.0117621 |

Because a change of - 9.22 percent in primary metering must be compensated for, the secondary metering must be increased 9.22%/1.5 or 6.15 percent (1.5 is the mixture-flow constant discussed above).

| Proposed Secondary Metering | Area (in.$^2$) |
|---|---|
| Original metering— | 0.0117621 |
| Plus 6.15% | + 0.0007234 |
| Proposed metering— | 0.0124855 |
| | |
| Secondary jet—0.135-inch— | 0.0143139 |
| Less proposed metering— | - 0.0124855 |
| Proposed secondary rod (power-tip) area— | 0.0018284 |

Using the formula $D = 2\sqrt{\dfrac{A}{\pi}}$ the actual power-tip diameter of the proposed secondary metering rod can be determined.

$$D = 2\sqrt{\frac{.0018284}{3.1416}} = 0.0482''$$

But, if you don't care to get out your calculator or square-root tables, refer back to Chart A to get a close idea of the power-tip diameter you'll be needing.

Now, look through a chart of available metering rods to find one with center-of-taper dimension equal to the original rod (0.0987-inch) and something close to the proposed power-tip dimension. The closest available rod has a power tip of 0.045-inch and is marked AD (PN 7033772). Although this is somewhat smaller than required, it is the only one available close to the required specification. Very seldom will your calculations exactly match an available rod so an intel-

ligent compromise is almost always required.

If no replacement secondary metering rods of the desired type are available, the power tip of the rods you have can be reduced in size. Chuck each rod (carefully!) in a lathe or drill press and use emery cloth or a pattern file to reduce the diameter. Continually check with a micrometer or caliper to make sure you are not removing more material than you had planned. This should be regarded as a temporary expedient for any rods to be used in a carburetor for a street/highway vehicle as the slightly hardened surface of the metering rod is removed when the diameter is reduced. Excessive wear can occur in a very short time as the metering rod rattles around in the sharp-edged stainless-steel secondary "jet" orifices.

**Power Mixture Cut-in Point**—As discussed on page 233, manifold vacuum provides the signal to control power-system fuel flow. This description also explains how to select a different power-piston spring/assembly to cause power-system operation to begin at a lower manifold vacuum than provided by the original parts combination. It also indicates similar results can be obtained by modifying the spring you now have.

Late-model cars are typically equipped with low numerical gear ratios (referred to as being high-geared) in conjunction with camshafts which cause manifold vacuum to fall quickly anytime the throttle is opened. The stock power systems feed fuel on medium-throttle light accelerations when the added fuel is not really required. Economy suffers. For economical operation it is a good idea to change the power system cut-in point to a lower vacuum setting so the driver will not be "into the power system" so much of the time during normal driving. By making this simple change, greater fuel economy is a foregone conclusion.

As with nearly any improvement, there are compromises. Reduction of power-system operating range may cause leanness during medium-throttle accelerations. This will seldom be objectionable except perhaps during engine warm-up following cold starts—typically for the first mile or so after the choke comes off. Personally, I will take a few weak tip-in sags for the first few blocks in exchange for added miles per gallon.

The best

power-system operating range is a 6-inch start with a 3-inch full-in calibrated spring assembly. This allows you to drive with normal traffic without using power mixtures and yet get added fuel required for heavy-throttle applications. If you are driving at high altitude, this calibration provides good drivability without compromises.

Q-jet power spring 7037851 has the 6—3 inch Hg characteristics if you would prefer to use a standard part instead of cutting coils off of the spring now in your Q-jet. If you want to try cutting coils, start with 1-1/2 coils, counting the end coil as a full coil. Two coils is the absolute maximum to cut off a standard Q-jet power-piston spring. When the spring has been cut, the piston must still return fully and decisively to the top of its bore when you push it down and release it. If the piston does not rise to the top of its bore fully and quickly, too much has been cut from the spring and the power system will never operate. It is a good idea to check the action of the power piston in its bore before disassembling the carburetor so you will know how it is supposed to act. Measure the height of the primary metering rod hanger above the gasket surface with the stock spring installed, then you will have a guide to check against when you install a modified spring.

If you are driving a car with a high numerical ratio in the rear end (low gear) you may not need to change the power-system spring because more throttle can be used without causing drastic reductions in manifold vacuum. This is especially true with mild cam timing. Careful observation of manifold vacuum during accelerations can tell you a lot about when you want the power system to operate.

**Spark advance**—This plays a large role in economy, too. On most late-model cars the spark is too lazy, requiring an overly long time to occur. This is built into the distributor's mechanical-advance mechanism. In many cases, vacuum advance is locked out by the emission-control devices until the car is in high gear. The *mechanical* spark advance should be quickened for better economy, but no additional *total* spark advance is usually needed. In general, the *vacuum* advance should be left alone. However, getting the mechanical advance in sooner may cause part-throttle pinging due to the built-in vacuum advance. If this occurs, the vacuum advance will have to be limited (required reducing the amount of travel in the mechanism), or the original mechanical-advance springs will have to be reinstalled.

Total advance (initial setting plus mechanical advance in distributor) should not be increased for any economy or drivability modifications.

**Camshaft**—Today's engines use camshafts designed for emission reduction when combined with low numerical axle ratios. As discussed further in chapter twelve, the combination pins your engine in an RPM range where it can never get much manifold vacuum. If the gear ratio were to be swapped for a higher numerical ratio (lower gear), the camshaft would probably work better as your engine might be able to take some advantage of the overlap characteristics. If you are seriously interested in economy, replace the stock camshaft with a reduced-overlap camshaft which will take full advantage of the existing low compression ratio. Racer Brown is one camshaft grinder who offers a number of camshafts specifically designed for smog-type engines. Installing one will jump the performance up to where it belongs and, if the cam is teamed with a high stream velocity manifold such as the Edelbrock Torker, chances are good the emissions will be better than the requirements imposed on the stock engine.

**Altitude**—Specially metered carburetors were once released as stock items for automobiles sold in high-altitude areas. But, because of the lean part-throttle and idle calibrations required for emission control, 1970 and later model GM cars and trucks—and perhaps some earlier than that—have been delivered with a single calibration for the model. Turn back to our discussion of altitude in chapter twelve, page 203, and read about how far off a sea-level calibration will be at your altitude. Unless you do most of your driving at low altitudes, say up to 2,500 feet, your car is wasting fuel. And, you are suffering a lack of performance considerably beyond that you would ordinarily expect from reduced air density at high altitude. The lack of performance is most noticeable during long passing maneuvers—say from 35 to 70 MPH—and if you are hauling a camper on your pickup or towing a trailer behind your car. If you live at high altitude and do nearly all of your driving there, make a few stopwatch

tests and be prepared to make several changes to your carburetor until you get it absolutely right. You'll be very glad you did because of two things: 1. You will be happy because your car performs better and is safer to drive. 2. Your pocketbook will be less deflated by gasoline bills.

Make stop-watch tests from 2,000 to 3,500 RPM while accelerating, or if you do not have a tachometer, make the tests from 35 to 70 MPH. Be sure to make the tests *before* making any modifications to the carburetor and write down the results so you will have a baseline from which to tune. Use the same level stretch of road and average your results from runs in both directions.

WARNING: An economy calibration designed specifically for high-altitude use can cause severe overheating and engine damage if the vehicle is used for hauling heavy loads or towing and/or if the vehicle is operated at lower altitudes. Remember this and recalibrate accordingly when you take your vehicle on a trip where you will be making the engine work hard and/or operating it at lower altitudes.

**Summary**—I have mentioned "thinking" several times throughout the book. It is the only way to stay in the winner's circle. Many short-term winners got there by copying with cubic dollars and luck. The dollars are definitely a great asset, but a full-time effort in copying generally only gets you equal to the copied—luck you can't depend on. So if you want to be a consistent winner you must out-think 'em.

With the preceding metering information and suggested approach to this very important phase of carburetion, you can hit perfection. The final specs have to be worked out on your machine.

In summary you use this metering approach in steps. In our exercise the part-throttle metering was changed first because the drivability was not good. Primary main metering jets were reduced, together with the primary metering rods. End result—primary WOT ends up slightly lean, which benefits economy. The reasoning is that if the part-throttle is more responsive, you will not be into the WOT mixtures nearly as often. When you do "lean on it" for passing and such maneuvers, the response will be very good. A slightly lean mixture generally gives crisp response.

The secondary

metering rods were changed to provide the added fuel required for extended WOT operation. This eliminates the possibility of frying pistons, valves and plugs on a long pull. Actually, this step could be skipped if the car was not going to be used for anything other than general transportation. However, it is a good safety measure to protect the engine in the event someone else should use the car and not know of the carburetion changes.

If your main concern is WOT drag or road race, the approach can be a bit different. Meter the primary WOT and secondary WOT fuel requirements first. If this ends up causing a part-throttle richness, then simply increase the part-throttle metering-rod area to reduce part-throttle fuel flow. You might also put in secondary metering rods with short power-tip ends so the intermediate opening positions are lean. . . . With this performance-priority metering you will undoubtedly compromise some economy potential for street driving.

## TWO-BARREL ECONOMY MODS

The modifications discussed here have been accomplished on two-barrel carburetors on Chevrolets, but the details are applicable to Buicks, Oldsmobiles and Pontiacs as well. To improve the economy and drivability it is generally correct to say the main-jet size should be increased one to two thousandths above that originally supplied. In the case of a 350 CID Chevrolet station wagon, this means changing the 50 jets used in production for size 51's. Or, if the car was being used in the extreme north where there is a lot of cold weather—or being used to pull a trailer or carry heavy loads—a size 52 might be even better.

When you make this change you have increased the total fuel available at the nozzle. Therefore, it is wise to reduce the power restriction area approximately 5%. Getting the power restrictions out so they can be replaced with smaller ones is somewhat difficult on late models with the APT feature because one of the plugs is larger and hard to remove. The two small plugs used on earlier models without APT are easy to get out. The photos and captions detail how to accomplish these modifications. Power restrictions are drilled brass cups or special power valves which are available from Doug Roe Engineering.

In the photos you will notice a small hole in the bottom of the bowl,

247

called the APT (adjustable part-throttle) feed hole. When you install larger main jets, it is important to plug this hole. The APT was factory-adjusted with a computer to provide a certain part-throttle air/fuel mixture with that particular carburetor when used for a specific application. You are not adding as much fuel with the larger main jets as you might suppose because you have eliminated that part of the main system fuel previously supplied by the APT.

Once the APT system is plugged you will have better metering consistency. The new combination will give better transition operation (off-idle to start of main-system fuel flow) and better tip-ins while cruising (light accelerations). Although dynamometer tests will not show any power difference between a carburetor with or without APT, driving the vehicle clearly shows better nozzle response (hence the better tip-ins and transitions) *without* APT. This is because all of the fuel comes directly through the main jets with no obstacle courses for the fuel to traverse before getting to the main well.

We also remove approximately three coils from the power-piston assembly spring so more throttle can be used without using power system fuel. See the preceding Q-jet economy description for more information on why this is important and how this modification will help economy.

Read the information on page 233 for details on how to determine where the power-piston assembly opens the power valve. In general, you want the power system to start feeding at about 6 inches Hg manifold vacuum and to be fully open by 3 inches.

Vega 2GV carburetors have a velocity-type power valve as described on page 234. The valve must be carefully opened by removing the cup at the top of the valve. Increase the weight of the valve by inserting a small lead ball into the recess in the top of the valve and punching it or epoxying it to stay in place. Replace the cup plug in the top of the valve body just even with the bottom of the screwdriver slot and secure it with epoxy for a good seal.

The compromise you have to accept with these economy modifications is detailed in the Q-jet economy tuning information.

Where our recommendations call for

Don't get excited when you first pull a two-barrel and find that only two bolts secure the carburetor to the manifold. That's all GM uses anymore. Add up the saving in capscrews and machine operations and you can see a big pile of dollars when you consider the volume.

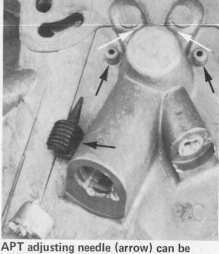

APT adjusting needle (arrow) can be eliminated. Cup plugs (arrows) are power-valve-channel restrictions which are driven through into main wells with a drift or 0.110-inch drill. White arrows show cup location between power-valve cavity and main-jet wells.

Top view shows APT feed hole (arrow) which will be plugged with lead ball. Drill in holder is used to open idle-tube restrictions. Power restrictions must be replaced with smaller ones or soldered shut and re-drilled for about 5% less area. This particular carburetor required 0.045-inch restrictions in place of the original 0.048-inch ones. Larger drill is used to knock out power restrictions so reduced-size ones can be installed. When the APT system is plugged, the main jets can be increased 0.001 or 0.002-inch. In this case the 0.064-inch jets were increased to 0.066-inch. Response can be improved by merely plugging the APT system inlet from the bowl and then installing 0.001 or 0.002-inch larger main jets. However, this will not provide the economy improvement which can be had by reducing the power-restriction size.

Epoxy the drilled plugs in place.

Sandpaper wrapped around a file is used here to eliminate casting flash in the venturis. Such smoothing can provide a considerable air-flow improvement, although this is of no help for economy tuning.

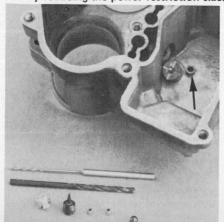

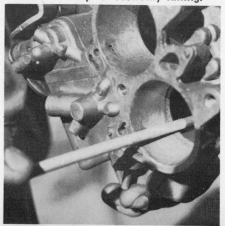

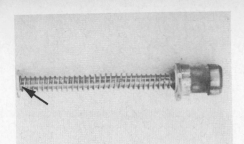

Three coils were cut from the end of the power-valve spring (arrow). This was done without taking the spring off the actuating rod. Power piston was removed from air horn in this case because actuation with a finger indicated erratic, sticky action. Piston and bore in air horn were polished with fine crocus cloth.

Cutting three full coils from 2G power-piston spring. Spring has been pulled up stem to allow easy coil counting. Good sharp diagonals are required to cut the spring.

Once spring has been cut, remaining spring is pulled out of the way with fingers and diagonals are used to unwrap cut-off coils from stem.

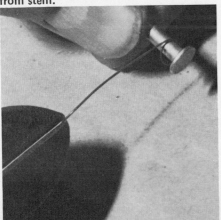

reducing the power-system fuel-flow area by 5% for economy, also reduce the power-valve channel restriction area an additional 1% for each 1,000 feet of altitude you live (and drive) above sea level. As an example, if you live in Denver at 5,000 feet altitude, reduce the power-system fuel-flow area by 5% for the economy reduction plus 5% (1% for each 1,000 feet) for altitude to make 10% total reduction. If the car seems sluggish during passing maneuvers, increase the power-system restrictions 1% in area at a time until you have obtained the best performance.

If you are driving a car with automatic transmission (Hydramatic), a too-lean power mixture will tend to make the engine hover at the RPM just before the 2-3 shift. It is a good idea to note how crisply the shift occurs before making any carburetor modifications—preferably by stopwatching several runs including this shift. Before making changes, be sure to run a baseline measurement as previously described—then make similar tests after each area change.

Where we suggest increasing the main jet one to two sizes, reduce the power-system restrictions and change the power-valve or power-system cut-in point *before* changing the main jets. You may find the stock jets to be "right on" once you have made the other changes and plugged the APT fuel supply.

Staking the power-piston assembly in place by pushing part of the metal over the edge of the retainer requires only a hammer and screwdriver.

## MONOJET ECONOMY MODIFICATIONS

In general, the Monojet has a strong and consistent idle system and changes to this part of the carburetor are seldom required. What is normally required to make one drive well and get better economy is to increase the main jet one or two sizes and plug the APT. This eliminates tip-in and nozzle-lag problems.

Additionally, cutting not more than two full coils from the power spring will allow the power system to come in at a lower manifold vacuum, thereby improving economy, as discussed in the Q-jet economy-tuning section earlier in this chapter.

In the case of the Vega Monojet, the velocity-actuated power system can be altered to come in at a higher road speed by increasing the weight of the power valve. Increase the valve weight to a total of 1.60 grams by dropping a lead ball inside the recess in the top of the power valve weight.

Also, be sure to read the references to altitude tuning in the preceding discussions of two-barrel and Q-jet economy modifications.

The compromise you have to accept with these economy modifications is detailed in the Q-jet economy tuning information.

To cure extreme tip-in sags or to obtain drivability for special uses, the idle-tube orifices can be opened 0.002 to 0.003-inch to obtain about a 5% increase in area. Such modifications will definitely destroy the engine's ability to pass emissions tests at any speed above curb-idle.

# Drag Racing

With the growing popularity of the AHRA showroom stock and NHRA stock classes which require retaining the original-equipment carburetor, it has become necessary to make do with what you're got.

In the case of almost all late-model, four-barrel-equipped GM V-8s and also some '70–'71 model 429-cubic-inch Fords, this means using the Rochester Q-jet. When prepared properly that isn't all bad. For years most drag racers have avoided the Q-jet as if it were some sort of contagious disease, but with the aid of the technical medicine made available in this book and some effort on the reader's part, your Quadrajet can become a weekend trophy grabber. As a matter of fact we will venture to say you will become partial to this versatile unit.

Blueprinting a carburetor is started in the same manner as blueprinting an engine, and this means complete disassembly and thorough cleaning.

During disassembly, care must be taken not to lose any of those tiny but ever-so-important parts that every carburetor is made up of. Probably the two most common parts inadvertently left out of the Quadrajet are the accelerator pump check ball and the small spring located beneath the power-valve piston.

**Inlet system**—We will attempt to explain the modifications to the Q-jet in a systematic manner, so let's start with the fuel entering the carburetor. The fuel must first pass through the paper or bronze filter located in the fuel-inlet housing on the main body of the carburetor.

For high-performance applications we suggest eliminating the factory-supplied filter in favor of one of the less-restrictive in-line type ahead of the carburetor.

After the fuel passes through the filter, it enters the float bowl via the needle valve and seat and at this point we offer some suggestions. Delco has three sizes of needle valves available for the Q-jet: 0.110, 0.125 and 0.135-inch diameter. While we do not automatically recommend discarding your needle valve in favor of the largest available, some applications might benefit by swapping the existing fuel inlet size for

Excitingly close-competitive vehicles have evolved from the popular "stock" drag race rules in the last few years. This Super Stock H Automatic is a strong contender on the West Coast. Owner-driver Jim Mehl does it himself without the aid of big-sponsor money.

Start the cleaning by doing a good job in the engine compartment. Frequent washing at the do-it-yourself car wash is good if you are a tinkerer.

Marine-engine filter 7029038 (arrow) is low-restriction nylon mesh in a plastic frame. It measures approximately 11/16-inch diameter and 1-1/8 inch long. Use this to replace the sintered-bronze or small pleated-paper filters shown here for comparison. This nylon-mesh filter can be used as a safety dirt catcher even when an in-line filter is used. It can even be used for drag racing and other competition applications. The nylon filter is available through Mercury Marine dealers as P/N 35-53336. Its filtering capability—80 microns (0.002–0.003 inch)—is the same as the sintered-bronze filters. Unlike the sintered-bronze and pleated-paper units which clog with use, the nylon mesh tends to be self-cleaning. Particles trapped by the mesh tend to fall off of the mesh to lodge in the bottom of the chamber surrounding the filter. The nylon filter is used on the 225 HP MerCruiser V-8 engine with Q-jet carburetion.

The stamped numbers and letters on primary metering rods are small. A spark-plug inspection light like this A-C unit is very useful in checking all small parts.

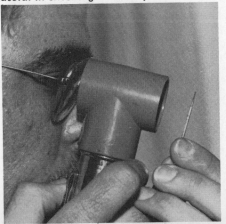

one of the optional units offered by Delco.

The neoprene-tipped needle is opened and closed by a closed-cellular plastic float that should be examined carefully for holes or other irregularities that could affect its operation. Having a spare one on hand is wise because they are subject to damage and at a competitive meet it's tough to get replacements. Carry such parts as these in separate protective containers, preferably a small box so the float's "skin" will not be dented or punctured.

**Primary Fuel Metering**—This is accomplished through the use of removable jets and metering rods. Rochester offers main metering jets in sizes from 0.066-inch to 0.078-inch and also a wide selection of metering rods.

The primary metering rods are identified with a stamped number and letter. Example: 42B. The number reflects the major metering area diameter in thousandths—0.042—and B rod indicates that the rod has a double taper. The small diameter at the bottom of the taper is 0.026-inch on all Q-jet primary rods except in GMC motor homes.

During part-throttle/high-vacuum conditions the larger diameter of the rod is restricting the fuel flow through the jet and thus part-throttle metering is accomplished. When a heavy-throttle condition arises such as those experienced in racing, manifold vacuum drops and the small spring beneath

the power-valve piston lifts the power valve and the attached metering rods so that the restriction in the jet is the small 0.026-inch diameter portion of the rod, thus allowing more metered fuel through the jet orifice.

We can conclude from this that significant metering changes for drag racing will have to be accomplished by changing the jets instead of the rods. You should be aware that part-throttle mixtures can become excessively rich and premature plug fouling could become a problem if the vehicle is subjected to slow driving in the pit and staging areas. In such cases a larger metering rod would be in order to afford proper part-throttle metering and yet retain the richer WOT fuel. Low-stall-speed automatic-equipped vehicles could benefit also by metering rod changes. They would be slug-

---

**Metering**—For WOT operation, meter the primary and secondary WOT fuel requirements *first*. If this causes part-throttle richness, increase the part-throttle metering-rod area to reduce part-throttle fuel flow. Use secondary metering rods with short power-tip ends so the intermediate opening power positions are lean. This performance-priority metering compromises street driving economy.

---

| h DIMENSION | IDENTIFICATION STAMP | h DIMENSION | IDENTIFICATION STAMP |
|---|---|---|---|
| .520 | B | .570 | L |
| .525 | C | .575 | M |
| .530 | D | .580 | N |
| .535 | E | .585 | O |
| .540 | F | .590 | P |
| .545 | G | .595 | R |
| .550 | H | .600 | S |
| .555 | I | .605 | T |
| .560 | J | .610 | U |
| .565 | K | .615 | V* |

*Only service hanger, P/N 7034522

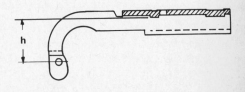

**Q-JET SECONDARY METERING ROD HANGERS**

Secondary rods are inserted with the bent ends pointing into the hanger. Secondary-rod changes can be made by taking out one screw which attaches the hanger.

Top view shows upper vertical passages. Tool is pointing to one, arrow indicates the other one.

Note fuel passage (black arrow) on the right side is 0.160-inch diameter up to intersection with vertical passage for the secondary air-bleed tubes. Small lower passage causes a restriction at the intersection (white arrow). For street use this is no problem; for high-performance applications, drill out the passages. Left passage has been drilled.

Secondary rod hangers vary in dimension to hang the rods at different height relative to the secondary metering orifices. Pages 243 and 251 provide additional information on the hanger and secondary rods.

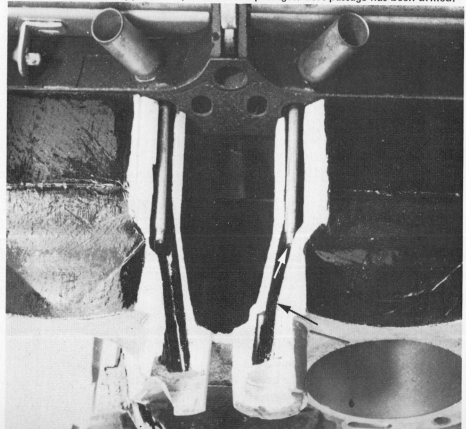

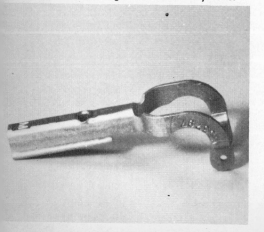

gish coming out of the hole if the low-end metering is rich.

Refer to the jet and rod-area information discussed in the economy/general performance section of this chapter for a better understanding of what the various metering rod and jet combinations can do to part-throttle and WOT metering.

**Secondary Fuel Metering**—This control is provided by multiple-stepped replaceable rods as the fixed-size 0.135-inch diameter orifices are not removable from the bowl.

Rochester offers a wide selection of secondary metering rods identified by two stamped letters. Example: CF. These rods and the sizes are listed elsewhere in this chapter.

The secondary metering rods are attached to what is referred to as a hanger. Rochester makes 20 types of hangers. The difference is the location of the hole where the secondary rods attach. These hangers are identified by a single stamped letter. (Hangers B through V have a hole location change of 0.005 per letter.)

Changing the hole location in the hanger changes the relative depth position between the rod and the secondary metering orifice, hence changing the restriction changes fuel flow. Delco offers only one replacement hanger 7034522 identified with the letter V. RPD uses the 20 different hangers to do specific model calibrating.

The rod hanger is one more tool for the fine tuner to help himself toward the finish line ahead of the competition. Again, Doug Roe Engineering can supply three different hangers for those who want an additional edge. Lean—0.520-inch; average—0.570; and rich—0.615-inch.

There are a few modifications explained below which can be made to the secondary fuel delivery system which will give quicker and more consistent engine response for heavy and WOT situations.

Secondary fuel passes through the metering orifices to a small chamber below the float bowl. At this point the fuel travels upward through two 0.160-inch-diameter passages which intersect with the 0.197-inch-diameter passages that lead to the nozzles. The two above-mentioned 0.160-inch passages sometimes offer an unwanted restriction to high fuel flow. This can be alleviated

The upper vertical passages are 0.197-inch which is large enough. Using a bit no larger than 0.200-inch, drill the upper passages to a new depth as illustrated. Note how close this gets to breaking out into the secondary venturi. Do not exceed 1-1/4-inch depth. If the bit breaks through to the venturi, it must be repaired. Some quality epoxy filler and a lot of care will take care of this.

Using the same bit, drill out the lower channels. To do this the fuel-chamber plugs (arrow) must be removed. Early-model Quadrajets have the cup-type plug as shown in cutaway. The later ones are thick metal plugs that resist removal. See next photos.

by drilling the passages out to 0.197-inch like the adjoining passages they feed. Refer to the accompanying photos to help with this modification.

With the main body of the carburetor inverted you will note two plugs side by side in the center of the carburetor. To gain access to the 0.160-inch passages referred to above, these plugs must be removed. On late-model Q-jets these plugs are die-cast and spun in and will require drilling out as shown. If these plugs are not die-cast they will be a brass cup plug which can be removed by prying them out with a good-quality ice pick or awl.

Care should be taken during this procedure not to do any unnecessary damage, but replacement plugs are available at most auto parts stores. And as these plugs have been a trouble spot due to leakage, we recommend using a good epoxy to cement the plugs into the well cavities.

Don't treat this lightly. This is the extreme bottom of your fuel reservoir and a leak here will let approximately a teacup of fuel into the engine.

Along with the passage modification one more step can be done to further alleviate this restriction. By inverting the top of the carburetor you will see four tubes, each approximately 1-1/4-inch long. (On a few Q-jets there will only be two.) The two smaller tubes, approximately 1/16-inch diameter each, are air bleeds to the secondary-fuel-delivery passages and when the carburetor is assembled the tubes extend into the intersection where the previously-discussed passages meet. Note the photo. By tapping these tubes further up into the air horn approximately 3/16-inch, the passage restriction can be reduced further. Their extended height will be 1 inch when driven in to the recommended depth. They do not move easily so fabricate a piece from aluminum or bronze (scrap) that will transmit driving forces to the shoulder without damaging the orifice at the tip.

If your Q-jet is equipped with the two 1/8-inch-diameter tubes, another modification is in order. These tubes act as a secondary accelerator pump in that they deliver a shot of fuel before air velocity is high enough to draw fuel out of the secondary-discharge nozzles. These tubes extend into wells at the rear of the float chamber (right and left side). The wells receive their supply of fuel through a 0.028-inch-diameter orifice located approximately 3/4-inch up from the floor of the float chamber.

When the secondary throttle blades start to open, a low-pressure area is created beneath the secondary air valve. Fuel then will flow from the well up the tubes and discharge from the 0.060-inch drilled orifice located just beneath the secondary air valve. On some carburetors this discharge orifice will be located just above the air valve, which means the fuel shot will be delayed until the air valve just starts to crack open creating a low pressure over the orifice.

---

**Performance rebuilding kits for the Q-jet—** Some kits have been marketed with the promise of better performance and more economy. Beware of such kits if they do not contain specific metering parts such as primary rods and jets and secondary rods. A new set of gaskets and a new pump may help in some isolated instances, but they cannot be relied on for additional performance. If in doubt, take a look at what the kit contains—or ask what it contains *before* you order it by mail. We don't know of any shortcuts. You have to look at what is in your carburetor and make metering changes based on what was there to begin with. We have explained explained exactly how to do that. A single kit with one metering configuration cannot fulfill all metering requirements for the entire Q-jet line comprising some 200 models with an enormous array of metering specifications.

---

On late-model Quadrajets the fuel-chamber plugs are driven in and then part of the die-cast material is pressed with a spinning tool until it is peened firmly over the plug as shown. This design is much less likely to leak than the early pressed-in-plug style.

Remove the retaining metal from over the plugs. Use a small chisel, a rotary file, a cutting stone and care.

*We always want—to those without patience, desires may never occur.*

—Doug Roe

Note the thickness of the removed plug being held by hand. The second plug is being drilled so it can be grasped for removal. A good epoxy filler will almost certainly be needed.

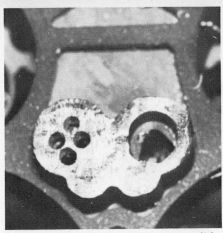

The battle has just begun. Rochester did not intend those plugs to fall out or be taken out.

Tomco makes these neat O-ring plugs for sealing these secondary fuel chambers. If you get the original plugs out without marking the walls of the plug wells, they will work fine. When in doubt—back it up with some epoxy. You do not want a leak of any sort at this point.

Early cup-type secondary-well plugs. Generally they can be removed by simply catching an inside wall with a sharp awl or punch and prying them out. This pair is secured with epoxy.

A piece of aluminum tubing can be drilled to 29/32-inch with a 3/32-inch or slightly larger drill. Then counterdrill an additional 1/4-inch with a 1/16-in. drill to provide clearance for the metering tip. This driver pushes against the tube shoulder to avoid damaging the tip.

These tubes have been driven to the preferred standing height of one inch as illustrated by the scale.

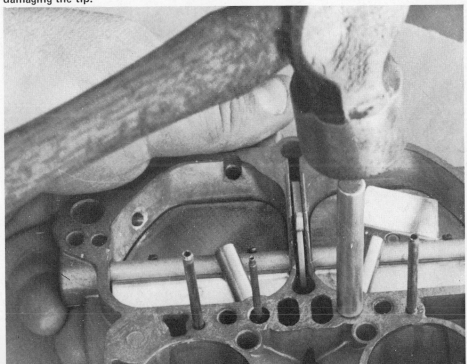

To create a smoother transition when the secondaries are opened the 1/8-inch tubes should be drilled as shown. Starting approximately 1/4-inch from the end of the tube drill four 0.030-inch diameter (1/32) holes approximately 3/32-inch apart straight through both sides of the tubes.

One more thing to keep in mind at this stage of the carburetor rework is that the above-mentioned wells require approximately 15 seconds to refill through the 0.028-inch drilled orifices. If after one or two hard burn-outs you encounter what seems to be a hesitation, or bog, then possibly the wells are not replenishing adequately. Increasing the size of the drilled orifice could be the answer.

Keep in mind the refill orifice serves as a metering restriction during extended secondary use. Secondary fuel metering could be affected drastically if this orifice is punched out too large. Make modest 0.001-inch to 0.002-inch changes.

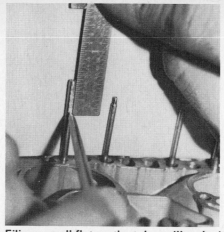

Filing a small flat on the tubes will make it easier to mark and prick-punch for drilling.

Holding the air horn as illustrated provides a good way to work on the tubes. A quality awl is an excellent tool for light prick-punching.

Most drill chucks will not clamp a small drill bit such as the 0.030-inch size used here. Note how a special chuck and shaft is clamped in the chuck of a drill motor. A drill press is preferred for this fine work but steady hands and care will get it done.

A hand-held drill job may net you some visible irregularities but if the hole sizes and spacing are right on, it is OK.

Note how precise the factory-drilled holes are. These tubes were standard in the 1970 Cobrajet Ford Quadrajet carburetors. Sorry—the tubes are not a replaceable item so cannot be purchased separately from Delco or your dealer.

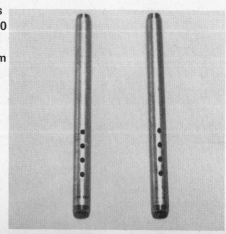

The auxiliary feed tubes get their fuel from reservoirs (black arrow). White arrow indicates 0.028-inch hole which fills the left side. A similar hole is on the right side.

Stop (arrow) can be modified to allow full 90° opening of the air valve. Some Q-jet air valves only open 70°, which can limit a 750 CFM unit to a mere 600 CFM. This opening may be sufficient to provide all the HP available from a modern-day emission-controlled engine. Remember to check this stop and modify it for full air-valve opening if you need all of the available capacity from your Q-jet.

Before you put your drills away there are a few more tricks to help put your "grocery-getter" in the trophy lane. The secondary air valve is opened by the low-pressure area created when the mechanically operated secondary throttle valves are opened. But as you probably have realized by now, the air-valve opening is not instantaneous and this is because Rochester had to keep the little old ladies happy as well as those of us who like to put our foot to the wood. If your vehicle has had the normal race-type preparation such as a little quicker advance curve, possibly a thinner head gasket and maybe a shorter set of gears, etc., you probably will benefit by letting the air valve open a little quicker.

Two major control mechanisms regulate the opening rate:

1. On the right side of the air-valve shaft an adjustable spring applies tension against the air valve. The factory settings are usually between 1/2 and 3/4-turn tension. By reducing this tension in steps of about 1/8-turn at a time, you will find which setting is best for your vehicle. The quicker you can open the air valve, the better for maximum performance. The limiting factor will be a lean sag at opening if it's too quick. Adjust to fit your application.

2. On the right front side of the 1967 and later Q-jet an enclosed vacuum diaphragm is attached to the air valve by a linkage arm. This diaphragm has a dual purpose: Number one, it serves as a vacuum break for the choke. Secondly, it slows the air-valve opening. This is what we will discuss here.

By all means do not throw the diaphragm away because this will cause the air valve to open too quickly. When this happens you have lost control—it will slam shut—bounce open —and lose you time.

If you remove the neoprene hose from the diaphragm, you will see a hole approximately 0.010-inch in diameter. By increasing this diameter in steps of about 0.005-inch at a whack, the damping effect on the air valve can be reduced progressively. All damping effect will be lost if the orifice is drilled much beyond 0.025-inch.

The 1965-66 Q-jets use a fuel-damped air valve. The fuel-damped piston slides loosely in a fuel chamber or cylinder. To tune this type for faster opening would require filing a flat on the edge of the piston,

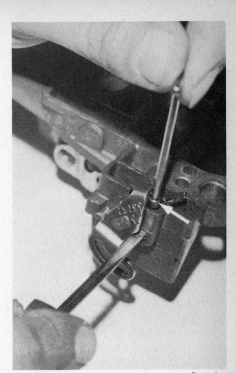

To adjust the secondary air valve—first insert a small screwdriver in the slot on the side and then loosen the allen screw (arrow). Now turn the conventional flat screwdriver counterclockwise until the air valve is closed with zero tension. If you turned the slotted shaft 1/2 turn to this zero tension, the setting was 1/2 turn. Before you move the slotted shaft it is a good idea to scribe a mark so you will know where it was set. This adjustment can be done with the carburetor on the vehicle but the allen screw will not be visible except with a mirror. Once you get onto it, the task is easy.

Tap the air valves slightly as you adjust the screw clockwise and counterclockwise. Provided there is no drag in the air-valve shaft, with a bit of practice you will find a zero-tension position easily. All setting references are from this point. For example, a 3/4 clockwise turn of the slotted shaft is a 3/4 setting. Remember to tighten the allen screw after each adjustment.

Do not turn the adjustment screw more than one turn from zero position because the secondary air-valve tension spring (shown here) will be distorted.

Sometimes the allen setscrew will be recessed as shown here (arrow). Air valve is in the open position.

Do not throw away the link connecting the diaphragm to the air valve (arrow). The air-valve dampening diaphragm described in the accompanying text refers to the plastic-type unit shown (white arrow).

The steel diaphragm-type unit illustrated here is difficult to alter because the dampening restriction is internal. Try to replace these with the plastic units before attempting to tune the air-valve opening rate.

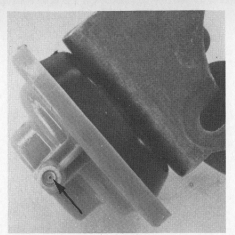

The size of this tiny orifice is critical to premium operation of the Quadrajet. Make small changes and careful performance checks.

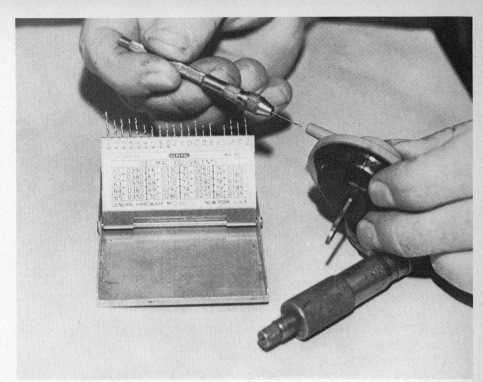

Wire drill index with drills from 0.0135 to 0.039-inch (Numbers 80 through 61) is an essential for the serious tuner. Pin vise holds the drill while finger power does the twisting—carefully! These drills and the pin vise are available at model shops, tool-supply houses and Sears.

starting with about 1/32-inch width. Still another possibility is drilling a 0.015-inch hole in the piston. The width of the flat, or the diameter of the hole can be increased until the desired damping action is achieved.

One further thing can be done to quicken secondary opening: Open the secondary throttles sooner by modifying the linkage from the primary throttles. This creates a pressure drop under the secondary air valve sooner so the secondary side can begin operating quicker. It is important to note this modification murders economy because the secondaries will be working at almost any road speed. Such modifications are not for street/highway-driven vehicles if economy is of any importance whatsoever. *Such modifications are strictly for the drag strip or road racing.*

To open the secondary throttles earlier requires modifying the secondary pickup lever. There are two types. In one case a slot must be deepened so a spring wound around the secondary throttle shaft causes the linkage to act quicker. In the other arrangement, the hole for the rod to the primary throttles must be relocated downward toward the secondary throttle shaft and moved toward the primary throttle shaft. Details for handling the first type mentioned are provided in the next section. It also explains how to obtain 90° or more secondary throttle position.

**Secondary Throttle Blade Angle**—Some Q-jets open the secondary throttles to a full-open 90-degree position. Others open the secondary throttles to a position somewhat less than 90 degrees. In either case, the carburetors will flow their full rated capacity.

**Secondary throttle opening vs. capacity**— Whether the secondary throttle blades open 5° short or 5° beyond the 90° position has little effect on air-flow capacity. But, this can affect distribution as explained on pages 14 and 261. Do not confuse throttle opening with air-valve opening.

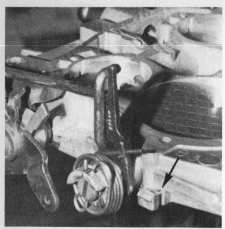

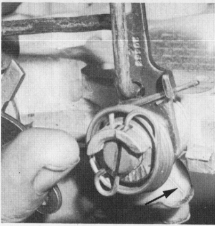

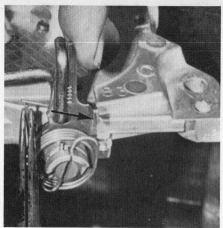

First step in getting lever off secondary shaft is to pry straight edge of inner spring (next to throttle body) off its ledge, then catch it with your finger and allow it to unwind its slightly-more-than-one-turn tension.

Next release tension on outer spring by releasing straight end of spring from its tang (arrow). Then hold secondaries closed as you rotate lever clockwise to allow spring to clear its slot in shaft end.

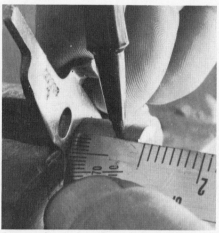

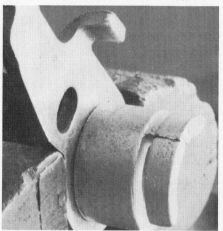

You must mark the lever to show original slot location before making any alterations. This gives a reference point. Mark by bluing and scribing or just file a mark.

A second way to mark the lever. Hold a straight edge against the end of the slot you are going to alter and draw a straight reference line with a scribe or pen.

Line marking end of original slot dimension. File notch at edge is insurance just in case the line gets rubbed out.

Do-it-yourselfer way of increasing slot depth with hacksaw blade. Because slot is a bit wider than blade, saw a little ways—then move blade over to keep original slot width. Saw only a few thousandths at a time, then reinstall lever to check throttle opening. Lever is slightly hard but a hacksaw blade will cut it.

How far should you saw the slot? The first time I sawed the slot 0.050-inch deeper which moved the 45° secondary pickup point to 38°. Another 0.020-inch moved it to 35° pickup point and a further clean-up got it to 30° pickup with the end opening angle of the secondaries slightly past a 90° position. I would not recommend any secondary pickup quicker than 30° or you will have them open all of the time.

Check yourself as you go with a protractor on the primary throttle shaft. Set it up with a wire pointer clamped to the throttle body so 0° indicates the closed-throttle position.

End point of the work. Protractor shows 30° primary throttle opening from closed and the primary lever linkage is starting to pick up the secondary throttle opening. The effort started with 45° primary opening before the secondary started to open, so we gained 15° at the pickup point with our modifications.

This photo graphically illustrates total movement of primary (and secondary) throttles does not exceed about 75°. Here primary side is open and protractor is indicating 75° travel. It looks like less in the photo because indicator is bent toward the 75°. Note primary throttle stop (arrow) which can be adjusted to allow a perpendicular position of the primary throttle blades relative to the throttle body. A few degrees blade angle does not affect carburetor capacity, but can affect mixture distribution. Secondary blades are over-center in this photo, showing successful lever modification to improve mixture distribution to front cylinders.

In some cases, notably Chevrolet V-8's, the front cylinders have a tendency to run lean. This may be caused by the secondary throttles not being fully opened by the stock throttle linkage. The lean tendency can be reduced or cured in many instances by re-working the linkage so the secondary throttles open to a position past 90 degrees—up to 5 degrees farther—so the throttle blades actually deflect the secondary mixture toward the front of the manifold.

The accompanying photos and captions describe how to accomplish the necessary changes in the linkage to get the secondary throttles to a full-open position or slightly beyond 90 degrees. It is important to note the need to keep the vertical portion of the rod striking the opening tang on the secondary lever as

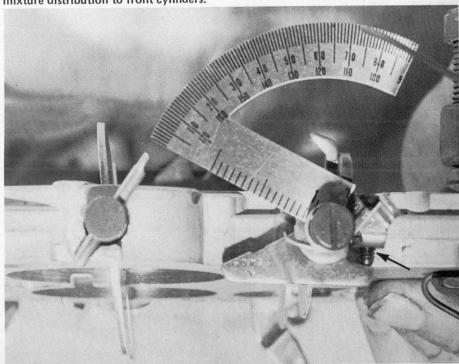

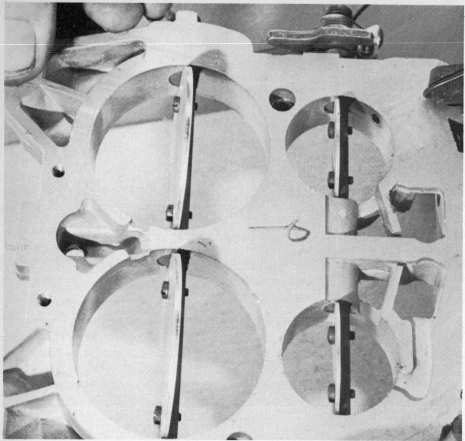

Underside of throttle body shows over-center secondary blade position as end result of modification.

Here's that dime-store protractor after cutting it in half and notching it for the screw to attach it to the primary throttle shaft.

Secondary closing spring must be installed first when replacing lever. Note wrap on outer, hooked end. Hook must be attached to lever, then inner straight end tensioned.

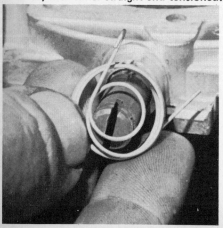

Spring has been hooked onto lever (arrow) and needle-nose pliers are being used to guide the straight end of spring past obstacles to attain slightly more than one turn wrap up when spring end is placed on ledge behind secondary shaft.

With inner spring installed and tensioned, hold throttles closed with one finger as you rotate lever clockwise to position the shaft slot and lever opening like so.

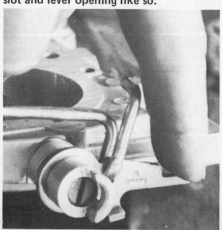

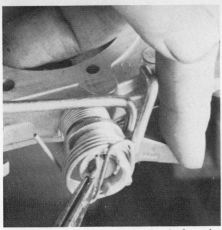

Install outer spring into slot in shaft end, pushing lever toward throttle body as you do so.

Outer spring "hairpin" end is installed in lever slot and shaft end slot and straight end of spring is being wrapped toward its tang.

Final twist on spring brings it behind tang as shown and lever is completely installed.

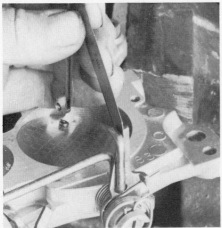

Checking air gap between actuating rod and secondary lever. It should be about 0.005 thousandths, as measured here with a feeler gage.

Adjustments of the vertical portion of the actuating rod can be made with large pliers. This is done to set air gap between top of rod and top of lever to ensure good opening characteristics. Extra leverage provided here "kicks" secondaries loose from their bores at the pickup point.

this provides additional leverage for opening the secondaries against the friction tending to hold them closed in their bores. And the secondaries must close easily and decisively when the work is completed. This is a function of the closing tang on the primary throttle lever and of spring tension on the secondary closing spring. Careful thinking and a studied approach is required to operate on either of the secondary opening mechanisms (there are only two types). Proceed slowly and methodically or you could find yourself laying out the long green for a new throttle body. The levers are not sold separately.

**Accelerator Pump**—Although the Q-jet accelerator pump is adequate for what it was intended it can be improved upon for drag-race applications. By using the inboard hole in the pump arm (the hole closest the pivot pin), you will get maximum design pump delivery, but this may fall short of your "fire breather's" demand. Refer back to page 229 where we've discussed getting more pump capacity.

---

**Don't bend the primary to secondary actuating rod** except at the vertical end to adjust for 0.005-inch clearance with the tang atop the secondary lever. It may be tempting to bend the rod to shorten it, but don't! If you do so, the secondaries may not close, or may hang up in the over-center condition. You don't want that to happen.

---

263

When the modification is completed, positive secondary closing is lost because the early pickup changes move the tang on the outer primary lever away from the inner secondary-actuating lever closing tang (indicated by screwdriver). Material must be welded onto the lever at point shown by arrow to close this gap to zero clearance at curb idle if the secondary closing feature is to be retained. Some Q-jets have teflon-coated shafts, so submerge the throttle body in water to keep the shafts cool and work quickly if any welding is done on the levers.

Where secondary lever is staked to throttle shaft, a different approach is required to get the throttles past the 90° position. Here several turns of mechanics wire are wound tightly around the lever and through the slot to reduce the slot length (arrow). Then clearance is reduced to 0.005-inch where the actuating rod's near-vertical portion strikes the opening tang at the top of the lever. This arrangement does not noticeably increase the secondary pickup beyond the usual 45° opening of the primary throttles.

**Carburetor Capacity**—When you start speeding up secondary air valve opening, then you better start thinking more capacity. Even before that, when you decide to make physical changes to venturis, etc., you better read the rule book. You want to be a winner, and if you work your way to that position, skeptical and/or curious people may want to check your skills against the rule book.

In closing, you might be wondering to yourself why these modifications and tuning tips have not been put in print previously. This was partially due to the lack of acceptance of the Q-jet by racers due to its seemingly-complex metering system and also the relative unavailability of metering

rods and jets. Study the economy and general performance section of this chapter for metering tips. The Quadrajet allows fine tuning for any purpose. The fact that Rochester has offered the Q-jet in only two sizes (750 and 800 CFM) may serve as a deterrent, because Holley offers a different CFM-rated carburetor for every day of the week and tremendous promotion to convince their buyers. Hopefully we have enlightened the drag-race enthusiasts to the fact that—when given proper preparation— a Q-jet-equipped engine can and does respond to the loud pedal with authority and either size unit will cover all demands up to its maximum flow capabilities.

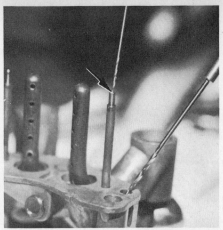

Small drill (arrow) is being used to open up idle orifice. Larger drill at right shows location of channel restriction in this two-barrel cluster.

## 2-BARREL PERFORMANCE METERING

We have already discussed the APT (adjustable part-throttle) system and the need for plugging it in our discussion of two-barrel economy. When you speak of performance it is essential to meter the fuel only through the main jet—as opposed to a combination of flows through the APT and the main jets.

If the car is being used for dual-purpose driving (street and strip), bring the power system in later and reduce the power-restriction areas as discussed in the economy section. Then do your performance metering with the main jets. Details on the power-system cut-in point and its importance to performance applications, especially with racing camshafts giving low manifold vacuum at idle, are provided on page 233.

Make main jet changes in 0.001-inch increments and keep track of the performance changes with a stopwatch or by the elapsed-time slips. If the carburetor will be used solely for strip or track applications, leave the power system metering restrictions alone: Use the ones which came in the carburetor.

There is no way to get a larger pump into the two-barrel unless it came equipped with the 0.730-inch cup. The pump bore is created by a very thin metal wall and there is not enough metal to bore it out as there is on the Q-jet. Some two-barrels have the large pump and these models would be our choice for performance applications where the rules permit.

When you are tuning a carburetor, taking the air horn on and off can be a frequent chore. It is easy to damage the gasket if it sticks to the air horn or carburetor body. To avoid this, oil the gasket so it stays pliable and easy to remove. If you have pulled a carburetor apart and will not be putting it back together immediately, this trick can save the gasket from drying out and becoming unusable.

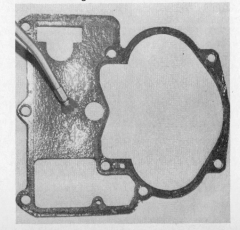

Very careful measurements of pump travel are required to ensure that the cup does not uncover the fill slot at the top of the well when the pump cup is raised to its uppermost position. Measure from the fill slot to the bottom of the well and make sure the pump-linkage geometry stacks up to give you exactly that amount of travel, ending up with the cup at the bottom of the bore. Otherwise, part of the pump shot will be pushed back into the bowl and ineffectual pump action will create an intolerable pump lag.

Make sure there is no slack or free play in the pump linkage with the throttle at idle position. Any movement of the throttle should start pump motion and only a few degrees of throttle movement should start a pump shot.

If you modify the idle feed restrictions to cure light-throttle maneuver weakness the two-barrel becomes prone to nozzle lag during medium-throttle accelerations. This is because anything you do to make the idle system feed stronger pulls down the fuel in the main well so the nozzle has to pull the fuel farther to get it to feed through the nozzle. The problem can be cured by setting a leaner idle mixture with the idle screw (thereby reducing the signal applied to the idle system so fuel is not pulled so far down in the main well). Or, a slightly larger main jet (one or two sizes) will help because the main well will be kept replenished even though the idle system is pulling strongly. This is also a problem with the Q-jet, although not to the same extent as on the two-barrels.

A Quadrajet carburetor supplies the air/ fuel mixture for the 455 CID Oldsmobile engine with Berkeley Jet in this Glastron-Carlson hull. Most of the jet and stern-drive power packages for pleasure boats are equipped with Q-jet carburetion.

## Off-Road Racing & Marine Tips

You probably are reading this section because you intend to do some off-road racing. I hope nothing discourages you; it can be the most rewarding thing you ever do. I have been involved in just about every kind of auto racing there is and each type has offered a challenge that can make you do strange things like: Work day and night on a car to make a race; spend money allotted for another good use; give up other fun things; and so on.

Even if it leads to this "squirrely" behavior, set your mind to going on and running some of the popular national events. Better yet, plan to run the Mexican Baja 500 or 1,000-milers. The purse isn't all that great but nowhere can you get away from the worldly pace like you can there. For these long races practice trips are essential to be competitive and they can amount to the best time of all. As I must go on with

the carburetor story it is time to stop PR for off-road racing.

Prepare the car well for off-road racing. Take time to blueprint the carburetor or it will act up in the rough-tough areas and that's 95 percent of the distance on many courses.

There is no carburetor I recommend higher for off-road use than the Quadrajet; so that is the one to be discussed here.

The Quadrajet will work equally well on boats. Although I have never raced boats personally, a good number of my friends do. The requirements are similar in some respects to off-road, so if you want to use a Quadrajet in your boat, read this section and the rest of this performance chapter. The carburetor can be blueprinted to take the abuse of a fast boat.

We are doing a special section on most popular types of racing be-

Another Q-jet-equipped Oldsmobile being used for something other than a Sunday afternoon drive. Photo made during 1973 Baja 500 race. 112 is same as car 167 shown later in this off-road section.

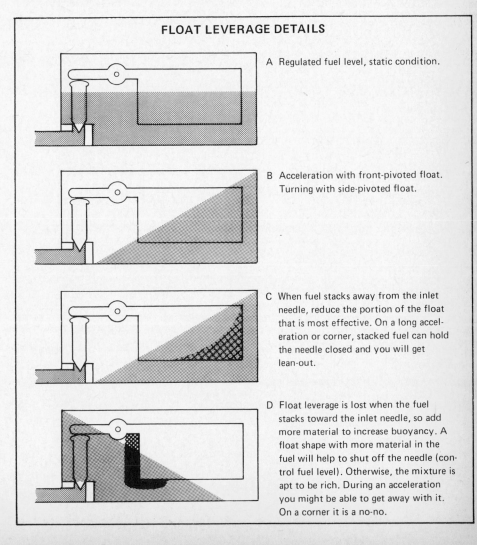

## FLOAT LEVERAGE DETAILS

A  Regulated fuel level, static condition.

B  Acceleration with front-pivoted float. Turning with side-pivoted float.

C  When fuel stacks away from the inlet needle, reduce the portion of the float that is most effective. On a long acceleration or corner, stacked fuel can hold the needle closed and you will get lean-out.

D  Float leverage is lost when the fuel stacks toward the inlet needle, so add more material to increase buoyancy. A float shape with more material in the fuel will help to shut off the needle (control fuel level). Otherwise, the mixture is apt to be rich. During an acceleration you might be able to get away with it. On a corner it is a no-no.

cause certain aspects of the systems become more or less important depending on use. I am going to start this off-road and marine carburetor section with the float and go on into a discussion of parts related to it. Controlling fuel level through the jolts, forces, and motions of a going off-road machine or boat is no light task. However, boats are not generally subjected to the high side loads or extreme angles which are commonly seen in off-roading.

**Floats**—You know, if I were an impatient sort and less aware of all the things that affect good carburetor performance, I would probably skip this section. The thought of a float discussion may not seem worthwhile. I admit, a number of years back a float to me was a float and it either stayed buoyant or got replaced. This happens to be now and I know the importance of these funny-shaped things and if you don't: Slow down to see what I have to say. This is about Quadrajets but many of the principles apply to any carburetor you want to relate them to.

First of all floats have many shapes. There are also size differences. Size is greatly influenced by the bowl and so is shape to some degree, but generally there is latitude for subtle variances. A small change may be all you need. See the illustrations.

Studying the float silhouettes shows you RPD put a great deal of thought into their design. The hinge pin is located forward on the Quadrajet and the bowl is quite narrow so right and left turns have little effect on level. When you brake suddenly, fuel will run forward toward the inlet needle. Rochester kept the forward part of the float quite thick for maximum buoyancy when fuel is forward and the long float leverage is thereby decreased.

In this case little can be done to improve on this because the visible angle is needed so the float can drop below center without the forward bottom edge of the float hitting the float chamber wall.

In the Quadrajet a well-engineered plastic spacer does the rest of the job by limiting the volume of fuel that rushes up around the inlet needle and away from the float during a forward brake stop.

Now let's suppose you have a mean dune buggy with tough tires and tractor gearing (high numerical) for traversing sand, rocks, ditches, etc. Each time you stab the throttle, fuel will stack up rearward and it would be easy to cause too efficient a shut-off leverage. Accelerating up a steep grade would add to the problem. If this causes lean-out I would get rid of buoyancy at the outer end of the float one way or another. I have gone float pontoon hunting and once the ultimate shape was found, it was integrated with the proper float arm. This practice is quite satisfactory when metal floats exist because they can be soldered in place if care is exercised.

The material used in most modern floats is a closed-cell plastic material and it is easy to alter but not likely to last for extended durations of time (meaning weeks and months). The material is easily cut and shaped and because of its cellular composition it will not leak—so it says in the book. This is true to a point but once the outside surface is disrupted you stand the chance

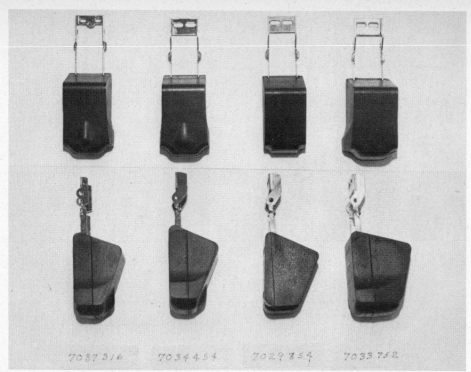

7087316    7034454    7029854    7033752

These few Q-jet floats show some of the variety of shapes used to fit particular applications.

Plastic spacer limits fuel volume which can rush away from the float during a forward brake stop. Float is centered between the primary bowls.

that some of the exposed outside cells will connect to inner cells. In this case fuel gets in and the float gets heavier—losing buoyancy.

If you are running in rough terrain and it's evident you need a new shape float: Cut it. If you anticipate such surgery have a small can of clear shellac with you. After the shaping is done, put a couple light coats of shellac over the new surface and let it dry thoroughly, and you are in business. Ambroid thinned sufficiently with acetone is perhaps an even more durable coating. Neither factory nor service dealers will recommend such practice because it can cause short life for the float. It is strictly a special-application, do-it-yourself exercise. Elsewhere in this section we suggest you carry an extra float or two. This is one of the reasons why.

Remember if you cut away much of the lower portion of the float, particularly at the outer end, appreciable buoy-

This Vega-powered test vehicle is shown getting the river-bottom treatment by author Doug Roe. A Quadrajet/turbocharger combination extracts over 200 HP from the 140 CID engine. With modest metering changes the same carburetor will handle a 472 CID Cadillac. Versatility plus!

ancy will be lost. Readjust with additional mechanical measurement from air horn gasket surface to the float to get back the liquid level you had before alterations. If you make small systematic changes when trying to beat one problem you are less likely to create another one elsewhere.

Remember to think it out, make subtle changes, and take the time to evaluate the total effect.

Now I want to discuss a situation that exists within the Quadrajet family of carburetors which can be important to off-roaders and boaters. The float-arm leverage will have one of two dimensions. The accompanying photos-illustrations show this difference. If you intend to compete with dogged determination to win, get a Quadrajet with the short fulcrum float arm. I have personally made both types work under the worst of conditions but the short fulcrum leverage makes the task easier and a margin better.

This sounds easier than it actually is because you cannot merely install the short-arm float into a carburetor which originally had one with the long arm. You actually have to switch the entire carburetor. Short- and long-arm floats cannot be interchanged.

There are currently approximately 200 models of Quadrajets. Most of them have subtle differences; some of the differences like this pivot are significant.

We can't possibly list every number and detail all dimensions but following is a guide that will narrow down the hunt as far as the float-arm fulcrums go.

| LONG FULCRUM | SHORT FULCRUM |
|---|---|
| Pontiac thru 1968 | 1969—1973 |
| Oldsmobile, all years | None |
| Cadillac, all years | None |
| Buick thru 1968 | 1969 to current |
| Chevrolet '66-'68— sporadic | 1966—68 sporadic 1969—1973 |

The 1965—66 and some '67 Quadrajets with the diaphragm type fuel inlet assembly had the long fulcrum. Because they date back a few years, parts to blueprint for your specific need will be limited. I recommend you avoid the use of these units for off-roading.

**Filter**—Providing you have scanned the fuel-system section, it is reasonable to assume you plan to use a good in-line filter just ahead of the carburetor. If not, you really should! Also plan to change it quite often, particularly if the carburetor sits days or weeks at a time between runs. Fuel systems can rust, gather condensation and harmful foreign contaminants when sitting. Like every other component, the filters have limitations you can exceed. Be on the safe side and change regularly: it's easy before the start flag comes down.

Here are two of the off-road-racing vehicles starting the 1972 Baja 1000 with Quadrajet carburetion. The Olds "Banshee" 213 was driven by actor/racer James Garner in 1970; Doug Roe in 1971; and Wally Dallenback in 1972. Olds Cutlass 167 was driven by James Garner and Doug Roe in the 1970 Baja 500 and has been driven in several Mexican and U.S. off-road events by world-famous drag racer and salt-flat record holder Paula Murphy. Chevy and Jimmy utility vehicles are common among trophy winners. Quadrajets do an excellent job in desert racing.

Short-fulcrum float at top compared with long-fulcrum float at bottom clearly shows difference in leverage. Short-fulcrum model has more closing force against needle valve. Differences and applications are described in text.

**Angularity**—Visualize the problem as you tip the carburetor. Consider how the fuel depth (or head) changes relative to the main jets, discharge nozzle and fuel-bowl vents when the carburetor is tipped. You might actually take a bowl with some fuel in it and observe these things. Note that the angle can affect the height of the discharge nozzle in relation to the fuel level. In some attitudes, a 40° tilt may cause fuel to drip or spill over from the discharge nozzle. This may require you to lower the fuel level. Increase the mechanical setting 1/32-inch to 1/16-inch per change. With the small inlet valve these changes are going to be more effective.

The problems with lowering the fuel level are: The power valve and main jets are uncovered sooner and nozzle "lag" may create a noticeable off-idle bog. Part of this lag can be offset with increased pump or richer metering which improves nozzle response. As with all other changes, make them in fine increments and don't go overboard.

## FLOAT SETTING CHANGE vs. INLET SIZE

| Fuel Pressure (psi) | Inlet Size (inch) | Fuel Level | Float Setting |
|---|---|---|---|
| 6.0 | 0.110 | 0.600 | 0.160 ± 0.010 to toe |
| 6.0 | 0.125 | 0.600 | 0.265 ± 0.010 to toe |
| 6.0 | 0.135 | 0.600 | 0.296 ± 0.010 to toe |

## RELATION OF INLET SIZE & FUEL PRESSURE TO FLOAT SETTING

| Carburetor (all 1973 models) | | Fuel Pressure (psi) | Inlet Size (inch) | Fuel Level | Float Setting |
|---|---|---|---|---|---|
| 7043244 | Buick | 4.5 | .125 | .750 | .479 ± .010 to toe |
| 7043243 | Buick | 5.25 | .135 | .655 | .405 ± .010 to toe |
| 7043515 | Chevrolet | 8.0 | .125 | .600 | .355 ± .010 to toe |
| 7043513 | Chevrolet | 6.0 | .125 | .600 | .265 ± .010 to toe |
| 7043274 | Pontiac | 6.0 | .135 | .675 | .390 ± .010 to toe |

**Inlet-valve size**—If you missed the considerable discussion on the inlet valve and how its size can affect your carburetion, take time to study it. This was presented earlier in this chapter.

To determine the best size for your application, carefully check performance with a fairly large valve. In most cases, use the production one. Do timed checks on a fairly smooth stretch of road. And do some seat of the pants tests on a rough course. The next step is to install an inlet seat with a smaller orifice. Repeat the same *exact* checks under as near the same conditions as possible. If your times are still good, drop the size still further. Continue dropping the size and testing, etc., until the accelerations fall off slightly. Now go up to the next size larger inlet valve and that part is settled. It is likely that the inlet valve will end up smaller for short spurts and rough stuff than for any other form of racing.

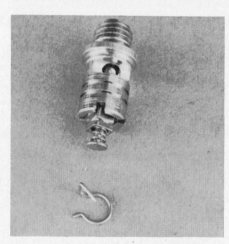

Assembled spring-loaded inlet valve assembly from Kaiser Jeep can be installed in any 2G carb. Exploded view shows relationship of the various parts.

**Off-road carbs/engines seldom run flat out**—Desert running or off-road racing is made up of a series of acceleration and braking maneuvers combined with tight turns and impossible-to-describe rough terrain. Occasionally a run will have fast sections but rarely should you tune solely for this.

Don't come back with an "of course not" because many engines for desert vehicles have been dyno-tuned at peak revs. Makes good ink but not good dust.

**Vent and pitot tubes**—Before we leave the float and fuel inlet area a few words need to be said about the fuel bowl itself. Once you get the fuel level pretty well controlled for any condition which you tested for, go ahead and try some running.

They say there is a rock, ditch or hole on every course with your name on it. Should these way-out unexpected slammers cause you problems the next thing to consider is extending your internal vent tubes upward. No harm in doing this before trouble has a chance to rear its ugly head.

Vent or pitot tubes should be extended upward as high as possible to handle sloshing fuel which can come out of the vent tubes under severe braking and/or bouncing. Off-road cars with remotely mounted air cleaners are sometimes equipped with appropriately sized hoses to extend the vents all the way up into the air cleaner. Do not extend them up against an air cleaner top so as to cause a restriction. Also remember to increase the size accordingly if you make the hoses exceptionally long. A restriction from length or obstruction could cause you real metering problems.

**Air cleaner**—Now and as long as you run the desert think about keeping your carburetor and engine clean. Don't cheat on the air cleaner.

The best possible air cleaner is one which completely encloses the carburetor, with the throttle operated by a cable. The cable housing can be sealed where it enters the baseplate for the air cleaner. More details on off-road air cleaners are in H. P. Books volume *Baja Prepping VW Sedans & Dune Buggies.*

Should you choose not to do a complete enclosure of the carburetor be sure you inspect the exterior for any possible leak. Silt and microscopic dust particles will do your engine in quick if you overlook the smallest crack.

Silicone rubber is really great for plugging holes. It should be used freely before any appreciable amount of running. The final time you install the cleaner or air duct adapter to the top of the carburetor seal it all the way around the gasket. Better yet, don't use a gasket. It can get cocked or creep away from the flange. A good uniform application of silicone rubber and a tight cleaner is assurance plus. It peels off easily if you do pull the

Air is filtered into the cab through three air cleaners on the back window of this off-road racing truck. Air cleaner on engine is supplied from the cab. Photo taken at 1972 Baja 1000 pits.

Here four thin filters with metal-mesh outside covering have been glued together with Silastic. Hood forms base for cleaner, which is ducted to top of carburetor.

Wrap-around polyfoam over a paper filter is often used for ultimate filtration. This combination really does the job!

Centrifugal-type filters are quite popular with off-roaders (and contractors!) because dust which collects during the day can be easily dumped to clean the unit.

Writer/racer John Thawley drove this tough beauty over parts of Baja California where goats and mules have trouble. The Quadrajet-equipped 350 CID V8 didn't whimper once. He did—many times!

Here the vent has been sealed with Silastic RTV compound. Choke assembly has been removed and the screw holes and vacuum passage sealed with the same material.

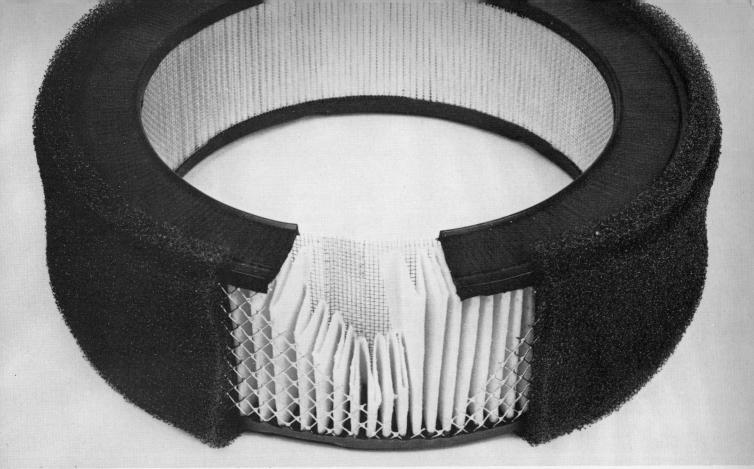

AC 222 C dual-stage filter is used on some Chevy Blazers and big trucks. It measures 5-3/4 x 12-1/4 inches and costs about $10 as Chevy 6421326. This filter is too big for most cleaner housings, so AC 223 C (only 2-7/8 x 12-1/4 inches) may be better for your application. It's cheaper at only $8.25 per. Ask for Chevy 6422680. These filters are nearly unpluggable because the foam outer covering keeps dirt from packing up inside the folded paper. When dirt builds up to a certain point, it just falls off, so the filter is self-cleaning to a certain extent. We highly recommend these filters for all off-road and desert type applications.

cleaner. Be sure to carry a tube. To a desert racer it's almost as valuable as racer's tape. **Throttle linkage**—The use of a cable drive is often recommended for off-road installations to transmit less vibration to the driver's foot. Many stock automobiles are now equipped with such arrangements and it is usually easy to obtain a cable throttle linkage with everything made to fit the job at hand. If it is light and small, carry an extra. The cable can get hung up with fine dust. Hydraulic linkages are also popular, but it is quite easy to overstress the throttle levers with such hook-ups. The hydraulic linkage should be installed so that the full travel of the actuating cylinder provides wide-open throttle and NO MORE. In some instances, the cylinder mounting has to be carefully thought out so that there is not an over-center condition during throttle operation. For very light vehicles you may decide the weight of these units is objectionable.

Sometimes a very simple open cable is the best way. Think about your own situation and decide.
**Vibration**—Floats take a real pounding in off-road and boat applications. Check their condition often.
**Dams and partitions**—Now and then an effective partition is fabricated into a location where it does some good. These are generally aspirin when you need a prescription, but sometimes worth the effort. Think about it in this light and see if you will benefit.

A short obstruction in the lower bowl can at best contain a few cc of fuel for a very short span of time. If you aren't able to replenish fuel to this area continuously during a potential problem maneuver or if enough fuel cannot be contained in the desired area for the entire span of the problem, of what value can this area possibly be?

Fuel stuffers as used in the Quadrajet

that prevent fuel from packing in useless areas are very beneficial. Deflectors high in the bowl directing fuel away from vents and downward toward metering orifices show merit in some cases. Direct all your first efforts to float, inlet needle, venting and metering. Then if you have problems, pursue other items. A nice neat dam blows the minds of your competitors who sneak a peek, so if one you have tried doesn't harm performance, leave it in.
**Metering**—Read the section in this chapter on general performance and economy. It gives you all the help needed to accomplish flawless metering of the Quadrajet for any purpose.
**Other RPD carbs**—The preceding recommendations for the Quadrajet will give you a very sound off-road or boating unit. Should you be working with another type Rochester, *all* of the same logic applies. However, basic carburetor design can make the task more difficult and perhaps less rewarding.

One type of float-assist spring. This is called a torsion spring arrangement. A similar type is used on some 2G and H carburetors. Another type has a float-arm tang which rides against a stretched spring.

**Fuel control problems**—The **4GC** floats and bowls are outboard which creates inherent fuel-control problems when angularity or forces are significant. This carburetor's float-assist spring does help to add stability to fuel level. This is of limited assistance when applied to a spread-out float and bowl arrangement such as this. If you must use one, think it through and do the best you can. It can be made to perform with 80—90 percent satisfaction.

The **H** carburetor found on Corvair engines has the same inherent problems—spread-out bowl chambers and long float arms. This is also a tough one even though it does have small inlet needles and float-assist springs. Spring-loaded needle valves were also used on some models, but this is strictly a junkyard exploring situation because the parts are no longer sold. Your best bet would be the 1963—64 era. The latter item is a real asset for washboard-type conditions but of little assistance to the problems that will get you—steep angles

and violently disrupted fuel from severe jolts and sudden turns.

The **2GC** offers you the most hope as an alternative to the Quadrajet. The entire float and arm is a bit too long but the bowl is a single chamber. Small inlet-valve assemblies are available which give you latitude for changing float length and shape as required. A spring-loaded needle-seat assembly is available from Kaiser Jeep as 7023895 or your Delco dealer as 30-95. Use it if your inlet-valve size requirement turns out to be approximately 0.100-inch. If you can get away with a smaller inlet such as the 0.086-inch 1971—72 Vega inlet valve (needle/seat) (Delco 30-3), the need for the spring-loaded assembly is minimized. As stated before, the spring-loaded assembly is effective in controlling fuel level over short choppy surfaces. Some models of the 2GC also have a float-assist spring. This is a good tool to work with when confronted with crazy fuel movement and unstable conditions in a carburetor bowl.

Let's say very steep grades and hard turns dictate you to build a float with 1/4 or 1/2-inch removed from the outer end. Quite likely the reduction in buoyancy and leverage would produce a flooder (lack of fuel control). Assuming the inlet valve is already minimum tolerable size, additional help is needed. Take off the float-assist spring and in increments of 1/32 per try, reduce the end-to-end tang measurement (free spring). Each time you reduce this measurement the float is more capable of shutting off the fuel. Like everything else it has limits—go too far and the float won't drop far enough to permit high volume fuel entry. I cannot give you a limit because I do not know your final float shape and leverage. Let's just say an 1/8-inch tang-to-tang change is a big one and it is probably nearing your limit. I have seen these 2G carburetors used on oval tracks, road races, hill climbs, slaloms and in the desert with good success. I wouldn't hesitate a minute to use them anywhere their capacity fills the bill.

Rich Sloma has been a successful competitor in West Coast SCCA B Production racing. He continued to use the Rochester Q-jet in 1973, even though non-stock carburetors were permitted by a rules change. 356 CID engine dyno-ed 402 HP at 6200 RPM before extensive modifications to the Q-jet. After changes to the carburetor the engine pulled 420 HP at 7500 RPM.

# Road Racing

Sitting snugly in a good handling machine with safety belts and helmet properly in place is when a driver can get great inner satisfaction. He knows the vehicle and the hours of skilled workmanship it took to make it what it is—a machine that will accelerate, brake, corner and haul down the straights whether they be short or long. He knows this assemblage of metal and miscellaneous materials is designed and built to respond to his own split-second reflexes.

How is all this tied to a carburetor? If you are going to run a circuit or series of sanctioned events, chances are the rules will classify vehicles by weight, HP, cubic inches, modified, production, etc. When words like *production, factory stock,* and *stock to the eye* apply to your car there is a good chance a carburetor is involved.

Take a production vehicle and install quick steering, wide tires, super brakes, special gearing, suspension, etc., it's a whole new game. A number of sanctioning organizations allow these changes to production-powered vehicles. Even though we say the Quadrajet is ex-

tremely versatile and lends itself to nearly any use, the production carburetor is no longer able to cope with the exotic performance machine surrounding it.

It's a rare production vehicle that exceeds a 0.5 G cornering capability and very few can approach an 0.8 G stopping ability. When a car is designed, built and run with over 1 G braking and cornering capabilities every accompanying piece must be up to that. There have been some production vehicles that do scat pretty quick but there is no reason for any one component supplier to exceed the performance parameters of the basic vehicle.

In road racing if each component of the unit you run gives anything less than 100 percent reliable performance, this is enough to put you second, third or further down the list from first. So take time to blueprint the Quadrajet and it too will become a competent part of your very special machine.

**Road-race tuning**—Road-race tuning is something else because these racer-mechanics go "deep" in their search for a state of "ultimate tune." We will tell you what these tuners do to get their carburetion "right on," but at the same time we have to warn you that this is not the sort of thing that Mr. Average Mechanic should undertake. This kind of tuning takes experience and the understanding that it is awfully easy to go "too far" in modifying parts. Some cannot be easily replaced . . . or put back in their original condition.

When tuning involves the idle-fuel systems, part-throttle and power-system metering, this gets you into carburetor modification which is close to carburetor engineering development. This is a tedious task requiring infinite care and patience—more than most mechanics would ever imagine.

Once the correct carburetor size has been selected, tuning proceeds like this: First the idle is set by turning the mixture screw in until the engine falters and then backing it out 1/8 to 1/4 turn. Monitor the idle RPM and adjust as necessary while setting the mixture. Next, the free-engine (no-load) RPM is increased slowly to 3,000 RPM to see if there is any point at which the engine stumbles or misses. The reason for doing this slowly is so the fuel from the accelerator pump will not confuse the lean condition which is being checked for. If missing or stumbling occurs, this indicates a lean condition. Back out the mixture screw slightly to see whether this helps the condition. If the screw has to be backed out more than 1/2 turn from the best idle setting which was previously established, open the idle tube 0.001 to 0.002-inch at a time. Do not exceed a size increase of over 0.006-inch.

If a miss still exists, drill out the channel restriction 0.001 to 0.002-inch at a time to a maximum increase of approximately 0.008-inch. See the general performance and economy section for details on altering this idle/off-idle system. Set curb idle after each change.
NOTE: Read the material on curb-idle throttle-plate positioning related to the transfer slots or holes as detailed for racing camshafts on page 236.

Once the idle tube and channel-restriction sizes are correctly established for the free-engine tests, the car is road-tested for surging at low constant speeds (the mixture is controlled by the idle tube from curb idle through 30 MPH steady speed or light-load conditions) and about 14 to 16 inches of manifold vacuum.

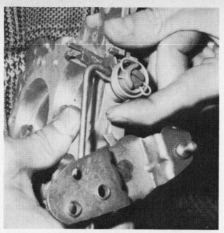

Any road racer should be meticulous about checking throttle openings and angles. Other material in this chapter details how throttle-plate angle can affect mixture distribution.

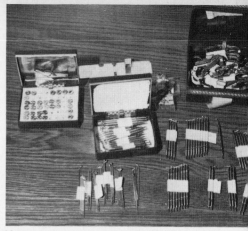

Here are the metering rods and jets used to tune Allan Barker's Corvette (pictured on page 1).

Metering—For WOT operation, meter the primary and secondary WOT fuel requirements *first.* If this causes part-throttle richness, increase the part-throttle metering-rod area to reduce part-throttle fuel flow. Use secondary metering rods with short power-tip ends so the intermediate opening power positions are lean. This performance-priority metering compromises street driving economy.

If the car "feels good" leave it alone. If it surges, a slight increase in the main jet will likely be enough to make the idle system feed more effectively just off curb idle.

Next, starting at 30 MPH, a series of "crowds" are made with the driver "crowding" a certain manifold vacuum as observed on a gage. These crowds are light accelerations made while keeping the manifold vacuum first at 12 inches, then at 10 inches, then at 8 inches, and so on—to the point of power-valve opening. If there is surging during these crowds, attempts to improve the condition are made by varying the area of the primary part throttle. If there is a lean surge, the primary jet size is increased until acceptable drivability is obtained. Incidentally, a rich condition is not usually found—except in some high-performance carburetors.

Next, wide-open throttle tests are made from 30 to 80 MPH in high gear or 20 to 60 in second or third. The WOT power fuel flow area is increased or decreased to get the best time, as determined by stopwatch. Study the details of area tuning in the general performance and economy section of this chapter. Continue with the stopwatch surge evaluations until your metering matches the engine and your needs. Having a good driving and performing road race machine makes it all better.

Detailed accelerator pump tuning procedures are described elsewhere in this chapter. Chances are the short pump arm position plus ensuring full 21/32-inch pump travel will be all that you need for road racing. If you feel more capacity is required, follow the pump-enlargement procedures we've provided.

Remember that conditions vary from hour to hour and day to day. Your fine tuning results will only relate to all things accomplished on the same day. Always take the time to rerun your baseline or best times to make sure that this has not changed because of variables such as engine condition, plug condition, tires, atmospheric pressure, etc.

# The Carburetor & Emission Controls

## The Carburetor Is Only Part of the Picture

Emission controls on 1973 Chevrolet Z-28 engine include air pump and distribution plumbing for AIR system, EGR (arrow), PCV, and ECS. The carburetor and distributor are carefully calibrated to provide the required emission performance with this equipment.

As can be seen in the chart below, each model year automobile requires a change to meet the tighter emission standards except from 1973 to 1974. All of these standards have been converted to true mass grams per mile for sake of comparison.

**A**ir pollution is a worldwide problem. This atmospheric condition is not limited to urban or industrial areas—it is now almost impossible to get away from pollution, regardless of where you might be on this earth. Growth in population and fuel use has been a major contributing factor. The magnitude of the problem—and the concern of the public—is manifested in state and federal legislation which has been enacted to regulate the amount of contaminants which can be put into the atmosphere by automobiles and light trucks. Standards for industrial emissions have also been set, but the major focus and enforcement has been in the area of new automobile manufacturers and makers of aftermarket (replacement) parts which directly affect emissions.

California, specifically the Los Angeles and San Francisco areas, has the greatest smog problem and hence this state has led in exhaust-emission-control regulations.

So, what is *smog,* anyhow?

It is a simple term for a complex happening. When unburned hydrocarbons (HC) and oxides of nitrogen ($NO_x$) combine in the atmosphere and are acted upon by sunlight, complicated chemical reactions occur to produce *photochemical smog.* Both areas mentioned have all of the necessary ingredients: HC + $NO_x$ + sunlight. These are aided in combining by the inversion layer over these areas—a dense layer of the atmosphere which prevents the escape of the ingredients into the upper atmosphere where they might be swept away by winds. The inversion layer is often likened to a lid on a pot and this "lid" holds the ingredients right there so that the reaction has plenty of time to take place—often several days at a time.

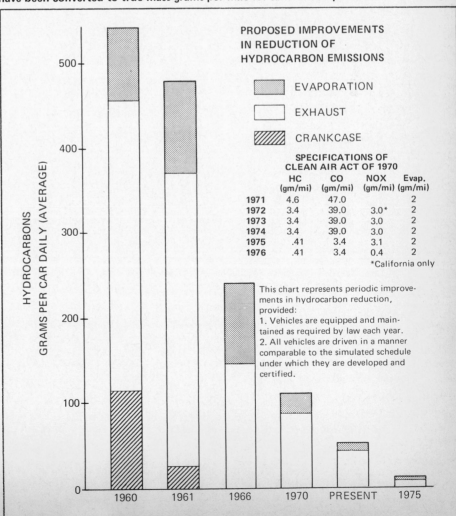

PROPOSED IMPROVEMENTS IN REDUCTION OF HYDROCARBON EMISSIONS

- EVAPORATION
- EXHAUST
- CRANKCASE

**SPECIFICATIONS OF CLEAN AIR ACT OF 1970**

| | HC (gm/mi) | CO (gm/mi) | NOX (gm/mi) | Evap. (gm/mi) |
|---|---|---|---|---|
| 1971 | 4.6 | 47.0 | | 2 |
| 1972 | 3.4 | 39.0 | 3.0* | 2 |
| 1973 | 3.4 | 39.0 | 3.0 | 2 |
| 1974 | 3.4 | 39.0 | 3.0 | 2 |
| 1975 | .41 | 3.4 | 3.1 | 2 |
| 1976 | .41 | 3.4 | 0.4 | 2 |

*California only

This chart represents periodic improvements in hydrocarbon reduction, provided:
1. Vehicles are equipped and maintained as required by law each year.
2. All vehicles are driven in a manner comparable to the simulated schedule under which they are developed and certified.

HYDROCARBONS GRAMS PER CAR DAILY (AVERAGE)

500 — 400 — 300 — 200 — 100 — 0

1960  1961  1966  1970  PRESENT  1975

Smog causes nose, throat and eye irritations. And, like carbon monoxide (CO), is extremely harmful to animal and plant life (including trees). Smog also causes deterioration of some plastics, paint, and the rubber in tires, seals, weatherstripping and windshield-wiper blades.

There are three major types of vehicle emissions to consider:
1. Crankcase
2. Exhaust
3. Evaporative.

In this section there are facts intended to acquaint you with the emission program equipment and how it affects you and your car. We must all become more knowledgable in this vital effort to help refute the black-magic fixes prescribed during shop-talk sessions. Theoretically, perfect combustion of gasoline and oxygen would produce only carbon dioxide and water as emissions. Unfortunately, the pressure and temperature of the combustion chamber cause several undesirable combustion products. Two primary ones are hydrocarbons and carbon monoxide. They are nothing more than partially burned air/fuel mixtures. Partial combustion is caused by a shortage of oxygen in the mixture, combustion-chamber shape, cam overlap and myriad other factors—all related.

Carbon monoxide forms whenever there is insufficient oxygen to complete the combustion process. Generally speaking, the richer the mixture, the higher the CO concentration. Even if the fuel/air mixture is chemically correct, CO cannot be reduced to zero because perfect mixing and cylinder-to-cylinder distribution is impossible to achieve.

Gasoline is composed of numerous and varied hydrogen and carbon compounds, hence the name *hydrocarbons*. Unburned hydrocarbons are just that—gasoline that did not get burned on its trip through the engine. There are several reasons why this can and does happen. Rich mixture is one. Fuel that does not get burned because of misfiring is another. So, either a lean or a rich A/F mixture can cause an increase in HC emissions. Other things affect HC concentration. Combustion chamber surface-to-volume ratio increase (high compression) increases HC. Spark advance. High vacuums under decelerations, etc.

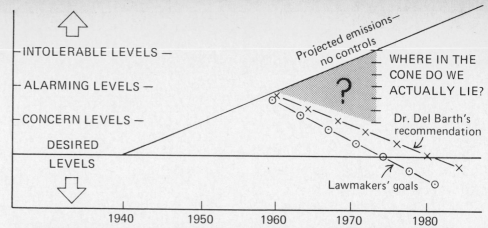

Nobody likes smog and air pollution. Starting in 1960 laws have been enacted, culminating in the Clean Air Act of 1970 which was greeted as the answer to all problems. One of the earliest qualified researchers in the field, Dr. Del Barth of Cal Tech, recommended a phased improvement program extending over the 1970-1980 decade. In the political arena this decade of orderly progress was arbitrarily shortened to only five years and the American public is now paying the price. In an atmosphere of panic, the Seven-Mode-Cycle shown below was devised and mandated even though widely considered to be unrealistic.

Meeting this accelerated time schedule and annual goals which are near-impossible forced car makers to resort to gadgetry and quick-fixes which are usually not understood by working mechanics. The results are upon us. Cars are more expensive and complicated; they consume up to twice as much gasoline; drivability is poor; it's difficult to get a proper tune-up. Authorized tuning adjustments are directed at reducing emissions instead of getting the engine to run satisfactorily.

Unless the public demands a pause in this pell-mell rush to disaster, it is likely to become worse. The government enforcers seem blind to the undesirable consequences of the Clean Air Act of 1970. There is no doubt that it accelerated the onset of the Energy Crisis, first recognized widely in 1973.

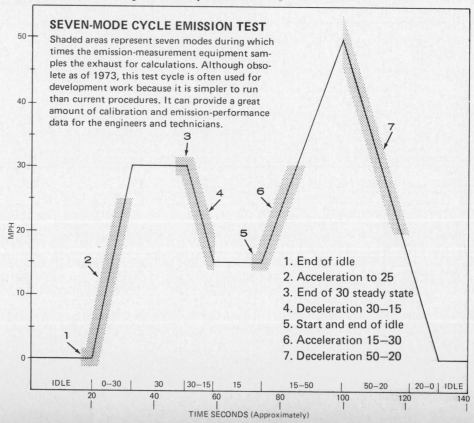

**SEVEN-MODE CYCLE EMISSION TEST**
Shaded areas represent seven modes during which times the emission-measurement equipment samples the exhaust for calculations. Although obsolete as of 1973, this test cycle is often used for development work because it is simpler to run than current procedures. It can provide a great amount of calibration and emission-performance data for the engineers and technicians.

1. End of idle
2. Acceleration to 25
3. End of 30 steady state
4. Deceleration 30—15
5. Start and end of idle
6. Acceleration 15—30
7. Deceleration 50—20

In the 1966 model year, California applied standards for tailpipe emissions of CO and HC, together with test procedures and sampling methods.

The Federal Government adopted the California test procedures and standards in 1968 and required—as California had earlier—all manufacturers to certify and prove that their vehicles met the prescribed standards.

A third product emitting from combustion chambers is oxides of nitrogen. These form when the normally inert nitrogen present in the air/fuel mixture combines with oxygen under high temperature and pressure. The tendency to form $NO_x$ is increased as the mixture is leaned because of the increased availability of free oxygen. Maximum $NO_x$ production occurs under the condition of approximately 16:1 A/F (0.062 F/A).

Starting in late 1970, legislation was passed which brought about tests for checking the HC, CO and $NO_x$. Prior to that only HC and CO were measured. A brief history is shown in charts on pages 279 and 280. Once test equipment became available and the test requirements and procedures were spelled out, manufacturers invented, designed, and tested many devices to meet the goals.

Before going into the engine hardware aspects of this, let's dwell on test schedules and comment on reported results.

## TEST PROCEDURES

The former test procedure consisted of seven duplicate cycles with varying driving modes. A tester drove the vehicle on chassis rolls as directed by audible signals and visual instruments. Exhaust samples were taken at certain intervals during every mode. The computer then analyzed each mode individually and compounded a total at completion. During development, separate modes were often run for expediency in testing. The numbers representing HC, CO and $NO_x$ in a certified test were always taken from a full seven-cycle run following a 12-hour minimum soak in a 75°F. room. The first two cycles encompassed emissions during choke operation. By the end of the second cycle the automatic choke was off and the remaining cycles represented warmed-up conditions. The fifth cycle was run per schedule but not sampled.

Many factors go into the computation of final results: percentages of certain cycles, vehicle weight, etc. All manufacturers and authorities work with the same regulations and formulas. No purpose can be served by writing pages of procedures most readers of this book will never get to exercise because the special equipment required is not readily available.

Starting in 1972, the actual dynamometer tests required to certify a car are part of a longer, more complicated procedure.

As in the seven-cycle tests, the car has to sit 12 hours in a controlled temperature of approximately 75°F. It is then pushed onto the emission chassis-dynamometer "rolls" and the exhaust is attached to the emission sensors. This procedure is required to be sure full choke operation is included in the results. The primary difference in exhaust-gas measurement between the seven-cycle and the current method is *total* sampling. The exhaust goes to a bag or chamber and is sampled continuously rather than only during certain modes. The engine is cold-started and idles 30 seconds in neutral. It is then put in gear and driven for 23 minutes in a series of acceleration "steady state' decelerations and idles. A maximum speed of 53 MPH is reached *once* during the test. Most of the remaining time is spent idling or at speeds ranging from 15 to 35 MPH. If you were to simulate this dynamometer cycle on the street, observant law-enforcement officials would certainly question your sanity.

It simulates a very heavy city-traffic situation being traversed by a driver who insists on chugging along in the wrong gear. In my opinion, this is *also* a very unrealistic test. With our fast highways and expressway systems the trend toward suburban and country living has spread. Longer holidays and vacations with the popularity of recreational-vehicle activity and earlier retirement puts a great majority of travel at 60–80 MPH. The criticism of these simulated schedules is not intended to be derogatory to those who put a great deal of sincere effort into this program. It is a most difficult task to simulate any aspect of the motoring public realistically; but with well over 2,000,000 highway and test miles to my credit, I feel entitled to express an opinion on the subject. Observed results of over

a decade of effort says it is an issue to consider further.

To be able to pass emissions tests and also to make the car reasonably drivable, the carburetor is calibrated *too-lean* at the lower speeds and *over-rich* at the higher speeds and for heavy throttle maneuvers, such as passing and climbing grades. In some cases this is particularly bad for altitude operation.

Let's not misunderstand this seeming criticism of what has been done to date to combat motor-vehicle emissions. It is for us, the motorists, that billions of dollars have been spent on this task. The question is if progress is being made in the *right* direction. Regulations are being met by the manufacturer. The question is: does all of this reduce emissions?

In addition to the emission tests required for exhaust-gas analysis, another test determines total hydrocarbons emitted from a non-operating vehicle.

A vehicle is placed in a sealed enclosure called a "bag." The evaporating hydrocarbon emissions are then measured as they accumulate in the enclosure. The test is in two parts:

**Simulated sun load**—Emissions increase when the vehicle is heated because this creates vapors. A parked vehicle being heated by sunlight is simulated by heating the gas tank at a linear rate. At the end of a one-hour period, hydrocarbon level in the enclosure is measured.

**Hot-vehicle soak**—This test simulates a parked vehicle which is still hot after being driven. The test vehicle is driven on the dynamometer until hot and parked in the enclosure. At the end of a one-hour soak, hydrocarbon level in the enclosure is measured again.

Total hydrocarbon losses are determined by summing the data obtained from the two vehicle tests.

Again, lengthy details of how these evaporative losses are trapped, measured and reported seem of little importance to those not directly involved in development or policing. As in the exhaust-analysis tests; procedures, testing and reporting are thorough through mass effort.

## VEHICLE EMISSION EQUIPMENT

We will now identify and show the purpose of the different emission-control devices introduced since 1959. These components will be illustrated and commented on so the novice and mechanic alike can understand what these mechanical units do.

## POSITIVE CRANKCASE VENTILATION (PCV)

Crankcase emissions were the first target of the lawmakers and automotive engineers. They were attacked first because approximately one third of all engine emissions come from this point. And anyone could look at the old road-draft vents—used on all cars and trucks through 1960—and see lots and lots of pollutants being spewed into the atmosphere. If you stand near an early car (pre-1965 in most of the United States) the smell of the escaping hydrocarbons and carbon monoxide is unmistakable.

The gasoline engine combustion process creates a highly corrosive gas. And, for every gallon of gasoline burned, more than a gallon of water is formed. Some unburned fuel and combustion products leak past the piston rings into the crankcase as *blow-by*. This blow-by must be removed before it condenses in the crankcase and reacts with the oil to form sludge. If this sludge is circulated with the oil it will cause corrosive and accelerated wear of pistons, rings, valves, bearings etc. And, because blow-by carries a certain amount of unburned fuel, oil dilution also occurs if it is not removed.

The first emission-control device—required by California in 1961—was subsequently required nationwide. This was a crankcase ventilation system including a metering valve plumbed between the crankcase vent and the intake manifold. These systems pulled air into the crankcase through the oil-filler cap's wire mesh, but this did not eliminate fumes escaping to the atmosphere, especially at idle and low engine speeds.

By 1968 the Positive Crankcase Ventilation (PCV) system was the standard crankcase ventilation system on all U.S. cars. It removes engine crankcase vapors resulting from normal engine blow-by. These are removed by using manifold vacuum to draw fresh air through the crankcase, sucking the undesirable corrosive gases and unburned fuel into the manifold so they can be burned in the engine.

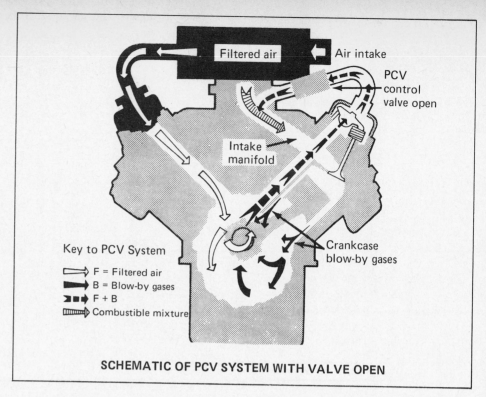

Key to PCV System

➡ F = Filtered air
➡ B = Blow-by gases
➡ F + B
➡ Combustible mixture

**SCHEMATIC OF PCV SYSTEM WITH VALVE OPEN**

The PCV valve varies the flow through the system according to the modes of operation: idle, cruise, acceleration, etc. This provides a proportioning of the additional vapors into the intake manifold so these vary in proportion to the regular air-fuel ratio being supplied by the carburetor. The valve has a spring-loaded large-area poppet and a fixed restriction. At periods of high manifold vacuum—as at idle or on the overrun—flow is only through the restriction, so very little flow occurs. As manifold vacuum drops, the spring opens the valve so PCV flow increases with engine speed.

As shown in the accompanying diagram, when the PCV valve is open air flows through the air cleaner, through the crankcase where it picks up vapor, through the

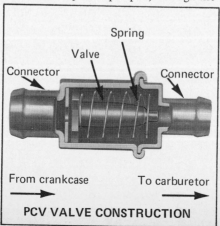

**PCV VALVE CONSTRUCTION**

PCV valve, and into the intake manifold.

At idle, or whenever the throttle is closed, the PCV valve also closes except for the small fixed passage so an excessive amount of air is not drawn into the manifold where it would lean out the idle mixture.

If blowby exceeds the ability of the PCV valve to extract it, then it will be extracted through two routes: the normal route through the PCV valve and also upward through the external air passage between the air cleaner and the engine.

Because additional air enters the intake manifold, carburetors used with the PCV system are calibrated to compensate for the air and blow-by gas entering the intake manifold from the crankcase.

Keeping the PCV valve and the vent tube from the air cleaner to the engine clean is essential to the correct operation of the engine because this is the only crankcase venting. A tube or threaded connection in the carburetor base allows at-

---

**PCV problems**—Although normally helpful in reducing HC emissions, PCV systems can increase CO emissions in two situations:
—1. A plugged PCV valve
—2. Engines with high blowby

tachment of the tube from the PCV valve. Because the vapors are reburned instead of escaping to the atmosphere, pollution is reduced. Because the blow-by components are positively removed from the crankcase, engine life is increased. A third benefit is that of added economy. A Chevrolet publication points out a gain in gasoline economy ranging from 2.3% at 50 MPH to 4.8% at 20 MPH and up to 15.4% at idle. This gain in economy is because the blow-by gas returned to the intake manifold is combustible and becomes fuel for engine operation.

There have been variations of the PCV system, but the one described is used on the vast majority of all engines in use today. Corvairs used a fixed restriction between the crankcase vent and the intake manifold vacuum line (instead of a variable-flow PCV valve).

The PCV valve should be checked at 6,000-mile intervals, and replaced at 24,000 miles. The easiest way to check this valve is to remove it from the rocker cover and shake it. If you can feel and hear it "rattle," it is still working.

The air filter should be checked at 6,000-mile service intervals and rotated 90°. Elements get dirtier in some spots because of air/dirt patterns. Replace at 12,000 miles or as recommended and more often when used under dusty driving conditions. The Vega has a 50,000-mile throwaway air cleaner. In dusty areas and some other conditions it must be changed more often.

## EXHAUST EMISSION CONTROLS

This area is more complex by far than the crankcase or evaporative ends of the problem. Exhaust emissions include unburned hydrocarbons, carbon monoxide and oxides of nitrogen.

There are various approaches to the problem of making engines and vehicles meet the emission standards. The first approaches were engine modification and air injection.

As of 1972, another engine modification was being used: exhaust gas recirculation or EGR. Subsequent modifications to the engine and exhaust system will involve thermal reactors and catalytic mufflers.

**Engine Modifications**—These include carburetor calibration and operational changes—along with spark-advance settings, curves and operational changes.

Carburetors are set up with leaner mixtures in the idle, off-idle and part-throttle ranges. Mechanical limiters on the idle-mixture adjustment screws prevent excessively rich idle mixtures. Dashpots and throttle retarders, plus special idle setting solenoids are used. And, choking mixtures are eliminated very quickly after the engine has been started. Inlet air is heated to ensure good fuel vaporization and distribution.

Decelerations create very high manifold vacuums unless special controls are used. With a closed throttle, so much exhaust is sucked back into the intake manifold that the A/F mixture is diluted (leaned) to the point of borderline firing, causing missing and consequent high emission concentrations of unburned hydrocarbons. Several controls can be used singly or in combinations to control deceleration emissions:

1. Shut off the fuel flow so that there will be no unburned hydrocarbons—because there is no fuel entering the manifold. This method has the problem of creating a "bump" when the fuel is turned back on near normal idling manifold vacuum.
2. Supply a richer mixture to ensure burning.
3. Retard throttle closing to avoid high vacuum buildup. Although this latter method is commonly used, it reduces the braking effect which would have been obtained from the engine during deceleration with a closed throttle.

Distributors are set up with more retard at idle and part-throttle and various switches, valves and other controls are used to provide advance or retard as required to meet emission requirements.

Additionally, basic engine modifications are being made, including reduction of compression ratio to reduce combustion pressures and temperatures, valve-timing variations and re-design of combustion chambers. Exhaust restrictions are also being used.

An additional engine modification, exhaust-gas recirculation, is discussed separately.

Sorting out which modifications are doing what is difficult. But, let's oversimplify and say that the retarded spark helps to reduce HC emissions by providing increased heat in the exhaust to ensure complete burning of any remaining hydrocarbons. By the same token, retarded spark does not let the engine develop peak pressures which it is actually capable of, so $NO_x$ is reduced simultaneously. Part of the combustion chamber alterations which are being made include the elimination of flat quench surfaces as another way to keep $NO_x$ down. Also, eliminating the cooler surfaces of quench areas reduces hydrocarbon emissions. The location of the rings on the piston has an effect on HC. If the rings are higher or the piston top is tapered toward the top ring—there is less volume in which unburned HC can "hide," thereby reducing HC still further.

Keeping the chamber hot also reduces deposits—which provide more places for unburned HC to "hide." These are all minimal things, but the game of reducing emissions is made up of a bunch of *small things*.

Keeping the compression ratio down reduces flame travel speed and keeps the peak pressures from becoming very high—reducing $NO_x$. Exhaust gas recirculation does much the same thing for the same gross effect.

Reducing emissions is a super balancing act—just like tightrope walking in many ways. The major complexity facing the engine designers is the internal-combustion engine's tendency to produce more oxides of nitrogen whenever hydrocarbons and carbon monoxide are being reduced—and vice versa.

Now let's examine the engine modifications in detail, system by system—various systems are interrelated in their effects: All are necessary to accomplish the desired end result of making the engine/vehicle combination pass the federal emission specifications.

## CONTROLLED COMBUSTION SYSTEM

This system increases combustion efficiency through leaner carburetor adjustments and revised distributor calibration. On the majority of installations, special thermostatically controlled air cleaners in conjunction with a heated air source piped over the exhaust manifold exterior are used to maintain intake air at approximately 100°F. or above.) The preheated air improves cold-start driveaways and prevents carburetor icing during mild ambients (30°–40°F.) and high humidity. The leaner mixtures can be tolerated with warmed inlet air because the warmer mixture gives improved distribution to the various cylinders.

The air-cleaner assembly includes a temperature sensor, a vacuum motor, a control damper assembly, and connecting vacuum hoses. The vacuum motor controlled by the temperature sensor operates the damper to control air flow, providing either preheated air from a shroud around the exhaust manifold or unheated underhood air—or a combination of the two.

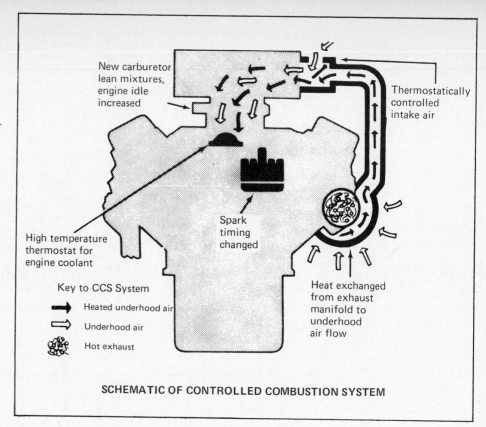

**SCHEMATIC OF CONTROLLED COMBUSTION SYSTEM**

Thermac control switch inside air cleaner is thermostatically actuated to open flapper valve to admit cooler air through snorkel.

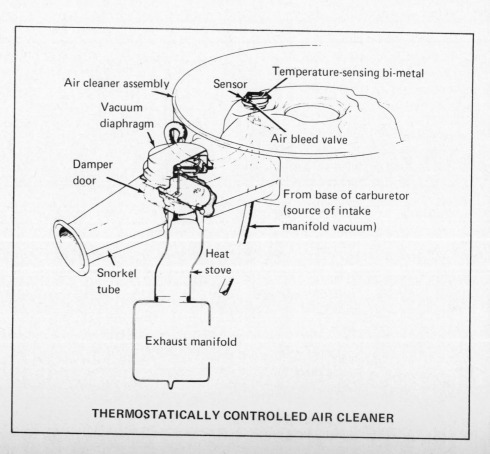

**THERMOSTATICALLY CONTROLLED AIR CLEANER**

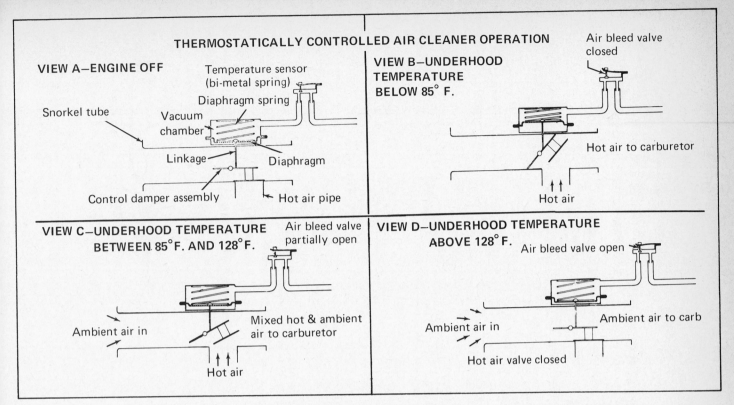

## THERMOSTATICALLY CONTROLLED AIR CLEANER OPERATION

**VIEW A—ENGINE OFF**
- Temperature sensor (bi-metal spring)
- Diaphragm spring
- Snorkel tube
- Vacuum chamber
- Linkage
- Diaphragm
- Control damper assembly
- Hot air pipe

**VIEW B—UNDERHOOD TEMPERATURE BELOW 85° F.**
- Air bleed valve closed
- Hot air to carburetor
- Hot air

**VIEW C—UNDERHOOD TEMPERATURE BETWEEN 85° F. AND 128° F.**
- Air bleed valve partially open
- Ambient air in
- Mixed hot & ambient air to carburetor
- Hot air

**VIEW D—UNDERHOOD TEMPERATURE ABOVE 128° F.**
- Air bleed valve open
- Ambient air in
- Ambient air to carb
- Hot air valve closed

## AIR INJECTION REACTOR SYSTEM

The first major use of air injection was on most 1966-67 California cars. GM calls the system AIR for *air injection reactor.*

Some engines have been equipped with AIR ever since the method was first introduced. Most of the cars equipped with these systems are those with high-performance engines. As of 1972, air-injection systems began to be used more widely to help meet increasingly stiff requirements for reduced HC and CO emissions.

Although this is a costly approach, the slightly richer mixture which can be used on engines which are thus equipped makes the car more drivable with less tendency toward surging in the mid-range.

The other side of the coin reveals a 5 HP loss to drive the AIR pump at high RPM.

This system actually adds air to continue burning of the unburned exhaust gases to reduce their HC and CO content. Air is drawn into an air pump where it is compressed and fed out through the diverter and check valves through the air pipes and into the exhaust manifolds (V8 engines) or cylinder-head exhaust ports (six-cylinder engines). When this compressed air mixes with the hot exhaust gases, further combustion occurs, burning most of any remaining HC and reducing CO in the exhaust before it leaves the vehicle's tail pipe. This increases the temperature of the exhaust gases.

Burning the last bit of exhaust gases is much like fanning dying embers. When the gases leave the cylinders they are still extremely hot and flammable if supplied with the other element of combustion, namely oxygen. Oxygen is in the air supplied by the pump. If the oxygen were to be added to the exhaust farther down-stream in the exhaust system, the gases could not be reignited to provide the desired HC and CO reduction. The primary reduction is in CO.

Because richer mixtures can be used for better drivability with the AIR system, engine overrun or deceleration produces very rich exhaust gases leaving the exhaust valves. If the AIR system was allowed to operate during such conditions, backfires would result as the air from the pump mixed with the rich vapors in the exhaust. A diverter valve eliminates this possibility. It is triggered by sharp increases in manifold vacuum, exhausting all of the pump output into the atmosphere so no backfire occurs. At high engine speeds excess air pressure is dumped to the atmosphere by a pressure-relief valve built into the diverter valve or AIR pump. This oc-

curs at about 55 MPH during cruising or light-throttle applications.

A timing valve in the valve stem controls how long the AIR pump output is diverted to the atmosphere. It bleeds off the manifold-vacuum signal to the diaphragm. Typically, four to six seconds elapse before pump output is again directed to the air pipes leading to the exhaust manifolds or cylinder-head ports.

Some early AIR systems (1966 U.S. and some later foreign GM cars) used an intake air bleed valve instead of a diverter valve. This also prevented backfires during deceleration. It added filtered air from the air cleaner to the intake manifold to lean out the air/fuel mixture. This added air caused the engine to run at a faster-than-idle speed, even with the throttle closed. Its function on deceleration was similar to that of an idle-stop solenoid as it leaned out the mixture by reducing intake manifold vacuum which would cause strong idle-system feed. It also ensures more complete burning of the mixture by reducing the excessive dilution of the mixture by exhaust gases on deceleration. Because the valve added a "gulp" of air to the manifold, it was often called a gulp valve.

Carburetors and distributors for engines with the AIR system are designed

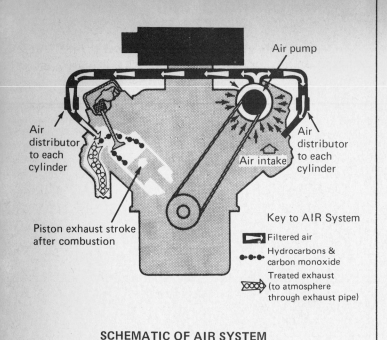

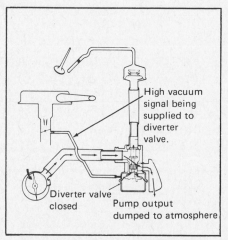

Air pump

Air distributor to each cylinder

Air intake

Air distributor to each cylinder

Piston exhaust stroke after combustion

Key to AIR System
- ▭ Filtered air
- •◄•◄• Hydrocarbons & carbon monoxide
- ◄◄◄◄ Treated exhaust (to atmosphere through exhaust pipe)

**SCHEMATIC OF AIR SYSTEM**

Air injection pump

Check valve

Diverter valve

Check valve

**AIR SYSTEM COMPONENTS**

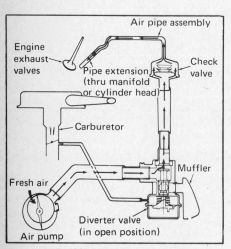

Air pipe assembly

Engine exhaust valves

Pipe extension (thru manifold or cylinder head)

Check valve

Carburetor

Muffler

Fresh air

Diverter valve (in open position)

Air pump

**AIR SYSTEM OPERATION**

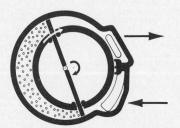

High vacuum signal being supplied to diverter valve.

Diverter valve closed

Pump output dumped to atmosphere

**DIVERTER VALVE OPERATION**

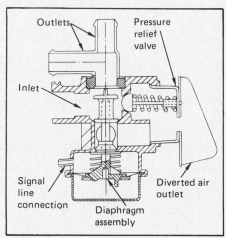

Outlets

Pressure relief valve

Inlet

Signal line connection

Diaphragm assembly

Diverted air outlet

**DIVERTER VALVE COMPONENTS**

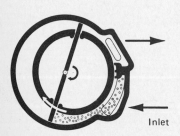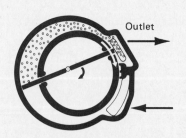

Inlet

Outlet

The vane is travelling from a small area into a larger area—consequently a vacuum is formed that draws fresh air into the pump.

As the vane continues to rotate, the other vane has rotated past the inlet opening. Now the air that has just been drawn in is entrapped between the vanes. This entrapped air is then transferred into a smaller area and thus compressed.

As the vane continues to rotate it passes the outlet cavity in the pump housing bore and exhausts the compressed air into the remainder of the system.

**AIR INJECTION PUMP OPERATION**

particularly for these engines. They should not be interchanged with or replaced by a carburetor or distributor designed for engines without the AIR system.

Exhaust-system components are also modified to get the best possible burning of any residual HC. As of 1973 vehicles, we are seeing enlarged exhaust manifolds to increase the heated area for further insurance that all possible residuals will be burned. It is expected that the exhaust-system design will play an ever-increasing role in emission reduction during the coming years.

This AIR system increases the temperatures applied to the exhaust-system components, necessitating special designs and materials.

A one-way check valve installed between the exhaust and air pump prevents exhaust gases from entering and damaging the air-injection pump. Without a check valve this could occur if exhaust pressure exceeded pumped pressure. A broken air-pump belt would be one cause for this condition to exist.

When correctly installed and maintained, the AIR system will effectively reduce exhaust emissions. However, if the AIR system does not func-

tion or is removed, CO goes abnormally high because the carburetor supplies a rich A/F mixture for pump use.

Because the state of engine tune has an effect on the amount of unburned gases in the exhaust, it has a direct bearing on AIR system operation. Be sure the engine is tuned correctly before analyzing related units.

The basic test to determine if the AIR is working is to loosen the hose clamp at the check valve. Pull the hose from the valve. Start the engine and hold your hand over the end of the hose. If the hose blows air with noticeable force, it is operating. Revving the engine should increase the pressure. The engine should not be run for extended periods with the hose disconnected as exhaust against the back side of the check valve will eventually damage it. If backfiring and popping noises are present when slowing to a stop, the diverter valve needs attention.

## COMBINED EMISSION CONTROL SYSTEM

In 1971 a new device was introduced on a number of GM Rochester-carburetor-equipped vehicles: the Combination Emission Control Valve (CEC). Chevrolet used the unit on six-cylinder engines sup-

plied to Buick, Olds and Pontiac, as did most 1-1/4 and 1-1/2-inch 2GV-equipped Chevrolet V8's. Chevrolet also used it on Q-jetted passenger cars and light-duty trucks. 1-1/2-inch 2GV and Q-jets on heavy-duty trucks did not use it. Vega also met emission requirements without CEC.

This emission-reduction device functions as follows.

When the valve is energized by the transmission, it acts as a throttle stop by increasing idle speed during high-gear operation of the engine. This controls hydrocarbons during deceleration by operating in the lean off-idle system of the carburetor. Keeping the lean mixture during deceleration is especially helpful in controlling HC. The limiting factor in this HC-reducing procedure is the amount of throttle opening a given vehicle can tolerate before objectionable deceleration rates occur. Do not set the CEC engine RPM above recommended specifications or your engine speed during deceleration could make braking difficult. The CEC valve also provides full-spark vacuum advance during high-gear operation but is de-energized in the lower gears and at idle for retarded spark timing during this period. The retarded spark further ensures

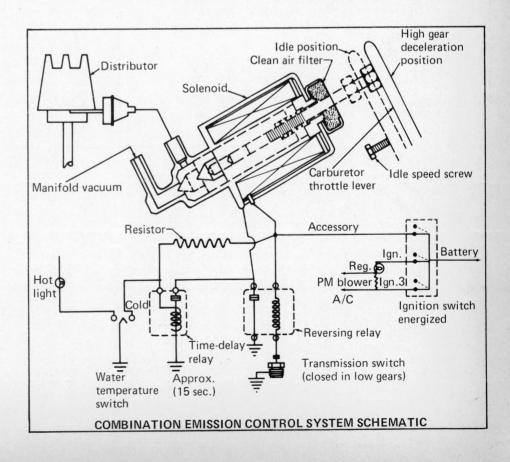

**COMBINATION EMISSION CONTROL SYSTEM SCHEMATIC**

complete burning to keep HC down. Normal idle-speed setting is made with the idle-stop screw or idle-stop solenoid.

With this system the distributor vacuum advance is eliminated in the low forward gears. Control of vacuum advance is accomplished by a solenoid vacuum switch which is energized by grounding a switch at the transmission.

When the CEC solenoid is in the non-energized position, vacuum to the distributor vacuum advance unit is shut off and the distributor is vented to the atmosphere through a filter at the opposite end of the solenoid. Air passage through the solenoid is provided by grooves molded in the spool and by the clearance between the adjusting screw and the plunger stop. When the solenoid is energized, the vacuum port is uncovered and the plunger is seated at the opposite end, shutting off the clean-air vent. Curb-idle speed and high-gear-deceleration throttle-blade setting must be performed with the solenoid in the non-energized position. With the vacuum line removed at the distributor and plugged, curb-idle speed is set with the idle-stop screw or idle-stop solenoid. The throttle-stop adjustment screw on the CEC solenoid is then pulled out to its most-extended position and adjusted to a specified RPM to provide the desired high-gear-deceleration throttle opening. Thus two separate throttle settings are: (1) curb idle and (2) deceleration hydrocarbon control.

The CEC solenoid is controlled by two switches and a time-delay relay. The solenoid is energized in the high forward gears and reverse on Hydramatics by a transmission-operated switch. On 1973 and later models, devices such as these must continue to provide emission control if they fail. Thus, a failed CEC could eliminate vacuum spark advance at all times.

A thermostatic water-temperature switch provides thermal override below 82°F. on models through 1972 and early '73 models. This is not allowed on 1973 models built later in the model year. The time-delay relay in the circuit energizes the CEC solenoid valve for approximately 15 seconds after the ignition key is turned on. Full vacuum supplied to the distributor (independent of engine temperature) for this time interval improves drive-away and helps to eliminate stall-after-start problems. Above 82°F. engine temperature, the distributor only receives vacuum when the transmission is in high gear.

Control of engine dieseling is achieved by using smaller throttle blade openings resulting from the lower curb-idle-speed setting. Additional benefits of the lower idle speeds are less transmission creep, less noise at idle and less heat rejection.

To reduce dieseling on air-conditioned vehicles with automatic transmissions, caused by the larger throttle angle required when setting idle with the air conditioning on, a solid-state time-delay device engages the air-conditioning-compressor clutch for approximately three seconds after the ignition is turned off. The compressor loads the engine to stop it sooner, thereby effectively reducing the tendency to diesel. Additional system features are:
1. Improved fuel economy due to lower idle speeds under city traffic conditions.
2. Improved reliability because of fewer assembly-plant operations required to install the system.
3. Reduced electrical loads because only one solenoid is required.

Broken or disconnected electrical connections resulting in failure of the solenoid to function will result in no vacuum advance and no high-gear deceleration throttle position. The valve is serviced as a unit replacement only.

## TRANSMISSION-CONTROLLED SPARK ADVANCE (TCS)

Transmission-controlled spark advance was first used on 1970 models. It prevents operation of distributor vacuum advance in the lower gears. On cars with automatic transmission, TCS allows distributor vacuum advance in high and reverse gears.

With the exception of 1971 and later Chevrolets (but including all Vegas), the TCS transmission switch *energizes* the TCS solenoid in the first and second gears, shutting off vacuum to the distributor. The solenoid is de-energized in high gear so the vacuum advance operates normally for best economy.

On 1971 and later Chevrolets (except Vega) the TCS transmission switch *de-energizes* the TCS solenoid to block vacuum advance in the lower gears.

Some six-cylinder TCS-equipped cars (prior to March 1973) included a time-delay device to allow distributor vacuum advance for about 20 seconds after starting a cold engine. This reduced stalls after starting. On small-block Chevrolet V-8's, the time delay delays distributor vacuum advance for 20 seconds after the transmission shifts into high gear.

Most emission-controlled Chevrolets (including Vegas) built prior to March 1973 had a hot or combination hot/cold temperature-override TVS switch which provided full vacuum advance in all gears at engine-coolant temperatures below about 63°F. and above 232°F. When engine-coolant temperatures are normal (63°–232°), vacuum advance is controlled by TCS-switch operation. Other vehicles using these Chevrolet engines used similar temperature-override features.

The accompanying chart for a 1971 Vega shows which transmission ranges allow/prevent vacuum to the distributor. TCS on other GM cars works similarly.

### DISTRIBUTOR VACUUM WITH TCS

| Transmission | Reverse | 2nd | 3rd | 4th |
|---|---|---|---|---|
| 3-speed | -- | -- | Vac | -- |
| 4-speed | -- | -- | Vac | Vac |
| Powerglide | -- | Vac | -- | -- |
| Turbo Hydramatic | Vac | -- | Vac | -- |

## TEMPERATURE VACUUM CONTROL SWITCH (TVS)

The retarded spark setting at idle and low speeds makes engines run hotter. Protection against overheating is accomplished on some models by a temperature-sensitive TVS switch which senses engine-coolant temperatures. The TVS switch allows the TCS switch to control vacuum to the distributor vacuum-advance mechanism when engine temperatures are normal. When engine coolant reaches about 210°F., the TVS valve opens against spring pressure to allow manifold vacuum to advance the distributor, regardless of the TCS switch position. When the engine cools to a normal operating temperature, the TCS switch controls the distributor vacuum-advance mechanism again.

## DISTRIBUTOR VACUUM-CONTROL SWITCH (DVS)

Some cars with air conditioning and heavy-duty cooling use a DVS switch. This combines the functions of the

TVS and TCS switches. The DVS solenoid vents the distributor vacuum signal to atmosphere in the lower gears and closes the vent in high gear to allow operation of the distributor-advance mechanism. When engine-coolant temperature rises to 210°F. or higher, full manifold vacuum is applied to the distributor vacuum advance until engine coolant cools down to normal operating temperatures.

## SPEED CONTROL SYSTEM (SCS)

Some GM cars control distributor vacuum advance according to road speed. Below 33-38 MPH the valve prevents vacuum advance. When road speed reaches 33-38 MPH the solenoid valve allows distributor vacuum advance according to the signal from the timed-spark port in the carburetor. The SCS sensor is actuated by the speedometer driven gear in the transmission. Some pre-1973 cars with SCS include a thermal-override valve which applies manifold vacuum to the distributor for full vacuum advance when an engine-water-temperature sensor indicates an overheat condition.

## EXHAUST GAS RECIRCULATION (EGR)

During the combustion process, nitrogen tends to combine with o oxygen at temperatures of about 2040°F., forming oxides of nitrogen ($NO_x$). Engine-combustion temperatures often exceed this figure. To reduce the formation of $NO_x$, an EGR valve meters inert exhaust gases into the intake manifold to displace a portion of the air/fuel mixture in the cylinders so peak-combustion temperatures and pressures are lowered. A portion of the exhaust gas is being recirculated into the combustion chamber, hence the term *exhaust gas recirculation.*

Although EGR does not affect vehicle operation at idle or wide-open throttle (where no EGR is occurring), engines do "object" to the recirculation of unburnable exhaust gas into the combustion chambers in amounts ranging up to 15% of mixture volume. The "objection" is made known by surging and uneven running during light-throttle/high-vacuum running at speeds from 25—60 MPH.

The EGR valve mounts on the intake manifold and connects to an exhaust manifold port on six-cylinder engines or the exhaust-cross-over passage on V-8's. The valve remains closed during idle and deceleration to pre-

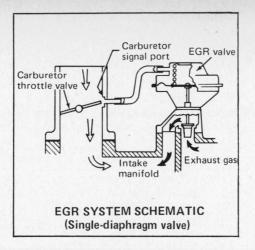

**EGR SYSTEM SCHEMATIC**
**(Single-diaphragm valve)**

vent rough running caused by excessive exhaust-gas dilution of the air/fuel mixture.

The EGR valve gets its opening signal from the carburetor. Vacuum from an off-idle port opens the valve when the throttle blade is opened past idle to uncover the EGR off-idle port, sometimes called a lower EGR port.

A second (upper) EGR port bleeds air into the vacuum channel to reduce the signal supplied by the lower port. This times the EGR valve for precise metering of exhaust gases depending on the location of the ports in the carburetor bore and the amount of throttle opening. As the throttle is opened farther into the part-throttle range, the upper port ceases to function as a bleed and is exposed to manifold vacuum to supplement the vacuum signal from the lower port so the EGR valve is maintained in the desired position. High vacuum opens the valve wide, low vacuum opens it slightly. Below 5 inches Hg vacuum or at wide-open throttle the valve closes.

In the GM line, only 1972 Buicks for California had EGR. By 1973 all GM passenger cars and light-duty trucks used the system. In 1974 some California models were equipped with dual-diaphragm EGR valves to provide improved metering of exhaust gases into the intake manifold according to engine speed and load.

As described, the single-diaphragm EGR valve is closed at both idle and wide-open throttle. It is closed at idle to avoid excessive dilution of the air/fuel mixture, recognizing that exhaust dilution already exists at idle speeds due to residual exhaust products in the cylinders. It is closed at wide-open throttle to avoid

reducing the available power from the engine when needed for passing maneuvers or climbing steep hills.

In effect, the single-diaphragm unit is responsive to throttle position and air flow through the engine.

A more sophisticated control makes use of a dual-diaphragm unit to allow the EGR valve to monitor and respond also to intake-manifold vacuum as an indication of engine load. The result is improved drivability with less surging at part-throttle.

There are two metal plates (or pistons) each sealed to the side of the EGR valve by a flexible diaphragm, and each attached solidly to the valve shaft.

The carburetor timed-port signal applied to the upper surface of the upper piston tends to open the valve against the closing force of the spring.

Intake-manifold vacuum is applied to the cavity between the two plates and *tends* to draw the two pistons toward each other. Because the upper piston is larger than the lower one, downward force is greater and manifold vacuum causes a net downward force on the valve shaft, tending to close the EGR valve.

Remembering that both pistons are firmly attached to the valve shaft, each of the metal pistons will transmit a force to that shaft according to the differential pressure across each piston. The larger upper piston compares the vacuum in the cavity between the pistons to the vacuum from the timed port. The net result is a force directed upward or downward, according to which side has the lowest pressure. Similarly, the lower piston responds to manifold vacuum in the cavity and ambient pressure on its bottom surface.

Manifold vacuum is also applied to the inner surfaces of the two flexible diaphragms which seal the pistons to the valve housing, and the vacuum also tends to pull the diaphragms closer to each other. Because the diaphragms are flexible they will assume a U shape and exert force both on the outer rim of the pistons and the walls of the housing.

The total force on a diaphragm will be divided about half and half between piston and housing. Force against the housing doesn't cause anything to move.

The small-area *diaphragm* around the large-diameter piston receives less total force from the vacuum and "wastes" less against the sides of the housing than does the large-area diaphragm.

Therefore the large-area *piston* receives more force from manifold vacuum than does the smaller piston and the net effect of manifold vacuum is to move the EGR valve toward closed.

An advantage of the dual-diaphragm control can be seen by assuming a worst-case condition: a car is traveling at medium speed with part throttle on a gentle downgrade. With a single-diaphragm unit, the throttle-valve would be positioned near the timed port and the EGR valve would be completely open resulting in too much exhaust dilution of mixture and surging or uneven running of the engine.

With the dual-diaphragm control, under the same road condition, timed-port vacuum tends to open the EGR valve but the high manifold vacuum tends to close it. The result will be a partially-open valve metering a tolerable amount of exhaust gas into the engine.

When engine load increases, pressure and temperature in the combustion chambers tend to favor the formation of $NO_x$. Manifold vacuum decreases, allowing the EGR valve to open farther. But, at wide-open throttle, both the single- and the dual-diaphragm EGR valves close so there is no exhaust-gas recirculation.

## EVAPORATIVE EMISSIONS

Here is the third major area for discussion. Evaporative emission includes hydrocarbon materials in fuel spilled or evaporated from fuel tanks and carburetors. You might consider this to be an insignificant part of the emission picture, but these emissions are estimated to equal the emissions which would occur if crankcase ventilation were uncontrolled.

The gas tank and carburetor were traditionally vented to the atmosphere—until 1970 in California and 1971 nationwide. Venting got rid of vapors and aided the cars' hot-starting capabilities. There was no real concern about emissions caused by spilled fuel during overfilling or gasoline sloshed out of vents during sudden maneuvers. In

EGR valve mounts on intake manifold to meter exhaust gas from the exhaust cross-over passage into the intake manifold. This reduces $NO_x$ formation by reducing peak pressures and temperatures in the combustion chambers. Exhaust gas recirculation amounts to as much as 15% of intake-mixture volume.

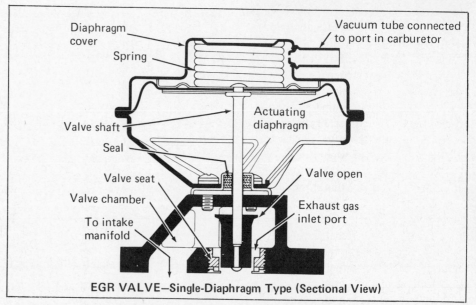

EGR VALVE—Single-Diaphragm Type (Sectional View)

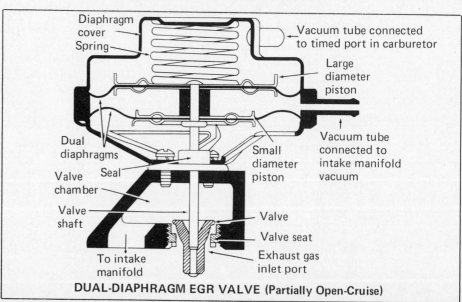

DUAL-DIAPHRAGM EGR VALVE (Partially Open-Cruise)

Engine surging at light load and part throttle is minimized by this dual-diaphragm EGR valve. Vacuum from carburetor timed port is urging EGR unit open while vacuum from intake manifold tends to close it. Result is partial opening of valve, metering exhaust gases in proportion to throttle position, air flow, and engine load. First use of these valves was on 1974 models sold in California.

1970-71 total evaporative emissions allowed from any car were six grams per hot soak. In 1972 the allowance was reduced to two grams per hot soak. See chart on page 279.

## EVAPORATION CONTROL SYSTEM (ECS)

This system reduces fuel-vapor emissions which would otherwise vent to the atmosphere from the gasoline tank and carburetor bowl. The evaporation-control system is one of the few emission-control devices which actually adds to the efficiency of the automobile. Anytime raw fuel can be burned instead of polluting the air, we are aiding car economy *and* ecology.

The obvious part of this system is the ECS charcoal canister which is usually found in the engine compartment. The ECS canister is a simple container for charcoal granules. The fuel vapor inlets and outlets are on one side of the charcoal and the purge air opening and filter are on the other. *Purge air* is the air drawn through the saturated charcoal to pick up the stored fuel vapors.

Vapors enter the canister from the carburetor bowl vent and the fuel tank liquid/vapor separator. The vapors are "stored" (adsorbed) onto the surfaces of the charcoal granules and stored in the air space in the canister. When the car is restarted, the fuel vapors and fresh air are sucked into the carburetor or intake manifold to be burned in the engine.

Canister purging methods vary widely with make and model.

**More about the canister**—Depending on car size, the canister contains either 300 or 625 grams of activated charcoal (carbon) in granules of 0.033 to 0.094-inch diameter. Each gram of charcoal can hold up to 35% of its weight in fuel vapor. This is because each gram offers a surface area of over 1100 square meters—over one-quarter acre! Thus, the 625-gram units have an internal surface area equivalent to 170 acres (165 football fields!). The fuel-vapor molecules are adsorbed (condensed and held) onto the surface of the charcoal. Adsorption is a process with a very weak attaching force so the fuel-vapor molecules are easily dislodged by flowing fresh air through the charcoal bed during the purge cycles.

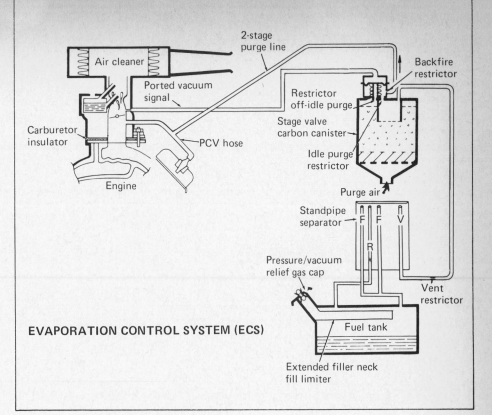

**EVAPORATION CONTROL SYSTEM (ECS)**

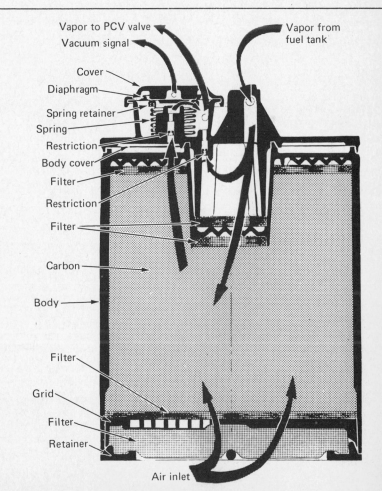

**THREE-TUBE PURGE-VALVE CANISTER**

Internal construction of three-tube purge-valve canister as used in 1970-71. Some 1970-71 models also directed vapors from the carburetor bowl to the canister. In 1972, all models had a carburetor bowl vent connected to the canister.

There are usually two calibrated orifices (either in the purge valve canister or carburetor throttle body) to return vapors to the engine. The smaller of the two (0.018–0.035-inch) connects directly to the intake manifold and passes fuel vapors and/or purge air any time the engine runs. This is called *constant-bleed purge.* The small orifice is sized according to how much vapor the engine can tolerate at idle.

A second orifice (0.055–0.065-inch), also connected to the manifold or carburetor, is controlled by a vacuum-operated valve on the canister and/or a timed port (off-idle) in the carburetor. This provides increased purging at speeds above idle when the engine speed is sufficient to accept the additional fuel/vapors.

In 1972 GM used a thermal-delay valve on some models so the canister is not purged until the engine attains operating temperature. Purging at idle or with a cold engine causes rough running and increased emissions because the additional vapor being added to the intake manifold can cause an over-rich mixture.

The air filter at the bottom of the canister requires replacement at 24-month or 24,000-mile intervals.

Certain wheel-splash conditions can plug the air filter in the vapor-canister base prematurely. The filter must pass air for the system to function. Inspect it periodically for mud and other contaminants.

ECS also uses sophisticated measures in the fuel-tank filling and construction areas. Thermal expansion is provided for by trapping up to three gallons of air during tank filling, then allowing this air to escape to the top of the main tank where it vents to the canister through a liquid/vapor separator. Any liquid trapped in the separator/s drains back to the tank and the vapor passes to the canister.

The gas cap is designed to contain the vapors in the tank and thus force them through the separator and into the canister. It takes approximately 1/2 psi to make the canister system work. Under extreme conditions, when the tank pressure goes above cap limit it starts venting to atmosphere. First-system caps would blow off at 1–2 psi pressure; varying by make and model. Starting in 1972 the fuel-tank caps have a higher blow-off pressure setting so the tank, straps and

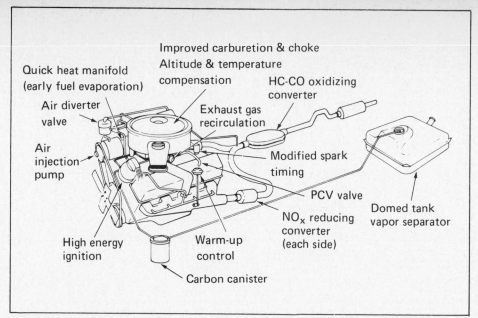

Complicated emission-control system expected to be needed to allow internal-combustion engines to meet the proposed 1976 federal standards. All of these controls and specialized components will increase the initial costs of the vehicle and will reduce fuel economy while causing the motorist to bear enormous costs for tune-ups and replacement components.

related components are stronger to withstand the increased pressure. The caps also include a vacuum vent to release any vacuum which forms in the tank as the fuel is used.

*Do not mix caps* between various makes and models. With large surface areas existing in fuel tanks it doesn't take much pressure to damage or burst them. Any emission parts which fail should be replaced only with original GM replacement parts to ensure correct performance.

## EMISSION SETTINGS

Currently all manufacturers are required to install a specification label in the engine compartment. It details the correct carburetor and distributor adjustments needed to maintain legal emission levels for that vehicle. No matter how effective or sophisticated the control system, a weak spark, fouled plug, bad plug wire, cracked distributor cap—or any of 200 other things—can wreck the system's efficiency. It only takes one of these things going wrong to cause the car to be-

come a pollutant emitter of a worse nature than a car with no controls at all.

A single bad spark plug is a good example. In SAE Paper 710069, "Exhaust-Emission Control for Used Cars," the authors pointed out that a single fouled plug increased the HC emission level six times: from 605 ppm to 3,609 ppm.

## WHAT'S IT COSTING?

The whole idea of emission controls is to clean up the environment—namely the air we breathe . . . and live in. But, as is always the case when you get down to the facts—there is "no free lunch." Anything you get—in this case cleaner air—costs *something.* That something includes reduced drivability, increased gasoline consumption, increased tune-up costs—and a car that weighs more.

To these obvious costs we must add increased complexity of the entire vehicle with associated cost increases in both engineering and manufacturing.

# Rochester Products' Test Facility

## ROCHESTER PRODUCTS "CLEAN ROOM"

The air we breathe, like the weather, has long been a subject that everyone talks about. But few people have done anything of real significance about the air until recently. Rochester Products Division of General Motors is doing something to improve our air and is continuing to set the standard of excellence in the field of exhaust emission control.

The twelve-million-dollar test facility, often called the "Clean Room," is technically known as Rochester Products' Carburetor Test Facility for Exhaust-Emission Control. But a "clean room" it is — in terms of what goes on inside — and what comes out of the facility as an end product.

The Test Facility is a unique manufacturing tool conceived, designed and built by Rochester Products personnel. It is a computer-controlled testing system providing the highest standards of assurance that Rochester carburetors will uniformly meet and generally exceed current federal and state exhaust-emission-control requirements.

The clean room's 128 computer-controlled test stands, test and set a total of 24,000 Rochester carburetors to close tolerance standards every day. Fifty-five man-years of programming and engineering effort went into the creation of the clean room facility. It took two years to build.

The facility gives Rochester carburetors heretofore unobtainable levels of exhaust-emission control. Besides constructing the most comprehensive carburetor-testing equipment ever devised for mass production, the overall quality of the carburetors being produced has been upgraded as a result. When you look at the clean room specifications, you begin to realize why all this has been possible.

A specially constructed, climate-controlled, 420- by 50-foot room is constantly maintained at 29.5-inches of mercury atmospheric pressure, 75-degrees Fahrenheit, and 50% relative humidity. Entry and exit for products and personnel

The heart of the Rochester Products' carburetor-flow test is located in this efficient computer room. Strategically located printers enable quality control and manufacturing supervision to be kept constantly informed.

On the test stands, all 95 of Rochester Products' current carburetor models are tested . . . covering the full range of their operation from idle to wide-open throttle.

The heart of the Rochester Products' carburetor-flow test is located in this efficient computer room. Strategically located printers enable quality control and manufacturing supervision to be kept constantly informed.

One of many GM chassis rolls used for exhaust analysis in emission development programs. These units and the accompanying computer-controlled equipment cost hundreds of thousands of dollars.

is through special air locks. A unique blending system controls the viscosity and specific gravity of the test fuel supplied to carburetors as they are wet-tested.

All 95 of Rochester Products current carburetor models are tested over a full range of operation from idle to wide-open throttle. Each test stand signals fuel flow, air flow and fuel pressure to the computer through electronic transducers repeatably accurate to within 0.025%.

A General Electric 4060 Process Computer controls and monitors the test facility. This computer system is one of the largest process computer systems in existence today. 6,700 digital channels and 420 analog channels connect the computer to the test stands and environmental controls.

The carburetor test is fully automatic and under constant computer control. Once the operator clamps a carburetor on the stand and initiates the cycle, the test is controlled by the computer. Accuracy of fuel and air flow is checked at each of the eight or more test points. Five servo motors automatically adjust metering screws

on the carburetor to bring fuel and air flow to an acceptable level. A code number on a digital read-out initiated by the computer indicates that the carburetor has passed the test — or the reason why it was rejected. Simultaneously, this information is stored in the computer memory for daily data reports on production output and rejection rates.

The computer periodically checks the accuracy and rejection rate of each test stand, environmental conditions in the clean room and its own operation. Out-of-limits "Alarm" conditions are typed out on automatic typewriters. "Alarm" conditions affecting the quality of acceptance automatically disconnect the test stands from computer control.

Because the computer works with test specifications for 95 carburetor models, model change-over is accomplished in less than a minute by feeding punched cards into the computer memory while minor mechanical adjustments are made at the test stands. Supervisors can obtain reports on how the test facility is working upon request from remote typewriters. Every hour a printer indicates how many carburetors were test-

ed, number rejected, reason for rejection and percent rejected. This current information on quality level allows taking immediate corrective action on the carburetor production lines if problems occur.

"Control" is the key word in talking about the Rochester Products carburetor test facility. This conscientious regard for control has created an unusually high quality level in carburetor manufacturing.

In addition to test facilities in Rochester, RPD maintains others at the General Motors Proving Grounds at Milford, Michigan; Mesa, Arizona and Pikes Peak, Colorado. Nearly every conceivable type of driving environment can be duplicated as Rochester engineers apply their talents to new models, new applications and experimental fuel systems of the future. In scores of rugged, over-the-road tests, they probe specific characteristics of power development, air and fuel flow, emission control, economy and all-around performance.

At Rochester Products, systems are tested all around before they are approved — and proved all around before they are produced.

# Index

## YES!

**THERE ARE A LOT MORE GOOD H. P. BOOKS** which you can get right away by using one of these order coupons. The big step toward mastery of *your own field of interest* is solid *factual* info written for you by experts. All H. P. Books focus directly on the real world. **How to do it. How things actually work.** You won't find any "Gee Whiz!" stuff or any limp warmed-over material in these Good H. P. Books. *You didn't find any in the book you just read, did you?*

Here are some mini-reviews to help you select the books you need:

*Want to be a carburetor expert?* You can't get there today by studying yesterday's carbs. Modern automotive carburetors are sophisticated instruments, but simple and easy to understand through the H. P. Book method of giving you words and pictures to make the function of each part perfectly clear. Carburetion of emission-controlled engines is a new ball game. Learn about *modern* practice in **Holley Carburetors** and **Rochester Carburetors**. Become an expert in tuning for performance AND economy.

*Want to pioneer a little?* Far-out stuff is done first by a few knowledgeable guys, then one day it becomes the hot set-up. If you have a few good ideas of your own, pioneering is fun. How 'bout hanging a turbocharger on your engine?

**How To Select & Install Turbochargers** lays it all out for you—cars or bikes. Or build a bullet-proof VW-based vehicle for bashing in the boonies. Take a look at **Baja-prepping VW Sedans and Dune Buggies.** All about the tough stuff.

*Tired of hiding in the weeds?* Arouse your Chevy with **How To Hotrod Small-Block Chevys** or **How To Hotrod Big-Block Chevys.** These *big books* are loaded with honest information—much of it factory-direct. Chevy heavy-duty parts numbers let you speak to the parts man in his native language. Straight scoop about the bolt-on equipment. With these good books, **you know what will happen before you make the mod.**

*Are H. P. Books really that good?* Roger Huntington, dean of American automotive writers says: "I rarely ever recommend specific technical books in my columns, but there is one company that is putting out such superior hop-up instruction manuals that I should give them a pat. It is Fisher Publications, Inc., 4058 North 14th Ave., Tucson, AZ 85705. This is Bill Fisher, who wrote the old "California Bill" hot rod books that some of you will remember from 20 years ago. He has brought a new breath of detail, accuracy and authority to this business of hop-up instruction manuals that has been sadly lacking in many cases. Most of his books are $5.00 each—but worth every penny. If you read it in a Fisher book, it works."

*Do it in the dirt?* The challenge of off-road motorcycling asks more of you than pluck and grit. It takes *know-how.* Get the benefit of Doug Richmond's years in the saddle by reading the entertaining **How To Select, Ride and Maintain Your Trail Bike.** Tells you everything the title promises and more, including the adventure of camping and touring in Baja. *What to take, what to do, how to fix it when it breaks.*

When you think you're pretty good at riding where others get off and push, you are ready to join the elite and ride Observed Trials. This booming off-road sport is rapidly attracting entire families because it's a bunch of fun. Carl Shipman's **How To Ride Observed Trials** takes you all the way from beginner to expert with photo-illustrated riding instruction for the *trick stuff*, bike modifications to help you do it better and advice on getting the best bike for you to ride. **Includes American Trials Association** official rules.

*We like librarians!* Your Friendly Local Library tries hard to have the books *people want to read.* They work to match the interests of young people, old people, and strange people. Outdoor types, indoor types, mechanical types, and philosophical types. Lots of libraries have H. P. Books on hand for your type and our type. If yours doesn't, why not suggest they get in touch with us and give them our address. **We have a good deal for libraries**, and that's a good deal for you too, dear reader!

*Meet the mighty mites!* If you drive a Datsun or VW you think you have a better deal than the lead sleds and gas gobblers. We tend to agree. Maybe you have thought about doing a little thing or two, nothing really humungus, to get a bit more performance—tailored to your personal driving preferences. We tend to agree. In fact we agree so much that we will be happy to send you **How To Modify Datsun Engines & Chassis: 510, 610 & 240Z.** Unless of course you drive a VW in which case **How To**

---

Please send:

**AUTOMOTIVE TITLES**

☐ Small-Block Chevy
☐ Big-Block Chevy
☐ Holley Carburetors
☐ Rochester Carburetors

☐ Turbochargers
☐ Datsun
☐ VW Engines
☐ Baja-Prepping
☐ Corvair

**MOTORCYCLE TITLES**

☐ Trailbike
☐ Tuning for Performance
☐ Observed Trials
☐ Minibikes

Enclosed is $ _____ ($5.30 per book including postage & handling; no COD's).

Remit in U.S. funds.          ☐ Check          ☐ Money order

Name _____

Street _____

City _____ State _____ Zip _____

**LIBRARIES, SCHOOLS, DEALERS—**
☐ Please send full information on volume purchase of H. P. Books.

Allow three weeks for delivery. Arizona residents, please add 15¢ sales tax per book.
**H. P. Books**, P. O. Box 50640, Dept. RO-1, Tucson, AZ 85703

---

Please send:

**AUTOMOTIVE TITLES**

☐ Small-Block Chevy
☐ Big-Block Chevy
☐ Holley Carburetors
☐ Rochester Carburetors

☐ Turbochargers
☐ Datsun
☐ VW Engines
☐ Baja-Prepping
☐ Corvair

**MOTORCYCLE TITLES**

☐ Trailbike
☐ Tuning for Performance
☐ Observed Trials
☐ Minibikes

Enclosed is $ _____ ($5.30 per book including postage & handling; no COD's).

Remit in U.S. funds.          ☐ Check          ☐ Money order

Name _____

Street _____

City _____ State _____ Zip _____

**LIBRARIES, SCHOOLS, DEALERS—**
☐ Please send full information on volume purchase of H. P. Books.

Allow three weeks for delivery. Arizona residents, please add 15¢ sales tax per book.
**H. P. Books**, P. O. Box 50640, Dept. RO-1, Tucson, AZ 85703

**Hotrod Volkswagen Engines.** Either way, you win. *Make a good impression!*

If you happen to be taking an auto technology or motorcycle course in school, *try walking in with an H. P. Book under your arm.* Your instructor will look twice. If you *are* the instructor, build *your* image the same way. Schools all over the country are adding interest and excitement to auto and motorcycle instruction, and science courses, by using H. P. Books as text or supplement. **We have a good deal for schools too.**

*No Sir! We didn't forget you.*

You with the classic Corvair. You know what you've got and you like it. Make it better by checking **How To Hotrod Corvair Engines** and do it right.

*How to kill two stones with one bird in the bush.* Need a gift for Uncle Harry or Cousin Pete? You don't know what they want. *Sly trick.* Give them something you want, like a good H. P. Book. They'll read it and enjoy it. Then you can borrow it and do the same. We do not advise actually keeping it when you borrow it back. Give your kid brother or sister a copy of **All About Mini- bikes** and you'll wind up reading every page.

*Remarkable Personal Book Collection Plan.* This is how professionals and advanced amateurs build a personal library of useful and interesting H. P. Books. Great for bad rainy days when you don't want to go out. Even greater for good days when you want to *fix or modify* your car, *ride or tune* your motorcycle, tune somebody's ailing carb. What you do is acquire H. P. Books singly or in bunches until you have them all. Smartest guy in your block . . . or town, maybe!

*Tune your bike.*

**Motorcycle Tuning For Performance** by Carl Shipman—contributing editor for Dirt Bike magazine—sold out the first edition and is now available in an expanded new Second Edition. Riding a bike in good tune is eight-hundred times as much fun as riding one which lurches, stumbles, and burps. Thoroughly covers motorcycle carbs, ignition, and tuning procedures for nearly any bike, street or dirt. Photo take-apart sequences for most carburetor brands and types of ignitions.

Not a recipe book, it gives you understanding of *how things work* and then *how to adjust them.* When you know the basics, you can work on engines you have never even seen before. Also includes practical info on engine modifications and improvements. More than repair manual—it begins where the repair manuals stop and helps you get best performance from your scooter.

SATISFACTION GUARANTEED. Strong words and we live by them—just as we have been doing since 1947 when we first started saying, "You must be satisfied or your money is refunded immediately." There are no advertisements in our books, just honest information you can trust—facts like the pros use. Our books are SUPERB.

*You better believe it!* We *promise* practical how-to-do-it info in our stuff and we *deliver.* Try this one in your computer. HOW TO ORDER H. P. BOOKS:

1. Cut out one of the adjoining coupons.
2. Make selection of all the books you need by putting a check mark in the box *by each title* you want to have.
3. Multiply number of books ordered by $5.30 and send along the total amount by your check or a money order which you can obtain at the post office or wherever you see the sign "money orders sold here." Make check or MO payable to H. P. Books. *If you live in Arizona* please add 5% to keep our government green.
4. Pitfall to avoid: *Do not send cash* through the mail.
5. Include *both* coupon and payment in an envelope, seal, put on stamp, *mail it yourself.* Use our correct address on the envelope with our zip.
6. Pitfall to avoid: Write your name and address on the coupon *so we can read it.*
7. What to expect: *Good books* in about three weeks. They will come to you parcel post, prepaid.
8. The big payoff: You will speak wisdom and perform mechanical miracles.
9. Handy Tip: Hurry.

CUT OUT THIS ORDER BLANK, PUT IT IN A STAMPED ENVELOPE WITH YOUR REMITTANCE, AND MAIL TO:

H. P. BOOKS
P. O. BOX 50640
TUCSON, AZ 85703

CUT OUT THIS ORDER BLANK, PUT IT IN A STAMPED ENVELOPE WITH YOUR REMITTANCE, AND MAIL TO:

H. P. BOOKS
P. O. BOX 50640
TUCSON, AZ 85703

**H.P. BOOKS**